METHODS IN MOLECULAR BIOLOGY

Series Editor
John M. Walker
School of Life and Medical Sciences
University of Hertfordshire
Hatfield, Hertfordshire, UK

For further volumes:
http://www.springer.com/series/7651

For over 35 years, biological scientists have come to rely on the research protocols and methodologies in the critically acclaimed *Methods in Molecular Biology* series. The series was the first to introduce the step-by-step protocols approach that has become the standard in all biomedical protocol publishing. Each protocol is provided in readily-reproducible step-by-step fashion, opening with an introductory overview, a list of the materials and reagents needed to complete the experiment, and followed by a detailed procedure that is supported with a helpful notes section offering tips and tricks of the trade as well as troubleshooting advice. These hallmark features were introduced by series editor Dr. John Walker and constitute the key ingredient in each and every volume of the *Methods in Molecular Biology* series. Tested and trusted, comprehensive and reliable, all protocols from the series are indexed in PubMed.

Bioinformatics for Cancer Immunotherapy

Methods and Protocols

Edited by

Sebastian Boegel

Mainz, Germany

Humana Press

Editor
Sebastian Boegel
Mainz, Germany

ISSN 1064-3745 ISSN 1940-6029 (electronic)
Methods in Molecular Biology
ISBN 978-1-0716-0326-0 ISBN 978-1-0716-0327-7 (eBook)
https://doi.org/10.1007/978-1-0716-0327-7

Cover Caption: Cover image made with biorender (www.biorender.com)

This Humana imprint is published by the registered company Springer Science+Business Media, LLC, part of Springer Nature.
The registered company address is: 1 New York Plaza, New York, NY 10004, U.S.A.

Preface

120 years ago Paul Ehrlich, a German physician and Nobel Prize winner, hypothesized that the immune system is capable of recognizing and killing tumor cells [1, 2]. With his postulate of immune surveillance in 1909, Paul Ehrlich identified the incredible potential to treat cancer by immunomodulation.

Cancer immunotherapy is a class of therapeutic strategies that use the host immune system as a means to control cancer by activating a patient's immune system against tumor cells [3]. 114 years after Ehrlich's immune surveillance hypothesis, cancer immunotherapy was acknowledged as "Breakthrough of the year 2013" [4] and the 2018 Nobel Prize was awarded for the discovery and therapeutic exploitation of T cell checkpoint receptors [5].

Two emerging concepts, that boost recognition of tumor cells by endogenous T cells are (1) cancer vaccines against neoantigens presented on the patient's human leukocyte antigen (HLA) class I and II alleles and (2) immunomodulatory agents, such as checkpoint inhibitors.

The process of neoantigen-directed personalized cancer immunotherapy approaches is reviewed in Chapter 1 and involves (1) identification of somatic mutations (Chapters 2–4), (2) determination of a patients HLA type (Chapters 5–8), (3) prioritization of variants according to their likelihood of being presented on HLA class I/II molecules and/or to elicit a T cell response (Chapters 9–12), and (4) modelling the (epitope-specific) T cell receptor (TCR) repertoire of a patient (Chapters 13–14).

The second concept, checkpoint inhibition, shows tremendous success in some tumors [6]; however, it is still not fully understood which patients benefit from this treatment. Several studies show that the number of mutations [7], the phenotype of tumor-infiltrating immune cells [8] (Chapters 15–19), the HLA class I type [9] and HLA expression [10] correlate with response to checkpoint blockade.

Furthermore, monitoring the activity of transcription start sites (TSSs) at single base-pair resolution across the genome in order to find active promoter and enhancer regions in cancer cells might have the utility to identify key factors to immunotherapy (Chapter 20).

Using high-throughput sequencing, it has become feasible to screen the entire exome and transcriptome of a cancer patient in short time and in addition recent progresses in mass spectrometry (Chapters 11–12) allows high-throughput discovery of HLA presented ligands. Many bioinformatic tools have been developed to address each of the steps mentioned above and also workflows integrating the whole process are available (Chapter 10). Thus, the focus of this volume is to gather a variety of in silico protocols of state-of-the-art bioinformatic tools and computational pipelines developed for neoantigen identification and immune cell analysis from high-throughput sequencing data for cancer immunotherapy.

This volume has been possible only through the contributions of the leading experts in the fields of bioinformatics and immunogenomics. I am grateful for the discussions throughout the process of assembling this book and their consent to share their expertise and knowledge with the scientific community.

I wish to thank Springer and the Series Editor Prof. John M. Walker for giving me the opportunity to assemble this outstanding collection of manuscripts and for their excellent support throughout the process of editing this book. Last but not least, I want to thank my former colleagues at TRON (Translational Oncology at the University Hospital Mainz, Germany) that gave me the chance to be part of this exciting journey in the field of cancer immunotherapy, as well as for teaching and inspiring me.

Mainz, Germany *Sebastian Boegel*

References

1. Ichim CV (2005) Revisiting immunosurveillance and immunostimulation: implications for cancer immunotherapy. J Transl Med 3:8
2. Rammensee H-G (2014) From basic immunology to new therapies for cancer patients. In: Cancer immunotherapy meets oncology. Springer, Cham, pp 3–11
3. Galluzzi L, Vacchelli E, Bravo-San Pedro J-M, et al (2014) Classification of current anticancer immunotherapies. Oncotarget 5:12472–12508
4. Couzin-Frankel J (2013) Breakthrough of the year 2013. Cancer immunotherapy. Science 342:1432–1433
5. Guo ZS (2018) The 2018 Nobel Prize in medicine goes to cancer immunotherapy (editorial for BMC cancer). BMC Cancer 18:1086
6. Ribas A, Wolchok JD (2018) Cancer immunotherapy using checkpoint blockade. Science 359:1350–1355. doi:10.1126/science.aar4060
7. Yarchoan M, Hopkins A, Jaffee EM (2017) Tumor mutational burden and response rate to PD-1 inhibition. N Engl J Med Mass Medical Soc 377:2500–2501. doi:10.1056/NEJMc1713444
8. Charoentong P, Finotello F, Angelova M, et al (2017) Pan-cancer immunogenomic analyses reveal genotype-immunophenotype relationships and predictors of response to checkpoint blockade. Cell Rep 18:248–262
9. Chowell D, Morris LGT, Grigg CM, et al (2018) Patient HLA class I genotype influences cancer response to checkpoint blockade immunotherapy. Science 359:582–587. doi:10.1126/science.aao4572
10. Roemer MGM, Redd RA, Cader FZ, et al (2018) Major histocompatibility complex class II and programmed death ligand 1 expression predict outcome after programmed death 1 blockade in classic hodgkin lymphoma. J Clin Oncol 36:942–950. doi:10.1200/JCO.2017.77.3994

Contents

Preface v
Contributors ix

1 Bioinformatics for Cancer Immunotherapy 1
Christoph Holtsträter, Barbara Schrörs, Thomas Bukur, and Martin Löwer

2 An Individualized Approach for Somatic Variant Discovery 11
Minghao Li, Ting He, Chen Cao, and Quan Long

3 Ensemble-Based Somatic Mutation Calling in Cancer Genomes 37
Weitai Huang, Yu Amanda Guo, Mei Mei Chang, and Anders Jacobsen Skanderup

4 SomaticSeq: An Ensemble and Machine Learning Method to Detect Somatic Mutations 47
Li Tai Fang

5 HLA Typing from RNA Sequencing and Applications to Cancer 71
Rose Orenbuch, Ioan Filip, and Raul Rabadan

6 Rapid High-Resolution Typing of Class I HLA Genes by Nanopore Sequencing 93
Chang Liu and Rick Berry

7 HLApers: HLA Typing and Quantification of Expression with Personalized Index 101
Vitor R. C. Aguiar, Cibele Masotti, Anamaria A. Camargo, and Diogo Meyer

8 High-Throughput MHC I Ligand Prediction Using MHCflurry 113
Timothy O'Donnell and Alex Rubinsteyn

9 In Silico Prediction of Tumor Neoantigens with TIminer 129
Alexander Kirchmair and Francesca Finotello

10 OpenVax: An Open-Source Computational Pipeline for Cancer Neoantigen Prediction 147
Julia Kodysh and Alex Rubinsteyn

11 Improving MHC-I Ligand Identification by Incorporating Targeted Searches of Mass Spectrometry Data 161
Prathyusha Konda, J. Patrick Murphy, and Shashi Gujar

12 The SysteMHC Atlas: a Computational Pipeline, a Website, and a Data Repository for Immunopeptidomic Analyses 173
Wenguang Shao, Etienne Caron, Patrick Pedrioli, and Ruedi Aebersold

13 Identification of Epitope-Specific T Cells in T-Cell Receptor Repertoires 183
Sofie Gielis, Pieter Moris, Wout Bittremieux, Nicolas De Neuter, Benson Ogunjimi, Kris Laukens, and Pieter Meysman

14 Modeling and Viewing T Cell Receptors Using TCRmodel and TCR3d 197
Ragul Gowthaman and Brian G. Pierce

15 In Silico Cell-Type Deconvolution Methods in Cancer Immunotherapy 213
Gregor Sturm, Francesca Finotello, and Markus List

16 Immunedeconv: An R Package for Unified Access to Computational Methods for Estimating Immune Cell Fractions from Bulk RNA-Sequencing Data 223
Gregor Sturm, Francesca Finotello, and Markus List

17 EPIC: A Tool to Estimate the Proportions of Different Cell Types from Bulk Gene Expression Data 233
Julien Racle and David Gfeller

18 Computational Deconvolution of Tumor-Infiltrating Immune Components with Bulk Tumor Gene Expression Data 249
Bo Li, Taiwen Li, Jun S. Liu, and X. Shirley Liu

19 Cell-Type Enrichment Analysis of Bulk Transcriptomes Using xCell 263
Dvir Aran

20 Cap Analysis of Gene Expression (CAGE): A Quantitative and Genome-Wide Assay of Transcription Start Sites 277
Masaki Suimye Morioka, Hideya Kawaji, Hiromi Nishiyori-Sueki, Mitsuyoshi Murata, Miki Kojima-Ishiyama, Piero Carninci, and Masayoshi Itoh

Index *303*

Contributors

RUEDI AEBERSOLD • *Department of Biology, Institute of Molecular Systems Biology, ETH Zurich, Zurich, Switzerland; Faculty of Science, University of Zurich, Zurich, Switzerland*

VITOR R. C. AGUIAR • *Department of Genetics and Evolutionary Biology, Institute of Biosciences, University of São Paulo, São Paulo, Brazil*

DVIR ARAN • *Bakar Computational Health Sciences Institute, University of California, San Francisco, San Francisco, CA, USA*

RICK BERRY • *PlatformSTL, St. Louis, MO, USA*

WOUT BITTREMIEUX • *Adrem Data Lab, Department of Mathematics and Computer Science, University of Antwerp, Antwerp, Belgium; Biomedical Informatics Research Network Antwerp (biomina), University of Antwerp, Antwerp, Belgium; Skaggs School of Pharmacy and Pharmaceutical Sciences, University of California, San Diego, CA, USA*

THOMAS BUKUR • *TRON—Translationale Onkologie an der Universitätsmedizin der Johannes Gutenberg-Universität Mainz gemeinnützige GmbH, Freiligrathstraße, Mainz, Germany*

ANAMARIA A. CAMARGO • *Molecular Oncology Center, Hospital Sírio Libanês, São Paulo, SP, Brazil*

CHEN CAO • *Department of Biochemistry and Molecular Biology, Cumming School of Medicine, University of Calgary, Calgary, AB, Canada*

PIERO CARNINCI • *Laboratory for Transcriptome Technology, RIKEN Center for Integrative Medical Sciences (IMS), Yokohama, Kanagawa, Japan*

ETIENNE CARON • *CHU Sainte-Justine Research Center, Montreal, QC, Canada; Department of Pathology and Cellular Biology, Faculty of Medicine, Université de Montré al, Montreal, QC, Canada*

MEI MEI CHANG • *Computational and Systems Biology 3, Genome Institute of Singapore, A*STAR (Agency for Science, Technology and Research), Singapore, Singapore*

NICOLAS DE NEUTER • *Adrem Data Lab, Department of Mathematics and Computer Science, University of Antwerp, Antwerp, Belgium; AUDACIS, Antwerp Unit for Data Analysis and Computation in Immunology and Sequencing, University of Antwerp, Antwerp, Belgium; Biomedical Informatics Research Network Antwerp (biomina), University of Antwerp, Antwerp, Belgium*

LI TAI FANG • *Bioinformatics Research and Early Development, Roche Sequencing Solutions, Belmont, CA, USA*

IOAN FILIP • *Department of Systems Biology, Columbia University Irving Cancer Research Center, New York, NY, USA; Program for Mathematical Genomics, Columbia University Irving Medical Center, New York, NY, USA*

FRANCESCA FINOTELLO • *Biocenter, Institute of Bioinformatics, Medical University of Innsbruck, Innsbruck, Austria*

DAVID GFELLER • *Department of Oncology UNIL CHUV, Ludwig Institute for Cancer Research, University of Lausanne, Lausanne, Switzerland; Swiss Institute of Bioinformatics (SIB), Lausanne, Switzerland*

SOFIE GIELIS • *Adrem Data Lab, Department of Mathematics and Computer Science, University of Antwerp, Antwerp, Belgium; AUDACIS, Antwerp Unit for Data Analysis and Computation in Immunology and Sequencing, University of Antwerp, Antwerp,*

Belgium; Biomedical Informatics Research Network Antwerp (biomina), University of Antwerp, Antwerp, Belgium

RAGUL GOWTHAMAN • *Institute for Bioscience and Biotechnology Research, University of Maryland, Rockville, MD, USA; Department of Cell Biology and Molecular Genetics, University of Maryland, College Park, MD, USA; Marlene and Stewart Greenebaum Comprehensive Cancer Center, University of Maryland, Baltimore, MD, USA*

SHASHI GUJAR • *Department of Microbiology and Immunology, Dalhousie University, Halifax, NS, Canada; Department of Pathology, Dalhousie University, Halifax, NS, Canada; Department of Biology, Dalhousie University, Halifax, NS, Canada; Beatrice Hunter Cancer Research Institute, Halifax, NS, Canada*

YU AMANDA GUO • *Computational and Systems Biology 3, Genome Institute of Singapore, A*STAR (Agency for Science, Technology and Research), Singapore, Singapore*

TING HE • *Division of Health Sciences Informatics, School of Medicine, Johns Hopkins University, Baltimore, MD, USA*

CHRISTOPH HOLTSTRÄTER • *TRON—Translationale Onkologie an der Universitätsmedizin der Johannes Gutenberg-Universität Mainz gemeinnützige GmbH, Freiligrathstraße, Mainz, Germany*

WEITAI HUANG • *Computational and Systems Biology 3, Genome Institute of Singapore, A*STAR (Agency for Science, Technology and Research), Singapore, Singapore; National University of Singapore Graduate School for Integrative Sciences and Engineering, National University of Singapore, Singapore, Singapore*

MASAYOSHI ITOH • *RIKEN Preventive Medicine and Diagnosis Innovation Program (PMI), Yokohama, Kanagawa, Japan*

HIDEYA KAWAJI • *Preventive Medicine and Applied Genomics Unit, RIKEN Center for Integrative Medical Sciences (IMS), Yokohama, Kanagawa, Japan; RIKEN Preventive Medicine and Diagnosis Innovation Program (PMI), Yokohama, Kanagawa, Japan; Tokyo Metropolitan Institute of Medical Science, Tokyo, Japan*

ALEXANDER KIRCHMAIR • *Biocenter, Institute of Bioinformatics, Medical University of Innsbruck, Innsbruck, Austria*

JULIA KODYSH • *Department of Genetics and Genomic Sciences, Icahn School of Medicine at Mount Sinai, New York, NY, USA*

MIKI KOJIMA-ISHIYAMA • *Laboratory for Transcriptome Technology, RIKEN Center for Integrative Medical Sciences (IMS), Yokohama, Kanagawa, Japan*

PRATHYUSHA KONDA • *Department of Microbiology and Immunology, Dalhousie University, Halifax, NS, Canada*

KRIS LAUKENS • *Adrem Data Lab, Department of Mathematics and Computer Science, University of Antwerp, Antwerp, Belgium; AUDACIS, Antwerp Unit for Data Analysis and Computation in Immunology and Sequencing, University of Antwerp, Antwerp, Belgium; Biomedical Informatics Research Network Antwerp (biomina), University of Antwerp, Antwerp, Belgium*

BO LI • *Lyda Hill Department of Bioinformatics, Department of Immunology, UT Southwestern Medical Center, Dallas, TX, USA*

MINGHAO LI • *Department of Biochemistry and Molecular Biology, Cumming School of Medicine, University of Calgary, Calgary, AB, Canada*

TAIWEN LI • *Lyda Hill Department of Bioinformatics, Department of Immunology, UT Southwestern Medical Center, Dallas, TX, USA*

MARKUS LIST • *Big Data in BioMedicine Group, Chair of Experimental Bioinformatics, TUM School of Life Sciences, Technical University of Munich, Freising, Germany*

CHANG LIU • *Department of Pathology and Immunology, Washington University School of Medicine, St. Louis, MO, USA*
JUN S. LIU • *Lyda Hill Department of Bioinformatics, Department of Immunology, UT Southwestern Medical Center, Dallas, TX, USA*
X. SHIRLEY LIU • *Lyda Hill Department of Bioinformatics, Department of Immunology, UT Southwestern Medical Center, Dallas, TX, USA*
QUAN LONG • *Department of Biochemistry and Molecular Biology, Cumming School of Medicine, University of Calgary, Calgary, AB, Canada; Departments of Medical Genetics and Mathematics & Statistics, Alberta Children's Hospital Research Institute, O'Brien Institute for Public Health, University of Calgary, Calgary, AB, Canada*
MARTIN LÖWER • *TRON—Translationale Onkologie an der Universitätsmedizin der Johannes Gutenberg-Universität Mainz gemeinnützige GmbH, Freiligrathstraße, Mainz, Germany*
CIBELE MASOTTI • *Molecular Oncology Center, Hospital Sírio Libanês, São Paulo, SP, Brazil*
DIOGO MEYER • *Department of Genetics and Evolutionary Biology, Institute of Biosciences, University of São Paulo, São Paulo, Brazil*
PIETER MEYSMAN • *Adrem Data Lab, Department of Mathematics and Computer Science, University of Antwerp, Antwerp, Belgium; AUDACIS, Antwerp Unit for Data Analysis and Computation in Immunology and Sequencing, University of Antwerp, Antwerp, Belgium; Biomedical Informatics Research Network Antwerp (biomina), University of Antwerp, Antwerp, Belgium*
MASAKI SUIMYE MORIOKA • *Preventive Medicine and Applied Genomics Unit, RIKEN Center for Integrative Medical Sciences (IMS), Yokohama, Kanagawa, Japan*
PIETER MORIS • *Adrem Data Lab, Department of Mathematics and Computer Science, University of Antwerp, Antwerp, Belgium; Biomedical Informatics Research Network Antwerp (biomina), University of Antwerp, Antwerp, Belgium*
MITSUYOSHI MURATA • *Laboratory for Transcriptome Technology, RIKEN Center for Integrative Medical Sciences (IMS), Yokohama, Kanagawa, Japan*
J. PATRICK MURPHY • *Department of Pathology, Dalhousie University, Halifax, NS, Canada*
HIROMI NISHIYORI-SUEKI • *Laboratory for Transcriptome Technology, RIKEN Center for Integrative Medical Sciences (IMS), Yokohama, Kanagawa, Japan*
TIMOTHY O'DONNELL • *Department of Genetics and Genomic Sciences, Icahn School of Medicine at Mount Sinai, New York, NY, USA*
BENSON OGUNJIMI • *AUDACIS, Antwerp Unit for Data Analysis and Computation in Immunology and Sequencing, University of Antwerp, Antwerp, Belgium; Antwerp Center for Translational Immunology and Virology (ACTIV), Vaccine and Infectious Disease Institute, University of Antwerp, Antwerp, Belgium; Department of Paediatrics, Antwerp University Hospital, Antwerp, Belgium; Center for Health Economics Research and Modeling Infectious Diseases (CHERMID), Vaccine and Infectious Disease Institute, University of Antwerp, Antwerp, Belgium*
ROSE ORENBUCH • *Department of Systems Biology, Columbia University Irving Cancer Research Center, New York, NY, USA*
PATRICK PEDRIOLI • *Department of Biology, Institute of Molecular Systems Biology, ETH Zurich, Zurich, Switzerland*
BRIAN G. PIERCE • *Institute for Bioscience and Biotechnology Research, University of Maryland, Rockville, MD, USA; Department of Cell Biology and Molecular Genetics, University of Maryland, College Park, MD, USA; Marlene and Stewart Greenebaum Comprehensive Cancer Center, University of Maryland, Baltimore, MD, USA*

RAUL RABADAN • *Department of Systems Biology, Columbia University Irving Cancer Research Center, New York, NY, USA; Program for Mathematical Genomics, Columbia University Irving Medical Center, New York, NY, USA*
JULIEN RACLE • *Department of Oncology UNIL CHUV, Ludwig Institute for Cancer Research, University of Lausanne, Lausanne, Switzerland; Swiss Institute of Bioinformatics (SIB), Lausanne, Switzerland*
ALEX RUBINSTEYN • *Department of Genetics and Genomic Sciences, Icahn School of Medicine at Mount Sinai, New York, NY, USA*
BARBARA SCHRÖRS • *TRON—Translationale Onkologie an der Universitätsmedizin der Johannes Gutenberg-Universität Mainz gemeinnützige GmbH, Freiligrathstraße, Mainz, Germany*
WENGUANG SHAO • *Department of Biology, Institute of Molecular Systems Biology, ETH Zurich, Zurich, Switzerland*
ANDERS JACOBSEN SKANDERUP • *Computational and Systems Biology 3, Genome Institute of Singapore, A*STAR (Agency for Science, Technology and Research), Singapore, Singapore*
GREGOR STURM • *Biocenter, Institute of Bioinformatics, Medical University of Innsbruck, Innsbruck, Austria*

Chapter 1

Bioinformatics for Cancer Immunotherapy

Christoph Holtsträter, Barbara Schrörs, Thomas Bukur, and Martin Löwer

Abstract

Our immune system plays a key role in health and disease as it is capable of responding to foreign antigens as well as acquired antigens from cancer cells. Latter are caused by somatic mutations, the so-called neoepitopes, and might be recognized by T cells if they are presented by HLA molecules on the surface of cancer cells. Personalized mutanome vaccines are a class of customized immunotherapies, which is dependent on the detection of individual cancer-specific tumor mutations and neoepitope (i.e., prediction, followed by a rational vaccine design, before on-demand production. The development of next generation sequencing (NGS) technologies and bioinformatic tools allows a large-scale analysis of each parameter involved in this process. Here, we provide an overview of the bioinformatic aspects involved in the design of personalized, neoantigen-based vaccines, including the detection of mutations and the subsequent prediction of potential epitopes, as well as methods for associated biomarker research, such as high-throughput sequencing of T-cell receptors (TCRs), followed by data analysis and the bioinformatics quantification of immune cell infiltration in cancer samples.

Key words Mutation, Cancer, Immunotherapy, Bioinformatics, T cell

1 Introduction

Cancer comprises a class of diseases caused by changes in the genome (i.e., mutations) of individual cells [1], allowing those cells to proliferate in an uncontrolled manner and eventually invade other parts of the body. These genomic changes include point mutations, small insertions and deletions, and larger structural variants, which are characterized by chromosomal rearrangements and which can result in fusion genes where the open reading frames of two genes are merged, giving rise to a novel gene product. All such mutations might engender new properties, enabling the cancerous behavior of the mutant cells.

T cells are a part of the adaptive immune system and respond to foreign antigens from, for example, infectious diseases as well as acquired antigens from cancer. During development, they undergo a process known as negative selection in the thymus. Negative selection prevents the recognition of self or autoantigens by nascent

Sebastian Boegel (ed.), *Bioinformatics for Cancer Immunotherapy: Methods and Protocols*, Methods in Molecular Biology, vol. 2120, https://doi.org/10.1007/978-1-0716-0327-7_1, © Springer Science+Business Media, LLC, part of Springer Nature 2020

T cells, while maintaining their ability to recognize nonself antigens, such as those from viruses or bacteria, or—in the case of cancer—so-called neoepitopes, which are caused by somatic mutations of the cancer but are essentially foreign to the body and its immune system.

The concept of personalized medicine extends stratified approaches to address individual patients' needs [2]. Personalized mutanome vaccines are a class of customized immunotherapies, the development of which is dependent on the detection of individual cancer-specific tumor mutations and neoepitope (i.e., mutated protein sequences accessible to the immune system) prediction, followed by a rational vaccine design, before on-demand production. Recent publications demonstrate the feasibility of this process [3–6].

The aforementioned neoepitopes (i.e., specific features caused by somatic mutations of cancer cells, which are foreign to the immune system and therefore can induce antitumor immune responses) are relevant not only to experimental treatments like personalized mutanome vaccines but also, more importantly, for immune checkpoint blockade therapies [7]. This class of treatments has demonstrated substantial promise for the treatment of several types of cancer, activating the immune system to combat cancer cells. Currently, the efficacy of these treatments has been demonstrated for more than ten types of cancers [8], with hundreds of additional clinical trials currently running [9]. Moreover, it has been shown that the number and features of mutations or neoepitopes of a given tumor is indicative for therapeutic response [10] of these highly relevant therapies. Factors involved in the immunogenicity of a neoepitope include the processing and transport of mutant proteins, the binding affinity to the major histocompatibility complexes (MHC) and the binding of a T-cell receptor (TCR) to the MHC-peptide complex.

The development of next generation sequencing (NGS) technologies allows large-scale, de novo detection of somatic mutations in extensive patient sample cohorts. Until recently, these studies focused primarily on the detection of recurrent events, that is, the detection of variants present in a significant portion of a patient population. This strategy allows a high count of patient cases to compensate for low detection sensitivities and specificities. Previous work has shown that short-read-based parallel NGS methods and the available analysis algorithms have a mediocre overall false discovery rate (50–70%) [11]. With regard to personalized medicine and mutanome vaccines, this lack of specificity becomes a severe challenge. In addition, the available software has a long running time, and the quality of the actual implementations varies drastically, which prohibits the use of some algorithms on high performance computing (HPC) clusters due to suboptimal software design. This interferes with the on-demand nature of personalized medicine and drug production, where a short turnaround time is crucial for patient healthcare. Therefore, common methods for improving the specificity, such as using multiple algorithms in parallel and using the consensus of the output, as applied to single-nucleotide variant (SNV) detection, are currently not feasible in a practical context due to long run times and are likely to reduce detection sensitivity [11].

As indicated, several steps are necessary to design personalized, neoantigen-based vaccines, including the detection of mutations and the subsequent prediction of potential epitopes. The accurate identification of neoantigens, both as treatment targets and as biomarkers, depends not only on mutation screening of NGS data but also on the prediction of the immunogenicity of the mutation-derived peptides [12]. Other procedures, which are rather useful for associated biomarker research, include the high-throughput sequencing of TCRs, followed by data analysis and the bioinformatics quantification of immune cell infiltration in cancer samples.

2 Mutation Detection

Mutation detection requires DNA extracted from tumor and healthy tissue (often blood) from the same patient, for comparison and correct classification of tumor-specific mutations. Precise mutation detection is often difficult with confounding factors such as tumor heterogeneity and contamination of healthy tissue by, for example, necrotic cells [13, 14].

In order to address this problem, bioinformatic approaches are based on the detection of mutant loci with statistically significance through short-read mappings provided by NGS, for example, by applying Poisson models to separate noise from real signals by use of the read-counts mapping to each locus. In addition to this, filters for known error sources or artifacts are employed using specialist knowledge. Examples of such filters are Mutect or Strelka [15, 16].

Besides approaches using statistics and filtering, new approaches using deep-learning methods are emerging. These currently encode the sequence read information in the form of images or tensors, rendering them usable by CNN (Convolutional Neural Networks). This eliminates the need for specialist knowledge of the underlying problems, which is always limited by how well these problems are actually understood. An example is DeepVariant, which trains and uses Neural Networks for SNP prediction to work around NGS read errors [17].

3 Epitope Prediction

Epitope predicton relies on the fact that mutations in tumor genomes can give rise to aberrant peptides, which in turn can facilitate tumor-specific epitopes. These neoepitopes are either presented by MHC class I molecules on antigen-presenting cells to $CD8^+$ T cells or by MHC class II molecules to $CD4^+$ T cells. In vitro immunogenicity testing involves evaluating whether blood samples from patients have an immune response to potential. However, many mutations do not facilitate spontaneous immune responses and are therefore difficult to assess for immunogenicity [4, 14]. In silico neoepitope-prediction is based on HLA (MHC) binding prediction and corresponding transcript expression.

The HLA type of a patient can be obtained computationally using NGS data; however, the HLA locus is highly polymorphic and shows high variability between individuals. Variant detection, which is necessary for accurate genotyping of said locus, is not easy. Generic short-read alignments of NGS-data to reference genomes are not able to account for the high variability. The use of sequence specialized alignment schemes and HLA gene database can help solve this issue. Examples of such alignment schemes are seq2HLA and HLA-HD among others [18, 19].

Binding prediction of potential epitopes to the patients HLA molecules can be done by a variety of algorithms, with NetMHC-pan being among the most popular [20].

Suitable neoepitopes are identified by comparing mutated and wild-type epitopes. Differences in MHC-binding scores or similarities in sequence between wild-type and mutated epitopes allow immunogenicity prediction [21–23]. The prediction of binding to MHC class II is particularly difficult as the binding core and position of the potential neoepitope is unknown [24].

Deep-learning approaches are also used in immunogenicity prediction, currently employing mass spectrometry data as opposed to binding assay data [25]. Convolutional and recurrent neural networks are the standard in these approaches, with peptide sequences and HLA alleles being used as input. Examples include DeepSeqPan and HLA-CNN [26, 27].

4 TCR Sequencing

Sequencing of TCR repertoires refers to the high-throughput sequencing and analysis of the CDR3 region from one of the TCR subunits [28], usually the TCR β chain. To successfully capture the CDR3 region, PCR primer pairs are often designed to match the V and J genes and to amplify either a sample's DNA or cDNA. TCR sequencing has an increasingly important role in clinical screening for biomarkers or response correlations [29–31], but analysis of the sequenced TCR repertoire is still not trivial. In total, it is estimated that the human body contains 10^{12} T cells. In theory, given the different combinatorial possibilities, 10^{15} different nonrandom TCRs are potentially possible. A young and healthy individual would have T cells comprising of 10^{7} different TCRs circulating in whole body.

Even with the use of NGS technology, the estimation of the TCR repertoire from a small blood sample remains a challenge. Hence, analysis of TCR repertoires is dependent on tools derived from ecological studies where population dynamics are studied from small samples. Thus, the general metrics to compare TCR repertoires are often richness, diversity, and repertoire dynamics [32].

Other tools have been developed to shed light on TCR sequence from a different angle. LYRA is a web-based tool to investigate the TCR structure from sequence [33]. However, even with the adoption of deep learning tools for the prediction of epitopes recognized by cytotoxic T cells [34], structural and functional predictions of TCRs still remain a challenge in immunoinformatics.

Technological developments led to further improvements of TCR sequencing: instead of only sequencing the TCR β chain of a bulk of T cells, new methods allow the sequencing of the TCR α and β chain pair in single cell resolution [35]. With the combination of TCR sequencing and transcriptomic profiling of the T cell, T-cell phenotypes and TCR sequences can be linked to investigate the function of specific T-cell subsets.

5 Immune Cell Quantification

Quantification of immune cells is an important process that may reveal any immune escape mechanisms employed by tumors. Immune cells infiltrate the tumor and are part of the tumor microenvironment, and knowledge about its composition may identify treatment options or aid in new drug discovery [36]. To assess cell populations in the tumor microenvironment, surface staining-based methods such as flow cytometry and IHC are the gold standard [37]. However, flow cytometry requires a large amount of tumor material, which is not always available, while IHC is usually applied on a single slice of tumor tissue, not accounting for heterogeneity within the tumor. Both methods also only apply a comparatively small sets of markers [38].

In silico methods (usually based on RNA-Seq data) can be divided into two different approaches, marker-gene based and deconvolution based. Marker-gene based approaches employ lists of genes characteristic for a cell type, derived from targeted transcriptomics. Cell types are independently quantified through expression values of the marker-genes. These expression values are then aggregated into abundance scores or statistically tested for enrichment of said genes [38]. Examples include MCP-Counter or xCell [39, 40].

Deconvolution based approaches employ a system of equations to calculate a weighted sum of the expression profiles of mixed cell types [41]. These are solved through linear least square regression, constrained least square regression, or support vector regression. Examples include TIMER, quanTIseq, and CIBERSORT [42–44]. We refer to Chapter 2 in this book about an in-depth review about immune cell quantification approaches.

6 Outlook

The past decade has benefited from various groups being remarkably active in building bioinformatics tools dedicated to supporting or enabling (individualized) cancer immunotherapy. Current treatments, however, focus on neoepitopes derived from SNVs or small insertions or deletions, which are appropriate for tumor entities with high mutational loads [5]. Other entities are characterized by far fewer mutations and often other mutation types [45]. For instance, structural variants or mis-splicing can result in new (fusion) transcripts and subsequently neoepitopes that are more foreign in terms of new amino acids compared with their wild-type counterpart than SNV-derived peptides. Although there are already diverse approaches to predict these mutations types (e.g., SvABA [46], Delly [47], MATS [48], or SpliceGrapher [49]) for implementation in the clinical setting, false discovery rates need to be lowered substantially. Moreover, it is more difficult to verify these alterations as truly tumor-specific as the genomic breakpoints of structural variants are often not covered by standard whole exome sequencing, and there is no representative germline RNA control to compare with alternative splicing events found in the tumor RNA. Thus, specificity is indirectly deduced, for example, from DNA mutations at splice acceptor or donor sites.

Once a broad set of putative neoantigens has been identified, the prediction of binding to the HLA molecules is the next step in refining the candidate list. Currently available tools, especially those designed for MHC class I-presented epitopes [12], address this challenge with confidence. A caveat here is that only a portion of the presented peptides will induce T-cell recognition or even anti-tumoral activity [4]. Modeling immunogenicity based on the peptide-MHC complex improves predictions compared with pure binding predictions (e.g., in references [50, 51]), but there is room for further improvement. Notably, the third variable part of the immunological synapse—the TCR—is currently rather neglected mostly due to the lack of sufficient data. More studies examining the triplet TCR-peptide-MHC complex are required to offer the underlying information to prediction algorithms that cover the whole process: processing and presentation along with recognition and immune activation.

References

1. Hanahan D, Weinberg RA (2000) The hallmarks of cancer. Cell 100(1):57–70
2. Britten CM, Singh-Jasuja H, Flamion B et al (2013) The regulatory landscape for actively personalized cancer immunotherapies. Nat Biotechnol 31(10):880–882. https://doi.org/10.1038/nbt.2708
3. Castle JC, Kreiter S, Diekmann J et al (2012) Exploiting the mutanome for tumor vaccination. Cancer Res 72(5):1081–1091. https://doi.org/10.1158/0008-5472.CAN-11-3722
4. Kreiter S, Vormehr M, van de Roemer N et al (2015) Mutant MHC class II epitopes drive therapeutic immune responses to cancer. Nature 520(7549):692–696. https://doi.org/10.1038/nature14426
5. Sahin U, Derhovanessian E, Miller M et al (2017) Personalized RNA mutanome vaccines mobilize poly-specific therapeutic immunity against cancer. Nature 547(7662):222–226. https://doi.org/10.1038/nature23003
6. Sahin U, Türeci Ö (2018) Personalized vaccines for cancer immunotherapy. Science 359 (6382):1355–1360. https://doi.org/10.1126/science.aar7112
7. Riaz N, Morris L, Havel JJ et al (2016) The role of neoantigens in response to immune checkpoint blockade. Int Immunol 28 (8):411–419. https://doi.org/10.1093/intimm/dxw019
8. Park Y-J, Kuen D-S, Chung Y (2018) Future prospects of immune checkpoint blockade in cancer: from response prediction to overcoming resistance. Exp Mol Med 50(8):109. https://doi.org/10.1038/s12276-018-0130-1
9. Darvin P, Toor SM, Sasidharan Nair V et al (2018) Immune checkpoint inhibitors: recent progress and potential biomarkers. Exp Mol Med 50(12):165. https://doi.org/10.1038/s12276-018-0191-1
10. McGranahan N, Furness AJS, Rosenthal R et al (2016) Clonal neoantigens elicit T cell immunoreactivity and sensitivity to immune checkpoint blockade. Science 351 (6280):1463–1469. https://doi.org/10.1126/science.aaf1490
11. Löwer M, Renard BY, de Graaf J et al (2012) Confidence-based somatic mutation evaluation and prioritization. PLoS Comput Biol 8(9): e1002714. https://doi.org/10.1371/journal.pcbi.1002714
12. Jurtz VI, Olsen LR (2019) Computational methods for identification of T cell neoepitopes in tumors. Methods Mol Biol 1878:157–172. https://doi.org/10.1007/978-1-4939-8868-6_9
13. Xu H, DiCarlo J, Satya RV et al (2014) Comparison of somatic mutation calling methods in amplicon and whole exome sequence data. BMC Genomics 15:244. https://doi.org/10.1186/1471-2164-15-244
14. Vormehr M, Schrörs B, Boegel S et al (2015) Mutanome engineered RNA immunotherapy: towards patient-centered tumor vaccination. J Immunol Res 2015:595363. https://doi.org/10.1155/2015/595363
15. Kim S, Scheffler K, Halpern AL et al (2018) Strelka2: fast and accurate calling of germline and somatic variants. Nat Methods 15 (8):591–594. https://doi.org/10.1038/s41592-018-0051-x
16. Cibulskis K, Lawrence MS, Carter SL et al (2013) Sensitive detection of somatic point mutations in impure and heterogeneous cancer samples. Nat Biotechnol 31(3):213–219. https://doi.org/10.1038/nbt.2514
17. Poplin R, Chang P-C, Alexander D et al (2018) A universal SNP and small-indel variant caller using deep neural networks. Nat Biotechnol 36 (10):983–987. https://doi.org/10.1038/nbt.4235
18. Kawaguchi S, Higasa K, Shimizu M et al (2017) HLA-HD: an accurate HLA typing algorithm for next-generation sequencing data. Hum Mutat 38(7):788–797. https://doi.org/10.1002/humu.23230
19. Boegel S, Löwer M, Schäfer M et al (2012) HLA typing from RNA-Seq sequence reads. Genome Med 4(12):102. https://doi.org/10.1186/gm403
20. Jurtz V, Paul S, Andreatta M et al (2017) NetMHCpan-4.0: improved peptide-MHC class I interaction predictions integrating

eluted ligand and peptide binding affinity data. J Immunol 199(9):3360–3368. https://doi.org/10.4049/jimmunol.1700893

21. Bjerregaard A-M, Nielsen M, Jurtz V et al (2017) An analysis of natural T cell responses to predicted tumor neoepitopes. Front Immunol 8:1566. https://doi.org/10.3389/fimmu.2017.01566
22. Ghorani E, Rosenthal R, McGranahan N et al (2018) Differential binding affinity of mutated peptides for MHC class I is a predictor of survival in advanced lung cancer and melanoma. Ann Oncol 29(1):271–279. https://doi.org/10.1093/annonc/mdx687
23. Duan F, Duitama J, Al Seesi S et al (2014) Genomic and bioinformatic profiling of mutational neoepitopes reveals new rules to predict anticancer immunogenicity. J Exp Med 211 (11):2231–2248. https://doi.org/10.1084/jem.20141308
24. Karosiene E, Rasmussen M, Blicher T et al (2013) NetMHCIIpan-3.0, a common pan-specific MHC class II prediction method including all three human MHC class II isotypes, HLA-DR, HLA-DP and HLA-DQ. Immunogenetics 65(10):711–724. https://doi.org/10.1007/s00251-013-0720-y
25. Abelin JG, Keskin DB, Sarkizova S et al (2017) Mass spectrometry profiling of HLA-associated peptidomes in mono-allelic cells enables more accurate epitope prediction. Immunity 46 (2):315–326. https://doi.org/10.1016/j.immuni.2017.02.007
26. Vang YS, Xie X (2017) HLA class I binding prediction via convolutional neural networks. Bioinformatics 33(17):2658–2665. https://doi.org/10.1093/bioinformatics/btx264
27. Liu Z, Cui Y, Xiong Z et al (2019) DeepSeqPan, a novel deep convolutional neural network model for pan-specific class I HLA-peptide binding affinity prediction. Sci Rep 9(1):794. https://doi.org/10.1038/s41598-018-37214-1
28. Woodsworth DJ, Castellarin M, Holt RA (2013) Sequence analysis of T-cell repertoires in health and disease. Genome Med 5(10):98. https://doi.org/10.1186/gm502
29. Wieland A, Kamphorst AO, Adsay NV et al (2018) T cell receptor sequencing of activated CD8 T cells in the blood identifies tumor-infiltrating clones that expand after PD-1 therapy and radiation in a melanoma patient. Cancer Immunol Immunother 67 (11):1767–1776. https://doi.org/10.1007/s00262-018-2228-7
30. Lin K-R, Pang D-M, Jin Y-B et al (2018) Circulating CD8+ T-cell repertoires reveal the biological characteristics of tumors and clinical responses to chemotherapy in breast cancer patients. Cancer Immunol Immunother 67 (11):1743–1752. https://doi.org/10.1007/s00262-018-2213-1
31. Jin Y-B, Luo W, Zhang G-Y et al (2018) TCR repertoire profiling of tumors, adjacent normal tissues, and peripheral blood predicts survival in nasopharyngeal carcinoma. Cancer Immunol Immunother 67(11):1719–1730. https://doi.org/10.1007/s00262-018-2237-6
32. Rosati E, Dowds CM, Liaskou E et al (2017) Overview of methodologies for T-cell receptor repertoire analysis. BMC Biotechnol 17(1):61. https://doi.org/10.1186/s12896-017-0379-9
33. Klausen MS, Anderson MV, Jespersen MC et al (2015) LYRA, a webserver for lymphocyte receptor structural modeling. Nucleic Acids Res 43(W1):W349–W355. https://doi.org/10.1093/nar/gkv535
34. Jurtz VI, Jessen LE, Bentzen AK et al (2018) NetTCR: sequence-based prediction of TCR binding to peptide-MHC complexes using convolutional neural networks. Preprint available on bioRxiv. https://doi.org/10.1101/433706
35. Han A, Glanville J, Hansmann L et al (2014) Linking T-cell receptor sequence to functional phenotype at the single-cell level. Nat Biotechnol 32(7):684–692. https://doi.org/10.1038/nbt.2938
36. Friedman AA, Letai A, Fisher DE et al (2015) Precision medicine for cancer with next-generation functional diagnostics. Nat Rev Cancer 15(12):747–756. https://doi.org/10.1038/nrc4015
37. Petitprez F, Sun C-M, Lacroix L et al (2018) Quantitative analyses of the tumor microenvironment composition and orientation in the era of precision medicine. Front Oncol 8:390. https://doi.org/10.3389/fonc.2018.00390
38. Sturm G, Finotello F, Petitprez F et al (2019) Comprehensive evaluation of computational cell-type quantification methods for immuno-oncology. Bioinformatics 35(14):436–445. https://doi.org/10.1093/bioinformatics/btz363
39. Aran D, Hu Z, Butte AJ (2017) xCell: digitally portraying the tissue cellular heterogeneity landscape. Genome Biol 18(1):220. https://doi.org/10.1186/s13059-017-1349-1
40. Becht E, Giraldo NA, Lacroix L et al (2016) Estimating the population abundance of tissue-infiltrating immune and stromal cell populations using gene expression. Genome Biol 17(1):218. https://doi.org/10.1186/s13059-016-1070-5

41. Finotello F, Trajanoski Z (2018) Quantifying tumor-infiltrating immune cells from transcriptomics data. Cancer Immunol Immunother 67(7):1031–1040. https://doi.org/10.1007/s00262-018-2150-z
42. Newman AM, Liu CL, Green MR et al (2015) Robust enumeration of cell subsets from tissue expression profiles. Nat Methods 12(5):453–457. https://doi.org/10.1038/nmeth.3337
43. Finotello F, Mayer C, Plattner C et al (2019) Molecular and pharmacological modulators of the tumor immune contexture revealed by deconvolution of RNA-seq data. Genome Med 11(1):34. https://doi.org/10.1186/s13073-019-0638-6
44. Li B, Severson E, Pignon J-C et al (2016) Comprehensive analyses of tumor immunity: Implications for cancer immunotherapy. Genome Biol 17(1):174. https://doi.org/10.1186/s13059-016-1028-7
45. Forbes SA, Beare D, Gunasekaran P et al (2015) COSMIC: exploring the world's knowledge of somatic mutations in human cancer. Nucleic Acids Res 43(Database issue):D805–D811. https://doi.org/10.1093/nar/gku1075
46. Wala JA, Bandopadhayay P, Greenwald NF et al (2018) SvABA: genome-wide detection of structural variants and indels by local assembly. Genome Res 28(4):581–591. https://doi.org/10.1101/gr.221028.117
47. Rausch T, Zichner T, Schlattl A et al (2012) DELLY: structural variant discovery by integrated paired-end and split-read analysis. Bioinformatics 28(18):i333–i339. https://doi.org/10.1093/bioinformatics/bts378
48. Shen S, Park JW, Huang J et al (2012) MATS: a Bayesian framework for flexible detection of differential alternative splicing from RNA-Seq data. Nucleic Acids Res 40(8):e61. https://doi.org/10.1093/nar/gkr1291
49. Rogers MF, Thomas J, Reddy AS et al (2012) SpliceGrapher: detecting patterns of alternative splicing from RNA-Seq data in the context of gene models and EST data. Genome Biol 13(1):R4. https://doi.org/10.1186/gb-2012-13-1-r4
50. Bjerregaard A-M, Nielsen M, Hadrup SR et al (2017) MuPeXI: prediction of neo-epitopes from tumor sequencing data. Cancer Immunol Immunother 66(9):1123–1130. https://doi.org/10.1007/s00262-017-2001-3
51. Kim S, Kim HS, Kim E et al (2018) Neopepsee: accurate genome-level prediction of neoantigens by harnessing sequence and amino acid immunogenicity information. Ann Oncol 29(4):1030–1036. https://doi.org/10.1093/annonc/mdy022

Chapter 2

An Individualized Approach for Somatic Variant Discovery

Minghao Li, Ting He, Chen Cao, and Quan Long

Abstract

Somatic variant callers identify mutations found within cancer genome sequencing data through mapping sequencing reads to a universal reference genome and inferring likelihoods from statistical models. False positives, however, are common among various tools as mismatches with the universal reference can also occur due to germline variants. Previous applications of personalized reference construction are not amenable with cancer genome analysis. Here, we describe an individualized approach for somatic variant discovery through the step-by-step usage of Personalized Reference Editor for Somatic Mutation discovery in cancer genomics (PRESM), a personalized reference editor for somatic mutation discovery in cancer genomes.

Key words Bioinformatics, Next-generation sequencing, Somatic mutations, Somatic variants, Cancer genomics, Personalized reference, PRESM

1 Introduction

Germline mutations are heritable alterations to the genetic sequence initially acquired in germ cells and subsequently carried over to all cells of the child organism and all ensuing descendants [1]. Somatic mutations, while not heritable, can arise through cell division and have widespread clinical implications in many different forms of cancer [2]. A key issue cancer researchers currently face is the challenge of confidently calling somatic mutations when analyzing cancer genomes. The difficulty lies mainly in two prominent factors: (1) the sparsity of somatic mutations, which has a broad range of 0.1–100 mutations per megabase depending on the type of cancer and (2) the inconsistent allele frequency of somatic mutations as a result of tumor heterogeneity [3–5]. The second point in particular is in contrast to germline mutations, which have a fixed frequency of 0.5 and 1.0 for heterozygous and homozygous variants, respectively [5]. As a result, despite the variety and abundance of somatic mutation callers published since 2010 [3, 6–18], high false positive (FP) rates continue to plague users [19].

Sebastian Boegel (ed.), *Bioinformatics for Cancer Immunotherapy: Methods and Protocols*, Methods in Molecular Biology, vol. 2120, https://doi.org/10.1007/978-1-0716-0327-7_2, © Springer Science+Business Media, LLC, part of Springer Nature 2020

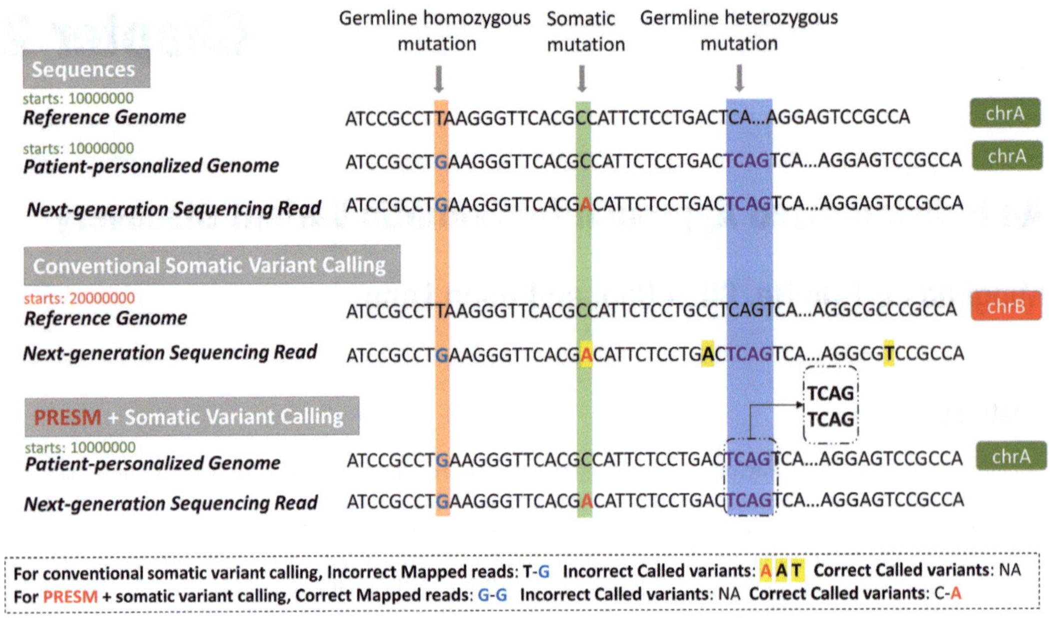

Fig. 1 Mismapping can be avoided by integrating germline mutations. The reference chromosome A (chrA), corresponding region in the patient's chrA, and a sequencing read from chrA containing a single somatic mutation. With conventional somatic variant calling, this read was mapped to reference chromosome B, resulting in four erroneous somatic variant calls. With the application of PRESM to interpolate the patient's reference genome, a somatic variant was called in the correct location with the correct mapping (Figure adapted from [26])

Established tools have an intuitive disadvantage as a reliance is placed on the quality of read mapping to the universal reference genome (e.g., hg19). The drawback of this approach comes from the abundance of germline mutations within the population. On average, an individual has 4.1–5.0 million germline variants distributed nonuniformly throughout the genome [20]. Based on this number, for read lengths of 100 bp, there is a 13–17% chance that a sequencing read carrying a somatic mutation will also carry a germline variant. Overall, this results in many mismatches with the universal reference which would then be incorrectly called as somatic variants, not to mention the potential for reads to be erroneously mapped to incorrect locations on the genome (Fig. 1).

Previous applications of customized reference genomes exist for RNA sequencing (RNA-seq), chromatin immunoprecipitation sequencing (ChIP-seq), and generalized variant calling [21–25], all with pronounced improvements in performance. In order to leverage personalized reference genomes for somatic mutation calling in cancer genomes, we present **P**ersonalized **R**eference **E**ditor for **S**omatic **M**utation discovery in cancer genomics (PRESM). PRESM is a user-friendly and parameter-free program aimed at using a dual reference genome approach (Fig. 2) to address the issue of both heterozygous and homozygous germline variants in

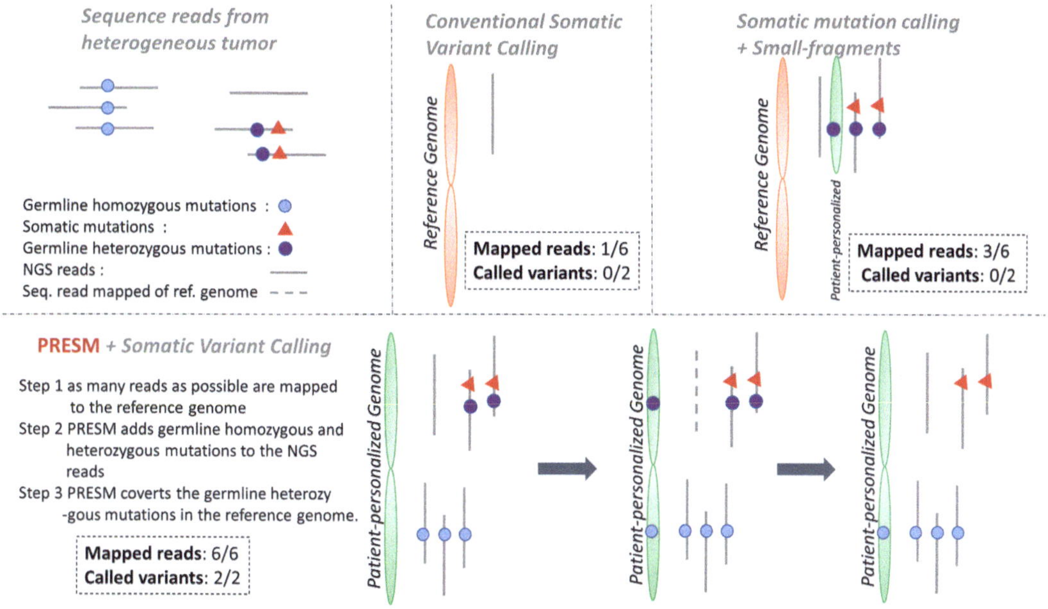

Fig. 2 Improvements in somatic variant calling with PRESM. Cells from heterogeneous tumor produce sequencing reads that may carry somatic mutations. Conventional somatic variant calling maps reads to standard reference genomes. Reads carrying more than a threshold of sequence differences will not be mapped. Naïve designs add alternative alleles and their flanking regions as small fragments but invalidate the variant likelihood score. With the PRESM strategy of dual personalized genomes, reads containing germline homozygous and heterozygous reads are more likely to be mapped to the correct locations in the personalized genome. Correct mapping is conducive to accurate calling of somatic variants (Figure adapted from [26])

cancer somatic mutation calling [26]. PRESM has been shown to greatly reduce the rate of FP calls in multiple somatic mutation callers across various datasets. The parameter-free aspect of the software also sufficiently addresses the gap in performance between authors and users observed frequently for bioinformatics tools.

In this chapter, we will describe the installation and usage of PRESM alongside its required dependencies in a detailed and beginner-friendly fashion. We will demonstrate its usage with data from the Dialogue for Reverse Engineering Assessments and Methods (DREAM) Genomic Mutation Calling Challenge, provided by the International Cancer Genome Consortium (ICGC) and The Cancer Genome Atlas (TCGA). The Broad Institute's (BI's) Mutect2 will be employed as the somatic mutation caller for demonstration purposes in the described workflow, although PRESM can be used with any somatic mutation caller and long-term support is offered for the software [26]. For readers wishing to use PRESM but consider themselves lacking in technical expertise, this chapter will serve as an excellent tutorial to get started.

2 Materials

PRESM takes the form of a Java JAR executable with a command-line user interface (CUI). Due to the platform independent nature of Java, PRESM is compatible with Windows, macOS, and Linux machines. The "batteries included" nature of JAR executables means that the tool is ready for immediate use, given that Java has been previously installed. It is recommended to use PRESM with Java 8 or newer releases. The Genome Analysis Toolkit (GATK) version 4, however, requires Java 8 specifically and does not offer support for Windows and newer Java releases. For researchers inexperienced with utilizing command-line tools, this subheading will provide some instructions on setting up the environment and getting PRESM up and running on Linux and macOS environments (*see* **Note 1**). We will use the percent sign ("%") to denote the prompt for shell commands. The backslash character ("\") will be used to allow for line breaks for the purpose of code formatting within this document. For actual usage, the backslash and line breaks are unnecessary, and the commands are meant to be inputted on a single line.

2.1 The Java Development Kit

If Java 8 has already been installed, readers should feel free to skip ahead to the following subheading. At the time of writing, Java 8 Update 201 is the latest available version and can be downloaded from the Oracle website (https://www.oracle.com/technetwork/java/javase/downloads/index.html). For readers referring to this document with future Java 8 releases, the only difference would be in the file names and version numbers. The following will detail some suggested best practices for Java installation.

The downloaded JDK installation file will be in either the GNU Zipped Tarball format (tar.gz; Linux) or the Apple Disk Image format (dmg; macOS). The first step is to extract the compressed files from archive.

1. Create a Java directory at a desired location to hold the Java installation:

   ```
   % mkdir -p /path/to/java_folder
   ```

   The "p" flag creates parent directories if they do not already exist.

2. Move the downloaded file to the newly created Java directory:

   (a) [Linux]

   ```
   % mv /path/to/download/jdk-8u201-linux-x64.tar.gz \
     /path/to/java_folder
   ```

(b) [macOS]

```
% mv /path/to/download/jdk-8u201-macosx-x64.dmg \
 /path/to/java_folder.
```

3. Verify that the checksum of the downloaded file matches the checksum provided by the download source. This step ensures the downloaded file is not corrupted; outdated; or worse, hijacked, and replaced by a malicious actor. In this instance, the Oracle website provides both the SHA256 and MD5 hash. As an example, we will demonstrate how to verify the MD5 checksum:

   (a) [Linux]

```
% echo "<MD5 checksum from website> \
 jdk-8u201-linux-x64.tar.gz" | \
 md5sum -c \
 /path/to/java_folder/jdk-8u201-linux-x64.tar.gz
```

   (b) [macOS]

```
% echo "<MD5 checksum from website> \
 jdk-8u201-macosx-x64.dmg" | \
 md5sum \
 -c /path/to/java_folder/jdk-8u201-macosx-x64.dmg.
```

4. Decompress the archive file:

   (a) [Linux]

```
% tar -xvzf \
 /path/to/java_folder/jdk-8u201-linux-x64.tar.gz
```

   The "xvzf" flags sequentially indicate extraction, verbosely list archive content, "gzip" utilization (for GNU Zip archives), and specifying the name of the file to be decompressed.

   (b) [macOS]

   Double click the file and follow the instructions on screen, setting the installation folder as the previously created folder.

   The next step is to make the Java program available from the command-line for the current user. Adding the Java binaries directory to the PATH environmental variable serves to accomplish this task. Assuming the default bash shell is being used in the terminal console command-line, we will do this within the ".bash_profile" file.

5. Using the text editor "vi," open (or create if it does not exist) a new text file named ".bash_profile" in the home directory:

```
% vi ~/.bash_profile
```

6. Press Shift + G to skip to the end of the document, then press the Insert key to enter "insert mode" in vi and then press the Enter key to insert a new line.
7. Type the following line at the end of the file:

```
export PATH="/path/to/java_folder/jdk1.8.0_201/bin:$PATH"
```

8. Press the Esc key to return to "normal" mode and then type ": wq" to save and exit the file. The colon character ":" precedes every vi command and "wq" refers to write and quit, respectively.
9. Read and execute ".bash_profile" to update the PATH environmental variable to include the Java binaries directory:

```
% source ~/.bash_profile
```

The final step is to verify the installation location and installed Java version to ensure that the user is not accidentally running a different installation of Java.

10. Verify that the correct Java binary is being located:

```
% if [[ $(which java) == \
    "/path/to/java_folder/jdk1.8.0_201/bin/java" ]]; \
    then echo "True"; \
    else echo "False"; \
    fi
```

11. Verify that the Java binary is the expected version:

```
% if [[ $(java -version 2>&1 | \
    grep "version" | \
    awk -F '\"' '{print $2}') \
    == "1.8.0_201" ]]; \
    then echo "True"; \
    else echo "False"; \
    fi
```

We use the "2 > &1" operator to pipe the standard error into the standard output as "java-version" returns information through the standard error.

The grep command searches for matching patterns, and the "F" flag for the awk command denotes the string delimiter character. As awk works with a 1-based index and the version number is wrapped with quotation marks in the "java-version" output, we have chosen the quotation mark character as the delimiter and selected the index 2 element (the element immediately after the first delimiter).

2.2 Downloading PRESM

The PRESM JAR executable is hosted on GitHub (https://github.com/precisionomics/PRESM). The repository can be cloned into a parent directory of choice.

1. Create a parent directory for PRESM at a desired location to hold the PRESM directory and change the current directory to the said parent directory:

```
% mkdir -p /path/to/PRESM/parent/dir && \
    cd /path/to/PRESM/parent/dir
```

2. Using "git," clone the PRESM repository from GitHub:

```
% git clone git@github.com:precisionomics/PRESM.git
```

   This will create a directory with the same name as the cloned repository within the parent directory, containing the repository files.

Henceforth in this document, for the sake brevity, the PRESM JAR file will be referred to as "**presm.jar**". The full file path will still need to be inputted whenever the file is utilized.

2.3 Anaconda

Anaconda is an open source Python distribution which, in conjunction with the conda package management system, hosts thousands of packages within the data science stack. Anaconda also hosts its own R ecosystem. Bioconda is a channel within the conda containing more than 6000 bioinformatics packages available for installation [27]. The following will briefly outline how to install Anaconda to your system. We recommend the full Anaconda distribution over Miniconda for ease of usage in the future as the default distribution includes many commonly used packages for data analysis.

1. Download the latest Anaconda installer (March 2019 version as of writing) appropriate for your system from the Anaconda website (https://www.anaconda.com/distribution/). Linux systems only have access to a CUI installer, while macOS users will also have access to a graphical installer. We recommend the Python 3 version.
2. Execute the file and follow the instructions on screen. Do not allow the installer to edit the ".bashrc" file (explaind below):
   (a) [Linux]

```
% sh Anaconda3-2019.03-Linux-x86_64.sh
```

   (b) [macOS]

```
% sh Anaconda3-2019.03-MacOSX-x86_64.sh.
```

3. Using vi, edit the ".bash_profile" and enable conda. Recall to the step doing the same for the Java binary folder, we must put the new line above the Java line:

```
% vi ~/.bash_profile
. "/path/to/anaconda3/etc/profile.d/conda.sh"
conda activate
<Java line here>
```

   The reason we do so is because the default Anaconda package list includes an older version of OpenJDK. Adding the Anaconda binary directory to the PATH variable after the Java binary directory would supersede our preferred Java executable with the Anaconda Java binary. Alternatively, to alleviate the issue altogether, uninstall the included OpenJDK package:

```
% conda remove openjdk
```

   ".bashrc" is also executed every time a new terminal console is opened while ".bash_profile" is only executed at login, therefore any modifications of the PATH variable within ". bashrc" would supersede ".bash_profile."

4. Add the bioconda and conda-forge repository channels to the conda package manager:

```
% conda config --add channels bioconda
% conda config --add channels conda-forge
```

2.4 The Genome Analysis Toolkit

GATK is a suite of tools provided by the BI designed for analyzing next-generation sequencing data with a focus on variant discovery [28]. As of version 4, GATK has also integrated Picard tools into its toolkit (https://broadinstitute.github.io/picard/). In addition to the software itself, BI also provides a set of "Best Practices" workflows for working with various GATK tools toward an assortment of genomic analyses [29, 30]. GATK also bundles Mutect2, the successor somatic mutation calling tool to MuTect [3]. The following subheading will detail the installation procedure for the aforementioned software [26, 31].

Download the latest version of GATK from the BI website (https://software.broadinstitute.org/gatk/download/) and unpack the archive. At the time of writing, GATK 4.1.1.0 is the latest version.

1. Create a GATK directory and move the downloaded Zip archive to the directory:

```
% mkdir -p /path/to/gatk_folder && \
mv /path/to/download/gatk-4.1.0.0.zip \
/path/to/gatk_folder
```

2. Using "unzip", decompress the archive:

```
% unzip /path/to/gatk_folder/gatk-4.1.0.0.zip
```

3. Create a conda environment named "gatk" using the included conda environment file:

```
% conda env create -n gatk -f \
    /path/to/gatk_folder/gatk-4.1.0.0/gatkcondaenv.yml
```

4. Activate the new conda environment:

```
% conda activate gatk
```

   Keep in mind that the "gatk" environment needs to be activated every time a new terminal prompt is opened as the default environment is still the base environment. To deactivate and return to the base, simply type:

```
% conda deactivate
```

   Alternatively, to set the "gatk" environment as the default, go to the ".bash_profile" file and change "conda activate" line to "conda activate gatk."

5. Some GATK tools have additional R dependencies, which are the packages "gsalib," "ggplot2," "reshape" (not reshape2), and "gplots." "ggplot2" is included within the "r-essentials package." Therefore, we need to install the R language and essential R packages from the R channel as well as the other three separately using conda within the "gatk" environment:

```
% conda install -c r r-base r-essentials
% conda install r-gplots r-gsalib r-reshape
```

6. Add the GATK directory containing the GATK binary to the PATH variable in ".bash_profile." This should be after the Anaconda binary PATH line and can be above or below the Java binary PATH line:

```
    % vi ~/.bash_profile
    <Anaconda line>
export PATH="/path/to/gatk_folder/gatk-4.1.0.0:$PATH"
```

2.5 Pindel, Burrows-Wheeler Aligner, and Samtools

Pindel is a program which uses a pattern growth algorithm to identify structural variants from short paired-end reads [32]. The PRESM workflow incorporates Pindel as a means to supplement GATK for capturing larger germline structural variants (*see* **Note 3**). The Burrows-Wheeler Aligner (BWA) is a sequencing read alignment tool based on the backward search and Burrows-Wheeler transform algorithms [33, 34]. BWA-MEM is an extension for the tool initially seeding alignments with maximal exact matches [35]. The tool delivers better heuristics and accuracy as standard short-read lengths expanded from 36 bp reads to ~100 bp reads [35]. Lastly, Samtools is a suite of programs aimed at working with SAM and BAM files.

All three programs are available through Bioconda, and conda will automatically install any required dependencies.

1. Ensure you are presently in the "gatk" environment:

```
% conda activate gatk
```

2. Using "conda," install Pindel, BWA, and Samtools:

```
% conda install pindel bwa samtools
```

2.6 The Genome Aggregation Database

The Genome Aggregation Database (gnomAD) is a collaborative research effort to aggregate sequencing data [36]. It is the successor to the Exome Aggregation Consortium (ExAC) and now includes genome-wide data in addition to exome-only data. For the purpose of somatic mutation calling, particularly with GATK4 Mutect2, gnomAD serves as the population germline variant resource and also provides information on allele frequencies. The most up to date version is gnomAD version 2.1.1, released in March 2019. gnomAD is currently built against Genome Reference Consortium (GRC) Human Build 37 (GRCh37; hg19); therefore, we are using GRCh37 as the universal reference for this pipeline. A GRCh38 version of gnomAD is currently under construction, and future readers of this document may opt to use that version and the corresponding universal reference instead. The following will detail how to download the gnomAD genome file, though the reader should keep in mind that the uncompressed file takes up a large amount of storage space (~3.5 TB). The downloads page is hosted on the gnomAD website (https://gnomad.broadinstitute.org/downloads).

1. Create a directory to hold the files:

```
% mkdir -p /path/to/gnomAD_folder
```

2. Download the all chromosomes Block GNU Zipped Variant Call Format (vcf.bgz) file for full genome samples. Go to the gnomAD downloads page, right click the download link, and click "Copy Link Location."
3. Using "nohup" and "wget," download the file without causing a hang-up in the terminal as the downloads proceed:

```
% nohup wget https://link/to/genome/file
    -P /path/to/gnomAD_folder &> \
    genome_download.log &
```

The "P" flag dictates the parent directory the downloaded file will be placed in.

"&>" pipes the nohup command output to the "genome_download.log" file, and the "&" indicates for the whole command to run in the background. The download progress can be periodically checked by peeking at the end of the log file:

```
% tail genome_download.log
```

If the download is interrupted at any point, the wget command above can be repeated with the "c" flag to continue.

4. Rename the ".bgz" portion of the file to ".gz" as to ameliorate its usage with downstream tools:

```
% mv gnomad.genomes.r2.1.1.sites.vcf.bgz \
    gnomad.genomes.r2.1.1.sites.vcf.gz
```

5. Using "gunzip," decompress the archive:

```
% nohup gunzip gnomad.genomes.r2.1.1.sites.vcf.gz &> \
    gnomad_gunzip.log &
```

6. (Recommended) Remove extraneous metadata. We have included this step because gnomAD includes a lot of information, such as the allele frequency of given ancestries, which are not required for this particular workflow. Removing the extra information greatly reduces the file size (3.5 TB to 15GB, multiple magnitudes of reduction!) and makes the file overall easier to work with:

```
% nohup awk 'OFS = "\t" {split($8, info, ";"); if ($1
    !~ /^#/) {if (info[3] !~ /^AF/) {info[3] = "AF=0";};
    $8 = info[1]";"info[3]; print $0;} else {print $0;}}'
    gnomad.genomes.r2.1.1.sites.vcf | \
    nohup sed '8d;10,574d' > \
    gnomad.genomes.r2.1.1.sites.less-metadata.vcf &
```

Using the semicolon character (";") as the delimiter, we are able to retain the allele frequency information from the "INFO" column.

The sed command-line deletes the 8th and 10th to 574th lines within the file. For the all chromosomes VCF file released under gnomAD version 2.1.1, these line numbers corresponds to the removed metadata as per the awk command. For future versions, please keep in mind the specific line numbers may be different.

7. Using GATK IndexFeatureFile, index the gnomAD file:

```
% gatk IndexFeatureFile \
    -F gnomad.genomes.r2.1.1.sites.less-metadata.vcf
```

Henceforth in this document, the modified gnomAD v2.1.1 VCF file will be referred to as "gnomad.vcf." Again, the full file path will still need to be inputted whenever the file is utilized.

2.7 Human Build 37

The GRC is an international effort aimed at improving reference genomes for various organisms. The final version of GRCh37/hg19 human reference is GRCh37 patch 13, released in June 2013. BI includes the University of California, Santa Cruz (UCSC) version of the hg19 FASTA file as part of the resource bundle hosted on their FTP server (ftp://gsapubftp-anonymous@ftp.broadinstitute.org/bundle/hg19) and can be accessed by leaving the password field blank. The universal reference can be downloaded with the following steps:

1. Create a directory to hold the files:

```
% mkdir -p /path/to/GRCh37_folder
```

2. Download the hg19 GNU Zipped FASTA and the corresponding MD5 file (fasta.gz; fasta.gz.md5) from the FTP server. Go to the FTP server page, right click the download link, and click "Copy Link Location." Repeat for both files.

3. Using "nohup" and "wget," download all files and verify the MD5 checksum:

```
% nohup wget ftp://link/to/file
    -P /path/to/GRCh37_folder &> \
    hg19_download.log &\
% md5sum -c /path/to/ucsc.hg19.fasta.gz.md5
```

   The downloaded MD5 files contain the full file path with regard to the FTP server. The reader will need to use a text editor to change the path to the file on the local disk.

4. Using "gunzip," decompress the archive:

```
% nohup gunzip file.gz &> \
    hg19_gunzip.log &
```

5. Using "sed," change the chromosome identifiers to match the gnomAD VCF file:

```
% sed -i -e 's/>chr/>/g' ucsc.hg19.fasta
```

   The "i" flag for the sed command refers to editing the file in--place.

   The "e" flag precedes a script command. The script 's/>chr/>/g' signifies substituting instances of ">chr" with ">"globally in the file. We are performing this step because the UCSC hg19 uses "chr#" for their chromosome identifiers, whereas gnomAD uses "#." We have elected to modify the reference file as opposed to the gnomAD file as it is far more computationally efficient to modify the header lines (e.g., 1–22, X, Y, M).

6. (Optional) Remove alternate loci. The BI provided UCSC hg19 file contains alternate loci (e.g., chr6_mann_hap4) which may complicate mapping. If this fits within the reader's use-case, this step can be ignored. Otherwise, it can be performed as follows:

```
% sed -n '>*_/q;p' ucsc.hg19.fasta > \
    ucsc.hg19.no_alt.fasta
```

   The "n" flag prevents implicit printing. The command signifies quit after recognizing the entered pattern, otherwise print explicitly. The printed output is then piped to our final file name.

7. Using GATK CreateSequenceDictionary, Samtools faidx, and BWA index, create the dictionary and index files for the reference fasta. We need to do this manually instead of downloading the files from the Resource Bundle because we have previously modified the FASTA file:

```
      % gatk CreateSequenceDictionary \
-R ucsc.hg19.no_alt.fasta \
-O ucsc.hg19.no_alt.dict
      % samtools faidx ucsc.hg19.no_alt.fasta
      % bwa index ucsc.hg19.no_alt.fasta
```

Henceforth in this document, the modified UCSC hg19 FASTA file will be referred to as "grch37.fa." Same concept applies with the full file path.

2.8 ICGC–TCGA DREAM Genomic Mutation Calling Challenge

The ICGC–TCGA DREAM Genomic Mutation Calling Challenge was comprised of seven separate minichallenge stages hosted from December 2013 to August 2016 (6 synthetic data, 1 real-world data). This set of data is considered the gold standard for assessing performance in variant callers and have 101 bp sequencing reads. The synthetic data were partially generated using BAMSurgeon and have coverage ranging from 30x to 40x [19]. The original PRESM publication utilized data from synthetic challenges 1, 2, and 3, which is what we will incorporate for usage.

Details on how to access the data can be found on the challenge website (https://www.synapse.org/#!Synapse:syn312572/wiki/402585).

3 Methods

The overall somatic variant calling workflow incorporating PRESM can be separated into four subheadings: data preprocessing, germline variant calling, personalized reference construction, and somatic variant calling. While PRESM itself is parameter free, other tools within the workflow are not (refer to **Notes 4–6** for a few more thoughts on this matter). The following will serve as a tutorial, detailing the tools and parameters utilized in generating the results as presented in the PRESM publication as well as showcasing the usage of PRESM with Mutect2, which departed beta status in GATK 4.0.2.0. GATK/Picard tools, in particular, have additional optional parameters (refer to **Note 2** if memory issues arise with the Java programs PRESM and GATK/Picard). It may be advisable for users to consult the online documentation for information on the additional parameters and see if they are suitable for a particular use-case.

3.1 Data Preprocessing

The DREAM Challenge synthetic datasets 1, 2, and 3 consist of single-end reads aligned using BWA-Backtrack [33], which has been succeeded by BWA-MEM [35] upon the release of the latter given performance improvements. As the datasets were provided in the form of BAM files, we will be extracting the sequencing reads and then following along the GATK Best Practices document (https://software.broadinstitute.org/gatk/best-practices/workflow?id=11165) to producing analysis ready BAM files for each of the datasets separately. Initially, the following steps should be repeated for all patient-matched normal sample BAM files individually. We will return later on to perform the same steps with the tumor sample BAM files twice, once each with our two constructed personalized reference FASTAs.

1. Using Picard SamToFastq, extract the sequencing reads into two FASTQ files:

```
% gatk SamToFastq \
    -I input.bam \
    -F /path/to/fq_folder/end_1.fq \
    -F2 /path/to/fq_folder/end_2.fq
```

2. Using BWA-MEM, map the reads to the universal reference:

```
% bwa mem -M -R <read_group_line> \
    grch37.fa \
    /path/to/fq_folder/end_1.fq \
    /path/to/fq_folder/end_2.fq > \
    /path/to/bam_folder/aligned_reads.sam
```

The "M" flag specifically makes BWA-MEM amenable for usage with GATK and Picard.

The "R" flag dictates the tab-delimited read group header line to be added to the final SAM file. An example could be as follows:

"@RG\tID:sample\tLB:sample\tPL:ILLUMINA\tPM:HISEQ \tSM:sample"

"sample" would be the user-specified sample name.

3. Using Picard SortSam, sort the SAM file by coordinate and create an output in the binary format (BAM):

```
% gatk SortSam \
    -I /path/to/bam_folder/aligned_reads.sam \
    -O /path/to/bam_folder/sorted_reads.bam \
    -SO coordinate
```

4. Using Picard MarkDuplicates, mark duplicates reads that can result from artifacts of the sequencing process:

```
% gatk MarkDuplicates \
    -I /path/to/bam_folder/sorted_reads.bam \
    -O /path/to/bam_folder/markdup_reads.bam \
    -M dup_metrics.txt
```

5. Using GATK, recalibrate base quality scores. For this step, we will use gnomAD for the database of known polymorphic sites. First, generate a recalibration table using GATK BaseRecalibrator:

```
% gatk BaseRecalibrator \
    -I /path/to/bam_folder/markdup_reads.bam \
    -R grch37.fa \
    --known-sites gnomad.vcf \
    -O /path/to/bam_folder/recal_data.table
```

6. Apply the recalibration using ApplyBQSR:

```
% gatk ApplyBQSR \
    -I /path/to/bam_folder/markdup_reads.bam \
    -R grch37.fa \
    -bsqr /path/to/bam_folder/recal_data.table \
    -O recal_reads.bam
```

7. Using Picard BuildBamIndex, generate an index for the final BAM file:

```
% gatk BuildBamIndex \
    -I /path/to/bam_folder/recal_reads.bam
```

At this point in the workflow, should storage space be a concern, the only files generated in this subheading which required downstream are the recalibrated reads BAM file and its associated index file.

3.2 Germline Variant Calling in Patient-Matched Normal Samples

Using the BAM files generated from normal (nontumorous) tissue, we will produce germline variants on a per sample basis. Leading downstream, this will be utilized by PRESM to construct the dual personalized reference genomes for each sample individual. In this subheading, we will use GATK and Pindel to produce three VCF files containing single nucleotide polymorphisms (SNP) and structural variants (denoted as INDEL). This part of the workflow will roughly follow along the GATK Best Practices document for calling germline variants (https://software.broadinstitute.org/gatk/best-practices/workflow?id=11145), deviating from it by skipping the consolidation step and applying hard filter parameters rather than quality score recalibration as we are working with an individualized approach and, therefore, on a single sample basis:

1. Using GATK HaplotypeCaller, call raw germline variants into a VCF file:

```
% gatk HaplotypeCaller \
    -I /path/to/bam_folder/normal_recal_reads.bam \
    -R grch37.fa \
    -O /path/to/vcf_folder/raw_var_recal.vcf
```

2. Using GATK SelectVariants, segregate SNPs and INDELs into separate VCF files:

```
% gatk SelectVariants \
    -V /path/to/vcf_folder/raw_var_recal.vcf
    -R grch37.fa \
    --exclude-non-variants \
    --remove-unused-alternates \
    -select-type SNP \
    -O /path/to/vcf_folder/raw_snps_recal.vcf
% gatk SelectVariants \
    -V /path/to/vcf_folder/raw_var_recal.vcf
    -R grch37.fa \
    --exclude-non-variants \
    --remove-unused-alternates \
    -select-type INDEL \
    -O /path/to/vcf_folder/raw_indels_recal.vcf
```

3. Using GATK VariantFiltration, apply hard filter parameters to the raw SNP and INDEL files:

```
% gatk VariantFiltration \
    -V /path/to/vcf_folder/raw_snps_recal.vcf
```

```
    -R grch37.fa \
    --filter-name "QD" --filter "QD < 2.0" \
    --filter-name "FS" --filter "FS > 60.0" \
    --filter-name "MQ" --filter "MQ < 40.0" \
    --filter-name "MQRS" --filter "MQRankSum < 12.5" \
    --filter-name "RPRS" \
    --filter "ReadPosRankSum < -9.0" \
    --filter-name "SOR" --filter "SOR > 4.0" \
    -O /path/to/vcf_folder/filtered_snps_final.vcf
% gatk VariantFiltration \
    -V /path/to/vcf_folder/raw_indels_recal.vcf
    -R grch37.fa \
    --filter-name "QD" --filter "QD < 2.0" \
    --filter-name "FS" --filter "FS > 60.0" \
    --filter-name "RPRS" \
    --filter "ReadPosRankSum < -20.0" \
    --filter-name "SOR" --filter "SOR > 10.0" \
    -O /path/to/vcf_folder/filtered_indels_final.vcf
```

The set of hard filters applied are the same values as used in the PRESM publication. Users could opt to use different values if desired.

4. Applying the Pindel pipeline will generate a third variant VCF file. First, using Picard CollectInsertSizeMetrics, obtain the mean insert size of the paired-end reads from the analysis ready normal tissue BAM file:

```
% gatk CollectInsertSizeMetrics \
    -I /path/to/bam_folder/normal_recal_reads.bam \
    -O /path/to/bam_folder/insert_metrics.txt \
    -H /path/to/bam_folder/insert_hist.pdf
```

While the DREAM Challenge BAM files are fairly large, an additional minimum percentage parameter ("M" flag) may need to be set to 0.5 for small files, as per the Picard documentation.

The mean insert size value will be within "insert_metrics.txt" and can be obtained using any text editor.

5. When working with BAM files using Pindel, a BAM configuration file is required denoting the file path, the average insert size, and a label for the sample name. Using vi, create a text file formatted as follows:

```
/path/to/file.bam <insert_size> <sample_name>
```

6. Using Pindel, generate a collection of variant files with the user-designed output prefix:

```
% pindel \
    -f grch37.fa \
    -i /path/to/pindel_folder/bam_config_file.txt \
    -o /path/to/pindel_folder/pindel_var
```

7. Using Pindel2VCF, consolidate all generated Pindel files and produce a VCF file containing all structural variants:

```
% pindel2vcf \
    -P /path/to/pindel_folder/pindel_var \
    -r grch37.fa \
    -R ucsc.hg19 \
    -d 20131208 \
    -v /path/to/vcf_folder/pindel_var.vcf \
    -e 4 -ir 2 -il 10 -pr 1 -pl 10 -as 250
```

The "P" flag collects all output files generated by Pindel and utilizes them as input, identifying the outputs from the previous step using the file prefix.

As with the GATK variant hard filters, the values shown here are the same values as used in the PRESM publication. Users could opt to use different values if desired.

3.3 Personalized Reference Construction

PRESM provides a comprehensive suite of functions for working with variant files and constructing personalized references. Table 1 contains a summary of the PRESM functions. For a more detailed overview, please refer to the user's manual.

Table 1
Summary of PRESM Functions

Function	Description
CombineVariants	Combine two VCF files into a single VCF file
MakePersonalizedReference	Generate a personalized reference FASTA file
MakePersonalizedVariantsDB	Generate a personalized variants database VCF
MapVariants	Map variants with coordinates on the personalized reference to the universal reference within a VCF file
RemoveOverlaps	Remove overlapping variants within a VCF file
ReplaceGenotype	Modify reads containing heterozygous alternative alleles to the reference allele within a SAM file
SelectGenotype	Generate a VCF file with only variants of the user-specified genotype
SomaticMutationsOnGermlineInsertion	Generate a text file containing relative coordinates of somatic mutations on germline insertions
SortVariants	Sort variants by coordinate within a VCF file
ViewFasta	Interactive; view a specified region of a reference FASTA file

In this subheading, we will use PRESM to generate two personalized reference FASTA files: one with all germline variants and the other with only homozygous germline variants:

1. Combine (twice), sort, and remove overlaps using the three VCF files. We also need to use an awk command to remove spanning deletions (https://software.broadinstitute.org/gatk/documentation/article?id=11029), which make up a small proportion of structural variants but can interfere with personalized reference creation:

```
% java -jar presm.jar -F CombineVariants \
    -variant1 /path/to/vcf_folder/gatk_snp.vcf \
    -variant2 /path/to/vcf_folder/gatk_indel.vcf \
    -R grch37.fa \
    -O /path/to/vcf_folder/gatk_var.vcf
% java -jar presm.jar -F CombineVariants \
    -variant1 /path/to/vcf_folder/gatk_var.vcf \
    -variant2 /path/to/vcf_folder/pindel_var.vcf \
    -R grch37.fa \
    -O /path/to/vcf_folder/germvars.vcf
% java -jar presm.jar -F SortVariants \
    -variants /path/to/vcf_folder/germvars.vcf \
    -R grch37.fa \
    -O /path/to/vcf_folder/sorted_germvars.vcf
% java -jar presm.jar -F RemoveOverlaps \
    -variants /path/to/vcf_folder/sorted_germvars.vcf \
    -R grch37.fa \
    -O /path/to/vcf_folder/germvars_rmol.vcf
% nohup awk 'OFS = "\t" {if ($5 != "*") {print $0;}}' \
    /path/to/vcf_folder/germvars_rmol.vcf > \
    /path/to/vcf_folder/germvars_final.vcf &
```

2. Construct a personalized reference FASTA with all germline variants, and normalize and index the file:

```
% java -jar presm.jar -F MakePersonalizedReference \
    -I grch37.fa \
    -germlinemutations \
     /path/to/vcf_folder/germvars_final.vcf \
    -O /path/to/ref_folder/pers_ref_all_custom.fa
% gatk NormalizeFasta \
    -I /path/to/ref_folder/pers_ref_all_custom.fa
    -O /path/to/ref_folder/pers_ref_all_germvars.fa
% gatk CreateSequenceDictionary \
    -R /path/to/ref_folder/pers_ref_all_germvars.fa \
    -O /path/to/ref_folder/pers_ref_all_germvars.dict
% samtools faidx \
    /path/to/ref_folder/pers_ref_all_germvars.fa
% bwa index /path/to/ref_folder/pers_ref_all_germvars.fa
```

3. Using PRESM SelectGenotype, create a VCF file with only homozygous germline variants:

```
% java -jar presm.jar -F SelectGenotype \
    -variants /path/to/vcf_folder/germvars_final.vcf \
    -genotype homo \
    -O /path/to/vcf_folder/germvars_hom.vcf
```

4. Create the second personalized reference with only homozygous germline variants and normalize and index the file:

```
% java -jar presm.jar -F MakePersonalizedReference \
    -I grch37.fa \
    -germlinemutations \
     /path/to/vcf_folder/germvars_hom.vcf \
    -O /path/to/ref_folder/pers_ref_hom_custom.fa
% gatk NormalizeFasta \
    -I /path/to/ref_folder/pers_ref_hom_custom.fa \
    -O /path/to/ref_folder/pers_ref_hom_germvars.fa
% gatk CreateSequenceDictionary \
    -R /path/to/ref_folder/pers_ref_hom_germvars.fa \
    -O /path/to/ref_folder/pers_ref_hom_germvars.dict
% samtools faidx \
    /path/to/ref_folder/pers_ref_hom_germvars.fa
% bwa index /path/to/ref_folder/pers_ref_hom_germvars.fa
```

3.4 Somatic Variant Calling in Tumor Samples

At long last, we arrive at the final stage of the workflow: somatic variant calling in tumor samples. This subheading will refer heavily back to Data Preprocessing (Subheading 3.1) as we will be repeating those exact steps with our personalized reference FASTAs. We will first remap the reads to the all germline variants personalized reference with, ensuring the reads are mapped to the correct locations. Next, we will replace heterozygous germline variants on sequencing reads with the reference allele and remap the reads once again to the homozygous germline variants only personalized reference. Lastly, we will utilize GATK Mutect2 to call somatic variants and map the coordinates back to the universal reference:

1. Using PRESM, produce a modified gnomAD VCF with personalized coordinates using the all germline variants VCF. Sort (with an awk command) and index upon completion:

```
% java -jar presm.jar -F MakePersonalizedVariantsDB \
    -I gnomad.vcf \
    -variants /path/to/vcf_folder/germvars_final.vcf \
    -O /path/to/gnomAD_folder/pers_gnomad.vcf
% nohup awk '{if ($1 ~ /^##contig/) {next}
    else if ($1 ~ /^#/) {print $0; next}
    else {print $0 | "sort -k1,1V -k2,2n"}' \
    /path/to/gnomAD_folder/pers_gnomad.vcf > \
```

```
    /path/to/gnomAD_folder/sorted_pers_gnomad.vcf &
% gatk IndexFeatureFile \
    -F /path/to/gnomAD_folder/sorted_pers_gnomad.vcf
```

2. Individually repeat **steps 1–6** in Subheading 3.1 using each of the DREAM Challenge tumor BAM files, "pers_ref_all_germvars.fa" and "sorted_pers_gnomad.vcf."
3. Edit sequencing reads in the remapped DREAM Challenge tumor BAM files to replace all heterozygous germline variants with the reference allele. Repeat for each file remapped tumor file and the original recalibrated patient-matched normal files. First, using Picard SamFormatConverter, convert the BAM file to SAM:

```
% gatk SamFormatConverter \
    -I /path/to/bam_folder/remap_recal_reads.bam \
    -O /path/to/bam_folder/remap_recal_reads.sam
```

4. Using PRESM ReplaceGenotype, replace heterozygous germline variants on the reads with the reference allele:

```
% java -jar presm.jar -F ReplaceGenotype \
    -I remap_recal_reads.sam \
    -germlinemutations \
    /path/to/vcf_folder/germvars_final.vcf \
    -genotype heter \
    -readlength 101 \
    -O /path/to/bam_folder/replace_germhet_reads.sam
```

The DREAM Challenge data we are working with have read lengths of 101 bp. The reader should change the value accordingly if following along with a different dataset.

5. Individually repeat **steps 1–6** in Subheading 3.1 using each of the "replace_germhet_reads.sam" files, "pers_ref_hom_germvars.fa," and "sorted_pers_gnomad.vcf."

   PRESM Replace does not alter the CIGAR and quality strings within the SAM file when replacing heterozygous germline variants. Thus, Picard tools within GATK4 will throw an error when parsing the file. As we only really need to extract the reads themselves, we can use an older version of Picard (v2.8.3, https://github.com/broadinstitute/picard/releases/tag/2.8.3) which is known to work with the altered SAM file:

```
% java -jar /path/to/picard.jar \
    -I=/path/to/bam_folder/replace_germhet_reads.sam \
    -F=/path/to/fq_folder/replace_germhet_reads_1.fq \
    -F2=/path/to/fq_folder/replace_germhet_reads_2.fq
```

6. Call somatic variants with personalized reference coordinates against the homozygous germline variants only personal

reference using Mutect2. gnomAD with personalized coordinates will serve as the population germline resource:

```
% gatk Mutect2 \
    -R /path/to/ref_folder/pers_ref_hom_germvars.fa \
    -I /path/to/bam_folder/tumour_final.bam \
    -tumor tumor_sample_name \
    -I /path/to/bam_folder/normal_final.bam \
    -normal normal_sample_name \
    --germline-resource \
    /path/to/gnomAD_folder/sorted_pers_gnomad.vcf \
    -default-af 0.00003125 \
    -O /path/to/vcf_folder/somatic_raw.vcf
```

The "default-af" parameter refers to the allele frequency of variants not within the population germline resource. As per the GATK4 Mutect2 tutorial (https://software.broadinstitute.org/gatk/documentation/article?id=11136), this value is calculated by 1/(2*num_samples). The gnomAD genomes file is produced from approximately 16,000 individuals.

7. Filter for confident somatic variant calls using GATK FilterMutectCalls:

```
% gatk FilterMutectCalls \
    -R /path/to/ref_folder/pers_ref_hom_germvars.fa \
    -V /path/to/vcf_folder/somatic_raw.vcf \
    -O /path/to/vcf_folder/somatic_final.vcf
```

8. Using PRESM MapVariants, map somatic variants back to coordinates on the universal reference using the homozygous germline variants file:

```
% java -jar presm.jar -F MapVariants \
-I /path/to/vcf_folder/somatic_final.vcf \
-germlinemutations \
/path/to/vcf_folder/germvars_hom.vcf
-O /path/to/vcf_folder/universal_somatic_variants.vcf
```

3.5 Interpreting Variant Call Format Files

We have briefly touched on the information contained within VCF files when removing extraneous metadata from the raw gnomAD file. VCF is a text file format beginning with metadata lines explaining the meaning of abbreviations and the type of value associated with each abbreviation. Following the metadata text chunk are the data lines displaying variant information in a tabular format, including chromosome, position, reference allele, alternate allele, and so on. Theoretically, the file can be treated as a tab-delimited file (*e.g.*, TSV) and opened in a spreadsheet program such as Microsoft Excel. However, as VCF files are usually very large, there is a real possibility of the spreadsheet program crashing when attempting to load the entire file to memory. Currently, the VCF specification is on version 4.3. For more information on the format itself, documentation is hosted on the Samtools GitHub website (https://samtools.github.io/hts-specs/VCFv4.3.pdf).

And that is it, thank you for reading and congratulations on making it to the end!

4 Notes

1. Overall, this pipeline is designed to be run on high performance computing clusters, which will generally be a Linux system. For smaller files, it may be possible to replicate the pipeline on beefy macOS workstations. For a primer on HPC clusters, please refer to the Materials section of [37].
2. If the Java programs are seemingly running out of heap memory, an option can be passed through the Java command to increase the maximum amount of memory allocated to the process:
   (a) For general Java applications, such as PRESM, this can be done by adding the "Xmx" flag:
      - `% java -jar -Xmx8g presm.jar [options]`
      - This allocates 8GB maximum heap memory to PRESM.

   (b) To do this with GATK, the "java-options" flag can be passed through the command:
      - `% gatk --java-options "-Xmx8g" <ToolName> [parameters].`
3. Pindel is a noted bottleneck in the entire pipeline. With the Dream Challenge data, it takes roughly 10 days to complete running. Users should decide whether they want to use Pindel, another structural variant calling tool that produces a VCF file, or bypass additional structural variant calling altogether.
4. The world of cutting edge software is fast moving. Be aware that changes can happen at any time and can break things (especially old pipelines) frequently and break things often, either through program restructuring or bugs. It is advisable to stick with "production ready" releases and avoid "nightly" versions, though that may not always be possible depending on the tool being used.
5. That being said, do not be afraid to seek out help when things do break. PRESM and GATK in particular have long-term support for major releases. Verbosity is a virtue when describing troublesome technical issues.

6. Tools come and go as standards change over time. PRESM's longevity lies in its versatility to be paired with any germline and somatic variant caller. We plan to offer frequent updates to PRESM to make the tool more amenable to new callers as they are released.

Acknowledgments

This work was supported by the Canadian Foundation of Innovation (grant #33605; QL) and a Natural Sciences and Engineering Research Council of Canada Discovery Grant (grant #RGPIN-2017-04860; QL). ML was funded through the Alberta Children's Hospital Research Institute (ACHRI) Graduate Scholarship. C.C. was funded through the ACHRI Postdoctoral Fellowship.

References

1. Wanger A, Chavez V, Huang RSP et al (2017) Chapter 13—Application of molecular diagnostics. In: Wanger A, Chavez V, Huang RSP et al (eds) Microbiology and molecular diagnosis in pathology, 1st edn. Elsevier, Cambridge
2. Zilberg C, Lee MW, Yu B et al (2018) Analysis of clinically relevant somatic mutations in high-risk head and neck cutaneous squamous cell carcinoma. Mod Pathol 31(2):275–287. https://doi.org/10.1038/modpathol.2017.128
3. Cibulskis K, Lawrence MS, Carter SL et al (2013) Sensitive detection of somatic point mutations in impure and heterogeneous cancer samples. Nat Biotechnol 31(3):213–219. https://doi.org/10.1038/nbt.2514
4. Gerlinger M, Rowan AJ, Horswell S et al (2012) Intratumor heterogeneity and branched evolution revealed by multiregion sequencing. N Engl J Med 366(10):883–892. https://doi.org/10.1056/NEJMoa1113205
5. Meyerson M, Gabriel S, Getz G (2010) Advances in understanding cancer genomes through second-generation sequencing. Nat Rev Genet 11(10):685–696. https://doi.org/10.1038/nrg2841
6. Christoforides A, Carpten JD, Weiss GJ et al (2013) Identification of somatic mutations in cancer through Bayesian-based analysis of sequenced genome pairs. BMC Genomics 14:302. https://doi.org/10.1186/1471-2164-14-302
7. Ding J, Bashashati A, Roth A et al (2012) Feature-based classifiers for somatic mutation detection in tumor-normal paired sequencing data. Bioinformatics 28(2):167–175. https://doi.org/10.1093/bioinformatics/btr629
8. Fan Y, Xi L, Hughes DST et al (2016) MuSE: accounting for tumor heterogeneity using a sample-specific error model improves sensitivity and specificity in mutation calling from sequencing data. Genome Biol 17:178. https://doi.org/10.1186/s13059-016-1029-6
9. Fang LT, Afshar PT, Chhibber A et al (2015) An ensemble approach to accurately detect somatic mutations using SomaticSeq. Genome Biol 16:197. https://doi.org/10.1186/s13059-015-0758-2
10. Goya R, Sun MGF, Morin RD et al (2010) SNVMix: Predicting single nucleotide variants from next-generation sequencing of tumors. Bioinformatics 26(6):730–736. https://doi.org/10.1093/bioinformatics/btq040
11. Hansen NF, Gartner JJ, Mei L et al (2013) Shimmer: detection of genetic alterations in tumors using next-generation sequence data. Bioinformatics 29(12):1498–1503. https://doi.org/10.1093/bioinformatics/btt183
12. Kim S, Jeong K, Bhutani K et al (2013) Virmid: accurate detection of somatic mutations with sample impurity inference. Genome Biol 14:R90. https://doi.org/10.1186/gb-2013-14-8-r90
13. Koboldt DC, Zhang Q, Larson DE et al (2012) VarScan 2: somatic mutation and copy number alteration discovery in cancer by exome sequencing. Genome Res 22(3):568–576. https://doi.org/10.1101/gr.129684.111

14. Larson DE, Harris CC, Chen K et al (2012) Somaticsniper: identification of somatic point mutations in whole genome sequencing data. Bioinformatics 28(3):311–317. https://doi.org/10.1093/bioinformatics/btr665
15. Rashid M, Robles-Espinoza CD, Rust AG, Adams DJ (2013) Cake: a bioinformatics pipeline for the integrated analysis of somatic variants in cancer genomes. Bioinformatics 29 (17):2208–2210. https://doi.org/10.1093/bioinformatics/btt371
16. Roth A, Ding J, Morin R et al (2012) JointSNVMix: a probabilistic model for accurate detection of somatic mutations in normal/tumor paired next-generation sequencing data. Bioinformatics 28(7):907–913. https://doi.org/10.1093/bioinformatics/bts053
17. Saunders CT, Wong WSW, Swamy S et al (2012) Strelka: accurate somatic small-variant calling from sequenced tumor-normal sample pairs. Bioinformatics 28(14):1811–1817. https://doi.org/10.1093/bioinformatics/bts271
18. Shiraishi Y, Sato Y, Chiba K et al (2013) An empirical Bayesian framework for somatic mutation detection from cancer genome sequencing data. Nucleic Acids Res 41(7): e89. https://doi.org/10.1093/nar/gkt126
19. Ewing AD, Houlahan KE, Hu Y et al (2015) Combining tumor genome simulation with crowdsourcing to benchmark somatic single-nucleotide-variant detection. Nat Methods 12 (7):623–630. https://doi.org/10.1038/nmeth.3407
20. Gibbs RA, Boerwinkle E, Doddapaneni H et al (2015) A global reference for human genetic variation. Nature 526(7571):68–74. https://doi.org/10.1038/nature15393
21. Rozowsky J, Abyzov A, Wang J et al (2011) AlleleSeq: analysis of allele-specific expression and binding in a network framework. Mol Syst Biol 7:522. https://doi.org/10.1038/msb.2011.54
22. Vijaya Satya R, Zavaljevski N, Reifman J (2012) A new strategy to reduce allelic bias in RNA-Seq readmapping. Nucleic Acids Res 40 (16):e127. https://doi.org/10.1093/nar/gks425
23. Stevenson KR, Coolon JD, Wittkopp PJ (2013) Sources of bias in measures of allele-specific expression derived from RNA-seq data aligned to a single reference genome. BMC Genomics 14:536. https://doi.org/10.1186/1471-2164-14-536
24. Yuan S, Qin Z (2012) Read-mapping using personalized diploid reference genome for RNA sequencing data reduced bias for detecting allele-specific expression. In: Proceedings - 2012 IEEE International Conference on Bioinformatics and Biomedicine Workshops, BIBMW 2012
25. Yuan S, Johnston HR, Zhang G et al (2015) One size doesn't fit all—refeditor: building personalized diploid reference genome to improve read mapping and genotype calling in next generation sequencing studies. PLoS Comput Biol 11(8):e1004448. https://doi.org/10.1371/journal.pcbi.1004448
26. Cao C, Mak L, Jin G et al (2019) PRESM: personalized reference editor for somatic mutation discovery in cancer genomics. Bioinformatics 35:1445–1452. https://doi.org/10.1093/bioinformatics/bty812
27. Dale R, Grüning B, Sjödin A et al (2018) Bioconda: sustainable and comprehensive software distribution for the life sciences. Nat Methods 15(7):475–476. https://doi.org/10.1038/s41592-018-0046-7
28. McKenna A, Hanna M, Banks E et al (2010) The genome analysis toolkit: a MapReduce framework for analyzing next-generation DNA sequencing data. Genome Res 20 (9):1297–1303. https://doi.org/10.1101/gr.107524.110
29. DePristo MA, Banks E, Poplin R et al (2011) A framework for variation discovery and genotyping using next-generation DNA sequencing data. Nat Genet 43(5):491–498. https://doi.org/10.1038/ng.806
30. Van der Auwera GA, Carneiro MO, Hartl C et al (2013) From FastQ data to high-confidence variant calls: the genome analysis toolkit best practices pipeline. Curr Protoc Bioinforma 43(1):11.10.1–11.10.33. https://doi.org/10.1002/0471250953.bi1110s43
31. Poplin R, Ruano-Rubio V, DePristo MA, et al (2017) Scaling accurate genetic variant discovery to tens of thousands of samples. bioRxiv. https://doi.org/10.1101/201178
32. Ye K, Schulz MH, Long Q et al (2009) Pindel: a pattern growth approach to detect break points of large deletions and medium sized insertions from paired-end short reads. Bioinformatics 25(21):2865–2871. https://doi.org/10.1093/bioinformatics/btp394
33. Li H, Durbin R (2009) Fast and accurate short read alignment with Burrows-Wheeler transform. Bioinformatics 25(14):1754–1760. https://doi.org/10.1093/bioinformatics/btp324
34. Li H, Durbin R (2010) Fast and accurate long-read alignment with Burrows-Wheeler transform. Bioinformatics 26(5):589–595. https://doi.org/10.1093/bioinformatics/btp698

35. Li H (2013) Aligning sequence reads, clone sequences and assembly contigs with BWA-MEM. arXiv. https://arxiv.org/abs/1303.3997v2

36. Lek M, Karczewski KJ, Minikel EV et al (2016) Analysis of protein-coding genetic variation in 60,706 humans. Nature 536(7616):285–291. https://doi.org/10.1038/nature19057

37. Long Q (2017) Chapter 15—Computational haplotype inference from pooled samples. In: Tiemann-Boege I, Betancourt A (eds) Haplotyping. Methods in molecular biology, vol 1551. Humana Press, New York, pp 309–319

Chapter 3

Ensemble-Based Somatic Mutation Calling in Cancer Genomes

Weitai Huang, Yu Amanda Guo, Mei Mei Chang, and Anders Jacobsen Skanderup

Abstract

Identification of somatic mutations in tumor tissue is challenged by both technical artifacts, diverse somatic mutational processes, and genetic heterogeneity in the tumors. Indeed, recent independent benchmark studies have revealed low concordance between different somatic mutation callers. Here, we describe **S**omatic **Mu**tation calling method using a **R**andom **F**orest (SMuRF), a portable ensemble method that combines the predictions and auxiliary features from individual mutation callers using supervised machine learning. SMuRF has improved prediction accuracy for both somatic point mutations (single nucleotide variants; SNVs) and small insertions/deletions (indels) in cancer genomes and exomes. Here, we describe the method and provide a tutorial on the installation and application of SMuRF.

Key words Somatic mutation calling, Next-generation sequencing

1 Introduction

Cancer is a genetic disease arising predominantly from accumulation of acquired (somatic) DNA mutations. Genomic instability and acquisition of somatic mutations are key enabling hallmarks of cancer initiation and progression. Somatic mutations are generated by mutational processes often involving defects in DNA replication or repair machinery. Environmental factors such as exposure to carcinogens can further promote mutagenesis. Mutations affecting critical genes or DNA regulatory elements can give the cell a selective advantage to survive and proliferate, leading to tumorigenesis. The identification of such driver mutations is crucial in our understanding of cancer.

Somatic mutation calling is the computational problem focusing on identification of the acquired genetic alterations in a given tumor. In most cases, this is achieved by comparing genomic sequences from a matched tumor and normal sample of a given cancer patient. The genetic alterations of interest can be grouped

Sebastian Boegel (ed.), *Bioinformatics for Cancer Immunotherapy: Methods and Protocols*, Methods in Molecular Biology, vol. 2120, https://doi.org/10.1007/978-1-0716-0327-7_3,

into single nucleotide variants (SNVs), short insertions and deletions (indels), and large-scale structural variants (SVs) that often involve complex chromosomal rearrangements. While SNVs and indels can be identified from sequences of short reads aligned to the genome, SV identification requires the analysis of discordant pair-end reads and split reads that are aligned to different locations of a genome. In this chapter, we will focus solely on identification and analysis of SNVs and indels.

A key goal in somatic mutation calling is to identify SNVs and indels with highest possible accuracy. Noise in the tumor sequencing data may arise from sequencing errors, alignment ambiguities, and tumor sample heterogeneity. Multiple SNV and indel callers have been developed, for example, MuTect2 [1], VarDict [2], VarScan2 [3], FreeBayes (ArXiv: https://arxiv.org/abs/1207.3907), and Strelka2 [4]. Benchmark studies have reported low concordance among existing variant callers [5–8]. Furthermore, the performance of individual variant callers is highly dependent on the specific variant calling workflow, depending on parameters such as alignment algorithm, use of local realignment, and variant caller configuration.

To address these issues, we developed a Somatic Mutation calling method using a Random Forest (SMuRF), an ensemble mutation caller that integrates the predictions from multiple callers using supervised machine learning (*see* Fig. 1). SMuRF SNV and indel models were trained on a set of gold standard somatic mutation calls that was manually curated by the International Cancer Genome Consortium (ICGC) from the matched tumor-normal WGS data of a chronic lymphocytic leukemia (CLL) patient and a medulloblastoma (MB) patient [9]. SMuRF accurately predicts

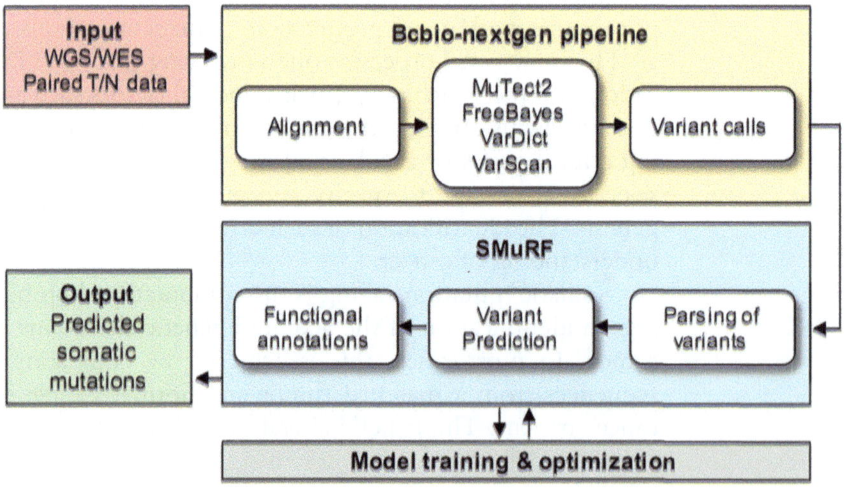

Fig. 1 Workflow of SMuRF

SNVs and indels at low allele frequencies and is robust when applied to multiple tested cancer types [10] (*see* **Note 1**).

In the next subheading, we describe the steps required to install and run SMuRF for somatic mutation calling and how to interpret the SMuRF output files.

2 Materials

2.1 Environment

SMuRF is implemented as an R package (requires R version > = 3.3.1) (https://cran.rproject.org) and runs on a standard computer (4 CPUs, 16GB RAM) (*see* **Note 2**). The source code of SMuRF is available at Github (https://github.com/skandlab/SMuRF).

To get started, download and install the package from Github with the following command in R:

```
install.packages("my/current/directory/smurf",
repos = NULL, type = "source")
```

SMuRF requires the following dependencies: *data.table*, *VariantAnnotation*, and *h2o-3.10.3.3*. These dependencies will be installed automatically the first time SMuRF is installed.

2.2 Input Requirements

SMuRF performs its prediction based on a set of input somatic mutations from the following four mutation callers: MuTect2, VarDict, VarScan, and FreeBayes (*see* **Note 3**). The required Variant Call Format (VCF) files from these four callers can easily be generated by the *variant2* bcbio-nextgen somatic mutation calling workflow (https://bcbio-nextgen.readthedocs.io). To ensure full compatibility with SMuRF, make sure you run the workflow with the following bcbio-nextgen YAML configuration parameters (Table 1).

The variant calls can alternatively be generated by independently running each of the four callers, but we recommend using *bcbio-nextgen* for ease of use and full compatibility with SMuRF and its input data requirements (Table 1).

Table 1
Bcbio-nextgen parameters required for SMuRF

bcbio-nextgen parameter	Value
mark_duplicates	True
Recalibrate	False
Realign	Gatk
Variantcaller	[mutect2,freebayes,vardict,varscan]

2.3 Test Dataset

A test dataset is installed when you install the SMuRF R package. The test dataset is partially derived from a CLL matched tumor and normal sample (European Genome-phenome Archive (EGA) accession number EGAS00001001539) [9]. Below, we will demonstrate the use of SMuRF based on analysis of this dataset.

3 Methods

3.1 Running SMuRF

Based on SMuRF's pretrained models, both SNVs and indels will be predicted using the "combined" function. SMuRF efficiently processes the data for both SNV and indel predictions concurrently. Run SMuRF on the test data provided in the package using the commands below:

```
library("smurf") #load package

mydir <- paste0(find.package("smurf"), "/data") #test data dir
setwd(mydir) #set target directory as working directory

myresults <- smurf(mydir, "combined") #run SMuRF
```

SMuRF expects one input VCF file (in the vcf.gz format) per caller: the vcf files should include one of the following keywords: "mutect," "freebayes," "vardict," and "varscan." For example, Test1-sample-mutect.vcf.gz, freebayes-sample2.vcf.gz.

For running SMuRF on multiple samples, simply run the function across the set of sample folders containing the respective input VCF files for each sample.

3.2 Retrieving Gene Annotation Information

The "cdsannotation" function annotates mutations found in the coding sequence (CDS) region using SnpEff [11] (generated from the bcbio-nextgen pipeline). For the "cdsannotation" function, VCF files need to be indexed with the following command (requires HTSlib; http://www.htslib.org/doc/tabix.html):

```
Linux:
tabix -p vcf myfile.vcf.gz
```

3.3 Interpreting SMuRF Output

SMuRF's output is stored as a list object under "$smurf_snv" and "$smurf_indel":

1. Variant statistics (stats)

   General statistics showing the total number of somatic mutations that passed each caller as well as SMuRF.

Table 2
Description for each column in the SMuRF output file

Column name	Description
Chr	Chromosome number
START_POS_REF/ END_POS_REF	Start and End nucleotide position of the somatic mutation
REF/ALT	Consensus Ref and Alt nucleotide changes of the highest likelihood
REF_MFVdVs/ ALT_MFVdVs	Reference and Alternative nucleotide changes from each caller; Mutect2 (M), Freebayes (F), Vardict (Vd), Varscan (Vs)
FILTER	Passed (TRUE) or Reject (FALSE) mutation calls from the individual callers
Sample_Name	Sample name is extracted based on your labeled samples in the vcf files
Alt_Allele_Freq	Mean Variant allele frequency calculated from the tumor reads of the callers
Depth ref./alt N/T	Mean read depth from the N/T sample for ref./alt alleles
SMuRF_score	SMuRF confidence score of the predicted mutation

2. Predicted mutations (predicted)

   *The predicted file is a data frame containing information on the predicted SNVs/indels, including their SMuRF scores (*Table 2*).*

3. Parsed-raw file (parse)

   The original list of somatic mutations passed by at least one individual mutation caller.

4. Predicted mutations with annotations (annotated)

   *The annotation file consists of 15 columns (*Table 3*). The information in the annotation file is extracted from SnpEff (http://snpeff.sourceforge.net/SnpEff_manual.html#input) annotations in the vcf files. Requires run with "cdsannotation" function.*

```
myresults <- smurf(mydir, "annotation") #run SMuRF and get
gene annotations
```

5. Time taken (time).

   Information on runtime.

3.4 Tweaking the Precision and Recall of SMuRF

The performance of SMuRF is calibrated based on the training and testing data. The user may want to adjust the SMuRF confidence score threshold to improve the precision or the recall (sensitivity) on their specific datasets. Lowering the score threshold will improve the recall but at the cost of precision, and vice versa.

The threshold can be readjusted easily from the parsed-raw file generated by SMuRF (Table 4). In the example (testdata-indel),

Table 3
Description for gene annotation features

Annotation info	Description
Allele	Alternate allele (ALT)
Annotation	Effect based on Sequence Ontology terms
Impact	Estimated level of impact (high, moderate, low, modifier)
Gene name	HGNC gene name
Gene ID	Gene ID
Feature type	Sequence type
Feature ID	Sequence ID
Transcript biotype	Ensembl biotypes (coding, noncoding, etc.)
Rank	Exon or intron rank
HGVS.c	HGVS DNA notation
HGVS.p	HGVS protein notation
cDNA.pos	Position in cDNA and transcript cDNA length
CDS.pos	Position and number of coding bases
AA.pos	Position and number of amino acids
Distance	Distance to nearest feature

Table 4
Readjusting thresholds using the SMuRF parsed-raw file

Chr	Start	End	Ref	Alt	Predict	Score
1	17,820,432	17,820,433	AT	A	TRUE	0870
1	32,639,063	32,639,064	TA	T	TRUE	0830
1	91,134,042	91,134,043	CT	C	TRUE	0565
1	81,654,021	81,654,022	CA	C	TRUE	0478
1	71,747,307	71,747,308	AT	A	FALSE	0391
1	238,440,063	238,440,065	GAA	G	FALSE	0358
3	181,270,479	181,270,483	AAAAG	A	FALSE	0267
13	22,442,400	22,442,401	AT	A	FALSE	0239
7	148,015,367	148,015,367	C	CGGCTG	FALSE	0228
9	95,191,597	95,191,599	AGG	A	FALSE	0195

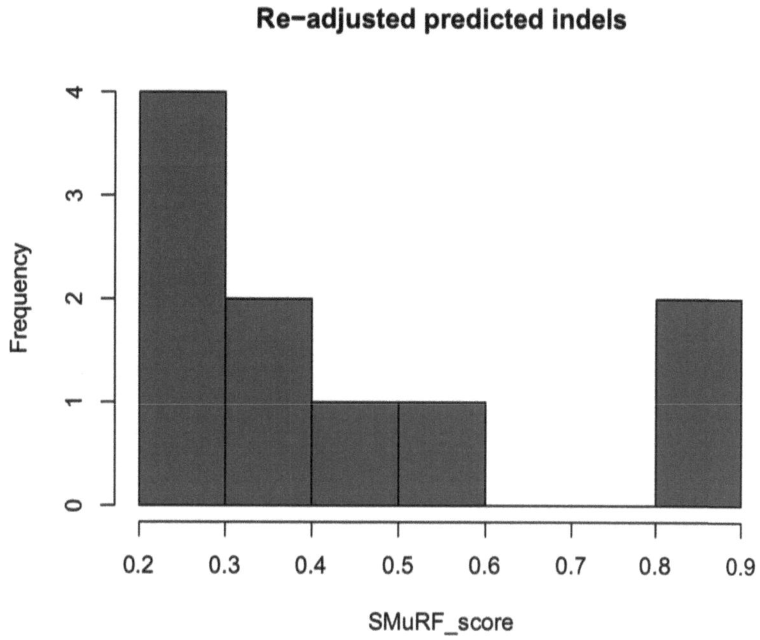

Fig. 2 Histogram of SMuRF scores of predicted indels

SMuRF predicts four indels at the default cutoff (Table 4, green). When the threshold for SMuRF score is lowered to 0.2, five additional indels (Table 4, blue, one indel omitted) are predicted. Rerun SMuRF with new cutoff "cut-off" to obtain gene annotations for the desired set of mutations.

```
#cutoff if not specified, will be at default
smurf("my/working/directory", "combined", indel.cutoff = 0.2)
```

3.5 Evaluation of SMuRF Predictions

The summary statistics for each run can be retrieved using "myresults$smurf_snv$stats_snv" and "myresults$smurf_indel$stats_indel" for SNVs and indels, respectively. A histogram can be plotted to visualize the distribution of the SMuRF scores of the predicted SNVs or indels (*see* Fig. 2).

```
#Adjusting the indel cutoff score
#myresults<-smurf(mydir, "combined", indel.cutoff=0.2)
myresults$smurf_indel$stats_indel
```

```
            Passed_Calls
Mutect2             1546
FreeBayes            339
VarDict              515
VarScan             2228
Atleast1            4343
Atleast2             244
```

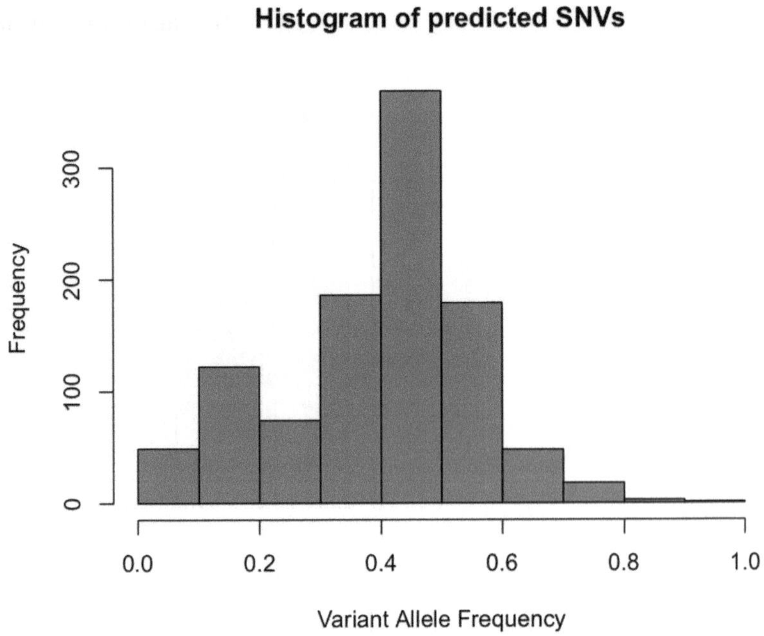

Fig. 3 Histogram of the predicted SNV variant allele frequencies

```
Atleast3              37
All4                   4
SMuRF_INDEL           10

#Plot histogram (Fig. 2)

hist(as.numeric(myresults$smurf_indel$predicted_indel
[,'SMuRF_score']), main = 'Re-adjusted predicted indels', xlab
= 'SMuRF_score', col = 'grey50')
```

In addition, SMuRF outputs the variant allele frequencies (VAF) of the predicted variants. SMuRF may in some cases be able to detect mutations at low allele frequencies (VAF < 0.1) (*see* Fig. 3) that are not called by the individual callers [10]. We can compare the distributions of VAFs of the passed calls and rejected calls by plotting the histograms using information from "parse_snv" and "parse_indel."

```
#Distribution of variant allele frequency in predicted SNVs
(Fig. 3)
hist(as.numeric(myresults$smurf_snv$predicted_snv[,'Alt_Allele_Freq']), main = 'Histogram of predicted SNVs', xlab =
'Variant Allele Frequency', col = 'grey50')
```

3.6 Saving Output

Use the following commands to save SMuRF indel output as tab-delimited files:

```
write.table(myresults$smurf_indel$stats_indel, file = "indel-stats.txt", sep = "\t", quote = FALSE, row.names = TRUE, na = ".")

write.table(myresults$smurf_indel$predicted_indel, file = "indel-predicted.txt", sep = "\t", quote = FALSE, row.names = FALSE, na = ".")

write.table(myresults$smurf_indel$parse_indel, file = "indel-parse.txt", sep = "\t", quote = FALSE, row.names = FALSE, na = ".")

write.table(myresults$smurf_indel_annotation$annotated, file = "indel-annotated.txt", sep = "\t", quote = FALSE, row.names = FALSE, na = ".")

write(myresults$time.taken, file = "time.txt")
```

4 Notes

1. We have benchmarked SMuRF to perform well on low tumor purity samples (up to 30% purity) and low variant allele frequency (F1 score of ~0.5 at VAF < 0.1).
2. SMuRF's runtime for a typical whole genome sequencing sample (~80–100× coverage) is ~10 min, while a whole exome sequencing sample (~200–400× coverage) takes ~3 min.
3. Although SMuRF is generally robust to variation in sequencing coverage, we recommend ensuring a tumor and normal coverage of at least 60× and 30×, respectively.

References

1. Cibulskis K, Lawrence MS, Carter SL et al (2013) Sensitive detection of somatic point mutations in impure and heterogeneous cancer samples. Nat Biotechnol 31:213. https://doi.org/10.1038/nbt.2514
2. Lai Z, Markovets A, Ahdesmaki M et al (2016) VarDict: a novel and versatile variant caller for next-generation sequencing in cancer research. Nucleic Acids Res 44(11):e108. https://doi.org/10.1093/nar/gkw227
3. Koboldt DC, Zhang Q, Larson DE et al (2012) VarScan 2: somatic mutation and copy number alteration discovery in cancer by exome sequencing. Genome Res 22 (3):568–576. https://doi.org/10.1101/gr.129684.111
4. Kim S, Scheffler K, Halpern AL et al (2018) Strelka2: fast and accurate calling of germline and somatic variants. Nat Methods 15 (8):591–594. https://doi.org/10.1038/s41592-018-0051-x
5. Hwang S, Kim E, Lee I et al (2015) Systematic comparison of variant calling pipelines using gold standard personal exome variants. Sci

Rep 5:17875. https://doi.org/10.1038/srep17875

6. Kroigard AB, Thomassen M, Laenkholm AV et al (2016) Evaluation of nine somatic variant callers for detection of somatic mutations in exome and targeted deep sequencing data. PLoS One 11(3):e0151664. https://doi.org/10.1371/journal.pone.0151664
7. O'Rawe J, Jiang T, Sun G et al (2013) Low concordance of multiple variant-calling pipelines: practical implications for exome and genome sequencing. Genome Med 5(3):28. https://doi.org/10.1186/gm432
8. Roberts ND, Kortschak RD, Parker WT et al (2013) A comparative analysis of algorithms for somatic SNV detection in cancer. Bioinformatics (Oxford, England) 29(18):2223–2230. https://doi.org/10.1093/bioinformatics/btt375
9. Alioto TS, Buchhalter I, Derdak S et al (2015) A comprehensive assessment of somatic mutation detection in cancer using whole-genome sequencing. Nat Commun 6:10001. https://doi.org/10.1038/ncomms10001
10. Huang W, Guo YA, Muthukumar K et al (2019) SMuRF: portable and accurate ensemble prediction of somatic mutations. Bioinformatics (Oxford, England) 35:3157–3159. https://doi.org/10.1093/bioinformatics/btz018
11. Cingolani P, Platts A, Wang le L et al (2012) A program for annotating and predicting the effects of single nucleotide polymorphisms, SnpEff: SNPs in the genome of Drosophila melanogaster strain w1118; iso-2; iso-3. Fly 6 (2):80–92. https://doi.org/10.4161/fly.19695

Chapter 4

SomaticSeq: An Ensemble and Machine Learning Method to Detect Somatic Mutations

Li Tai Fang

Abstract

A standard strategy to discover somatic mutations in a cancer genome is to use next-generation sequencing (NGS) technologies to sequence the tumor tissue and its matched normal (commonly blood or adjacent normal tissue) for side-by-side comparison. However, when interrogating entire genomes (or even just the coding regions), the number of sequencing errors easily outnumbers the number of real somatic mutations by orders of magnitudes. Here, we describe SomaticSeq, which incorporates multiple somatic mutation detection algorithms and then uses machine learning to vastly improve the accuracy of the somatic mutation call sets.

Key words Somatic mutations, Sequencing, Bioinformatics, Machine learning, Ensemble method

1 Introduction

To discover somatic mutations in a cancer genome, the tumor and its matched normal tissues (commonly blood or adjacent normal tissue) are typically sequenced side-by-side, with the normal acting as a control to filter out germline variants. However, the whole human genome consists of over 3 billion base pairs, and the coding regions alone make up over 30 million base pairs. There is a plethora of modern algorithms developed by different research groups to detect somatic mutations in such data sets [1–11]. However, they usually produce more false positive calls than actual somatic mutations.

SomaticSeq achieves higher accuracy by [12]:

1. Combine the call sets from these somatic mutation callers that were incorporated into the SomaticSeq workflow.
2. For each somatic variant call (*see* **Note 1** for SomaticSeq's definition of a unique variant call), extract genomic and sequencing features from the tumor and normal BAM files.

Sebastian Boegel (ed.), *Bioinformatics for Cancer Immunotherapy: Methods and Protocols*, Methods in Molecular Biology, vol. 2120, https://doi.org/10.1007/978-1-0716-0327-7_4,

3. Deploy an adaptive boosting [13] machine learning classifier to separate the false positives from the true mutations. The classifiers can be trained from semisimulated data sets, which we will describe in Subheading 3.4.

2 Materials

SomaticSeq is freely available under BSD 2-Clause open source license. The source code is located at https://github.com/bioinform/somaticseq. SomaticSeq Docker images can be found at https://hub.docker.com/r/lethalfang/somaticseq.

2.1 Software

To run SomaticSeq, the following tools and packages must be installed in a Linux or Unix environment:

1. SomaticSeq was developed in *Python 3* under Linux environment. We recommend using *Python* 3.5 or newer. In addition, Python libraries of *NumPy* (v1.13 or newer), *SciPy* (v1.0 or newer), and *pysam* (v0.13 or newer) are also required. Currently, SomaticSeq's latest stable version is v3.3.0, and the protocols in this book represent the version 3 branch of SomaticSeq. There have been substantial improvements in features and stability since SomaticSeq was first published in 2015.
2. SomaticSeq also uses *BEDTools* [14] (v2 or newer) to manipulate bed file inputs, that is, regions to include and/or exclude in the workflow.
3. *R* (v3.2 or newer) and *ada* (v2.0.5 or newer) library were implemented as the machine learning algorithm.
4. At its core, SomaticSeq combines and then filters the results of multiple somatic mutation detection algorithms based on many sequencing features. Generally speaking, at least one compatible caller needs to be run to generate a list of mutation candidates for SomaticSeq to evaluate. It is compatible with the following callers: the original MuTect/Indelocator as well as GATK4's Mutect2 [1], VarScan2 [2], JointSNVMix2 [3], SomaticSniper [4], VarDict [5], MuSE [6], LoFreq [7], Scalpel [8], Strelka2 [9], TNscope [10], and Platypus [11].
5. *Docker* [http://www.docker.com] is a container technology that can be used to package software and their dependencies in a portable Docker images, which can be used to execute a workflow reproducibly across different platforms and environments. SomaticSeq does not require Docker per se, but we have created Docker images of it, along with a number of compatible somatic mutation callers to make life easier for new users. The advantage of using container technologies like Docker is that one does not necessarily have to create the right software

environment with the correct dependencies for every software in a workflow, for example, by creating a Docker image for MuTect2, the users can simply use that Docker image for MuTect2 tasks. Otherwise, they must make sure to have the correct Java version and other dependencies to run MuTect2, and that those dependencies do not conflict with other software that may need different Java versions.

2.2 Download and Install SomaticSeq

Source code of SomaticSeq is available via Github repository: https://github.com/bioinform/somaticseq. The latest source code can be cloned with the following git command:

```
git clone https://github.com/bioinform/somaticseq.git
```

Alternatively, a fixed version may be downloaded and unpacked, for example,

```
wget https://github.com/bioinform/somaticseq/archive/v3.3.0.
tar.gz tar -xvf v3.3.0.tar.gz
```

Installation is not required to run SomaticSeq. You may specify the full path of all the SomaticSeq scripts. Nevertheless, the simplest way to place SomaticSeq executables in your $PATH is to install it:

```
cd somaticseq
./setup.py install
```

To have fully functional SomaticSeq, you must also install BEDTools and R (and ada library) as specified in Subheading 2.1 and place them in your execution $PATH.

3 Methods

3.1 Running SomaticSeq in Tumor-Normal Paired Mode

To see global parameters for either tumor-normal paired or tumor-only single mode,

```
somaticseq_parallel.py --help
```

To see input parameters specific for tumor-normal paired mode,

```
somaticseq_parallel.py paired -help
```

The following is an example command to run SomaticSeq's core algorithm after completing some (or all) of the compatible somatic mutation caller(s). This command will invoke the default consensus mode (Fig. 1a). Keep in mind that *not* all the input VCF

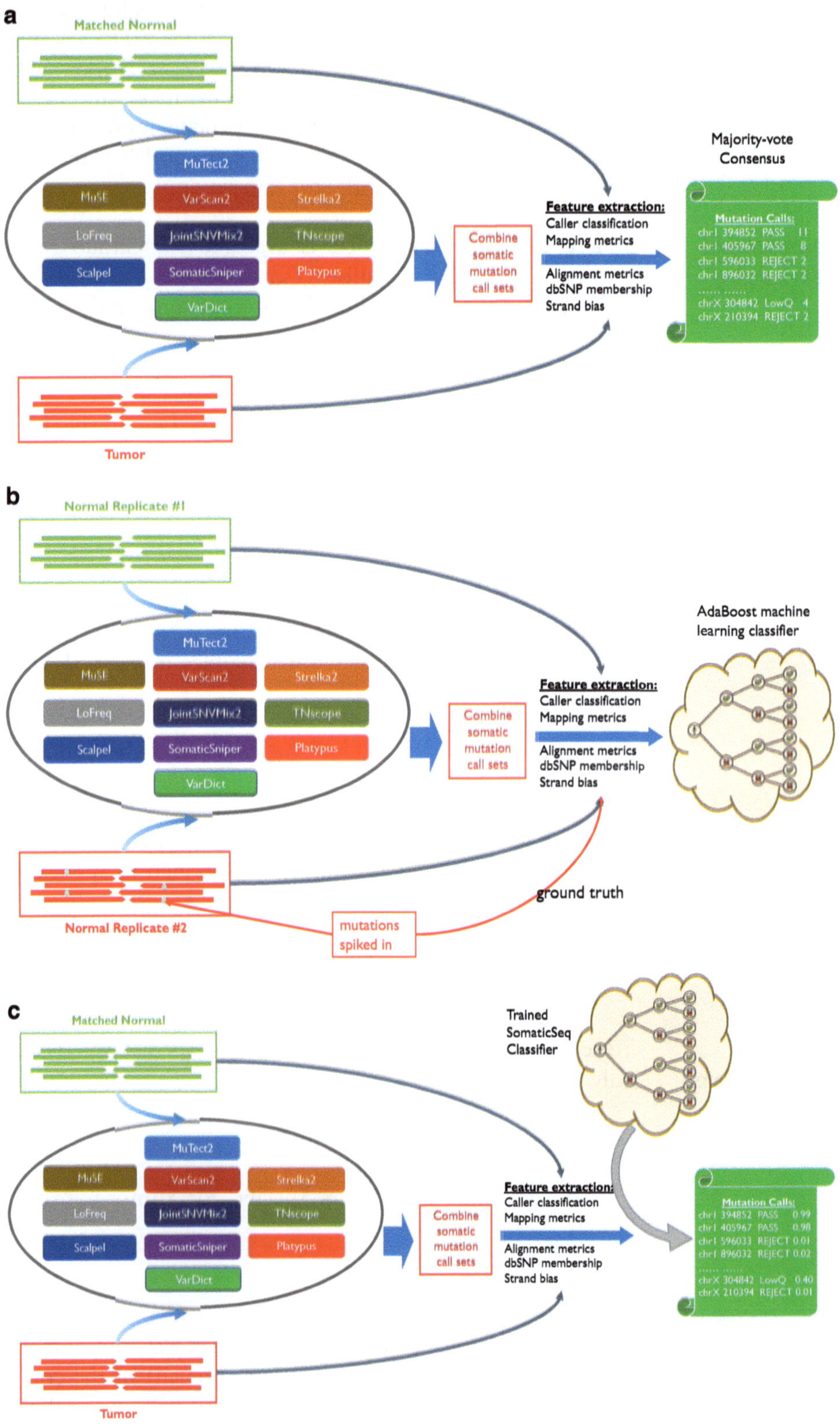

Fig. 1 The three modes for SomaticSeq

files from all the somatic callers are required. What somatic mutation callers you want to run is your choice based on how well they work on your data sets and the cost and availability of compute resources:

```
somaticseq_parallel.py \
--output-directory   OUTPUT_DIR \
--genome-reference   GRCh38.fa \
--inclusion-region   genome.bed \
--exclusion-region   blacklist.bed \
--threads             12 \
paired \
--tumor-bam-file           tumor.bam \
--normal-bam-file          matched_normal.bam \
--mutect2-vcf          MuTect2.vcf \
--varscan-snv          VarScan2.snp.vcf \
--varscan-indel            VarScan2.indel.vcf \
--jsm-vcf               JointSNVMix2.vcf \
--somaticsniper-vcf  SomaticSniper.vcf \
--vardict-vcf          VarDict.vcf \
--muse-vcf                   MuSE.vcf \
--lofreq-snv           LoFreq.somatic_final.snvs.vcf.gz \
--lofreq-indel         LoFreq.somatic_final.indels.vcf.gz \
--scalpel-vcf          Scalpel.vcf \
--strelka-snv   Strelka/results/variants/somatic.snvs.vcf.gz \
--strelka-indelStrelka/results/variants/somatic.indels.vcf.gz
\
--tnscope-vcf          TNscope.filtered.vcf.gz \
--platypus-vcf         Platypus.vcf
```

If you have SomaticSeq classifiers that you want to use to evaluate/score/classify the mutation candidates (Fig. 1b), point to them *before* the *paired* option, that is,

```
--classifier-snvEnsemble.sSNV.tsv.ntChange.Classifier.RData \
--classifier-indelEnsemble.sINDEL.tsv.ntChange.Classifier.
RData \
```

If this is a training data set for which you want to create SomaticSeq classifier (Fig. 1c), make sure to have your VCF files containing only the true variants, one for SNVs and one for indels, and place them *before* the *paired* option along with the *--somaticseq-train* flag. Every variant call in the inclusion region but not in the truth set will be considered a false positive, that is,

```
--truth-snv          TruePositive.snv.vcf \
--truth-indel   TruePositive.indel.vcf \
--somaticseq-train \
```

3.1.1 Inputs and Parameters

The *paired* argument puts SomaticSeq into tumor-normal paired mode. The caller outputs (VCF files) and the tumor-normal BAM files are placed **after** the *paired* argument. Everything else that is agnostic of paired or single sample mode goes **before** the *paired* argument, for example, genome reference, ground truth VCF files, inclusion/exclusion regions, and resource files such as dbSNP, COSMIC, and so on.

Arguments placed *before* the *paired* argument

- *--output-directory*: the path to output directory. Default is the current directory.
- *--genome-reference*: genome reference file in fasta format (typically .fa or .fasta extension). Always required, and *also* required are the existence of the index file (.fa.fai) and the dict file (.dict).
- *--truth-snv*: a VCF file containing true positive SNVs. When included, every SNV call in this VCF file will be labeled a true positive, and everything else a false positive. This is required in training mode. (If an inclusion region file is specified, then only calls within the inclusion regions may be considered. If there is an exclusion region file, then calls inside the exclusion regions will be ignored.)
- *--truth-indel*: same as above, but for indels.
- *--somaticseq-train*: flag to invoke training mode. When invoked, will create SomaticSeq classifiers if truth-snv and/or truth-indel files are specified.
- *--classifier-snv*: the trained SomaticSeq SNV classifier (.RData). When this file is specified, it will automatically invoke the prediction mode.
- *--classifier-indel*: same as above, but for indels.
- *--pass-threshold*: in prediction mode, this is the threshold (between 0 and 1) above which a call will be labeled PASS. Default is 0.5.
- *--lowqual-threshold*: in prediction mode, this is the threshold above which (but below the PASS threshold) a call will be labeled LowQual. Default is 0.1.
- *--homozygous-threshold*: a variant allele frequency (VAF) threshold, above which the GT field will be labeled 1/1. Default = 0.85.
- *--heterozygous-threshold*: a VAF above which (but below the homozygous threshold) the GT will be labeled 0/1. Default = 0.01.
- *--minimum-mapping-quality*: the minimum mapping quality score to count the reads. Default = 1.
- *--minimum-base-quality*: the minimum base-call quality score to count the base. Default = 5.

- --*minimum-num-callers*: the minimum number of caller(s) to call a variant candidate to be included in the combined call set. Default is 0.5, which means it will include some calls that are only called LowQual by a caller.
- --*dbsnp-vcf*: dbSNP VCF file. If included, can be used as a feature in machine learning.
- --*cosmic-vcf*: COSMIC VCF file. If included, it is only used for annotation of the variants. COSMIC is *not* a SomaticSeq feature.
- --*inclusion-region*: if included, then only variant calls in these regions will be considered.
- --*exclusion-region*: if included, then variants in these regions will be excluded. If a call is in both inclusion and exclusion regions, it will be excluded.
- --*threads*: number of threads. It will split the job into equal-sized regions (based on inclusion BED file or the .fa.fai file) to execute each thread. Default = 1.
- --*keep-intermediates*: a flag to tell SomaticSeq not to delete any intermediate files, for debugging purposes.

Parallel processing is achieved by splitting the inclusion BED file into a number of temporary BED files of equal sizes (i.e., same number of base pairs per BED file), named 1.th.input.bed, 2.th.input.bed, ..., n.th.input.bed. Then, each process will be run using each temporary BED file as the inclusion BED file. If there is no inclusion BED file in the command argument, it will split reference genome's index file (e.g., GRCh38.fa.fai) instead.

Arguments placed *after* the *paired* argument

- --*tumor-bam*: sorted and index tumor bam file. Required.
- --*normal-bam*: sorted and index normal bam file. Required.
- --*tumor-sample*: tumor sample name to place in the header of the output VCF file. Default is TUMOR.
- --*normal-sample*: normal sample name to place in the header of the output VCF file. Default is NORMAL.
- --*mutect-vcf*: VCF file output from the original MuTect v1. (Optional)
- --*indelocator-vcf*: VCF file output from Indelocator. (Optional)
- --*mutect2-vcf*: VCF file output from GATK4's Mutect2 and then FilterMutect-Calls. (Optional)
- --*varscan-snv*: the snp VCF file output VarScan2. (Optional)
- --*varscan-indel*: the indel VCF file output VarScan2. (Optional)
- --*jsm-vcf*: JointSNVMix2's output modified into a VCF file by SomaticSeq. (*See* Subheading 3.3; Optional)

- *--somaticsniper-vcf*: VCF file output from SomaticSniper. (Optional)
- *--vardict-vcf*: VCF file output from VarDict. (Optional)
- *--muse-vcf*: VCF file output from MuSE. (Optional)
- *--lofreq-snv*: somatic snv VCF file output from LoFreq. (Optional)
- *--lofreq-indel*: somatic indel VCF file output from LoFreq. (Optional)
- *--scalpel-vcf*: VCF file output from Scalpel. (Optional)
- *--strelka-snv*: somatic snv VCF file output from Strelka2. (Optional)
- *--strelka-indel*: somatic indel VCF file output from Strelka2. (Optional)
- *--tnscope-vcf*: VCF file output from Sentieon TNscope caller. (Optional)
- *--platypus-vcf*: VCF file output from Platypus. (Optional).

SomaticSeq supports any combination of the somatic mutation callers we have incorporated into the workflow. SomaticSeq will run based on the output VCFs you have provided. It will train to create SNV and/or indel classifiers if you provide the true-Positives.snv.vcf and/or truePositives.indel.vcf file(s) and invoke the *--somatic-seq-train* option. Otherwise, it will fall back to the simple caller consensus mode.

3.2 Running SomaticSeq in Tumor-Only Single Mode

SomaticSeq also supports tumor-only mode, in which case the *paired* option is replaced with *single* option. It has the same set of global input parameters as the paired mode, but a different set of arguments/options to be placed after the *single* option. To see what they are, you may run the following:

```
somaticseq_parallel.py single -help
```

3.2.1 Inputs and Parameters for Tumor-Only Mode

Arguments placed before the *single* option is same as the ones described in Subheading 3.1.1. The following options are placed *after* the *single* option:

- *--bam-file*: the BAM file for the tumor sample. Required.
- *--sample-name*: sample name to place into the VCF file. Default is TUMOR. (Optional)
- *--mutect-vcf*: VCF file output from the original MuTect v1. (Optional)
- *--mutect2-vcf*: VCF file output from GATK4's Mutect2 and then FilterMutect-Calls. (Optional)

- --*varscan-vcf*: VCF file with both SNVs and indels from VarScan2. (Optional)
- --*vardict-vcf*: VCF file from VarDict. (Optional)
- --*lofreq-vcf*: VCF file with both SNVs and indels from LoFreq. (Optional)
- --*scalpel-vcf*: VCF file from Scalpel. (Optional)
- --*strelka-vcf*: VCF file with both SNVs and indels from Strelka. (Optional).

3.2.2 Interpreting the Output Files

SomaticSeq will output a number of TSV and VCF files as results. The SNVs and indels are in separate files. The TSV files contain all the genomic and sequencing features extracted from the BAM files and/or some field of the individual callers' output. If the truth is VCF files are supplied, then each variant candidate will also be labeled true positive or false positive. The header of the TSV file describes each column. Missing values, for example, tBAM NM Diff (difference in average edit distances between variant-supporting and reference-supporting reads) when there is no reference-supporting read will be "nan."

The VCF files and the TSV files contain the same variants, though the VCF files only record a subset of the SomaticSeq features recorded in the TSV files. The descriptions are in the VCF headers, but we will describe them here. The QUAL column is a standard column in VCF format. This column contains the Phred-scaled SomaticSeq score (i.e., probability) if the VCF file was generated during Prediction mode. By default, variants with probability $\geq$0.5 (Phred-scaled QUAL $\geq$3.01) are labeled PASS. Variants with 0.1 $\leq$ probability $\leq$0.5 (0.458 $\leq$ QUAL $\leq$3.01) are labeled LowQual, and those under that threshold are labeled REJECT. In Consensus code, however, the QUAL column will always be zero, and the PASS/LowQual/REJECT labels are determined by the consensus of the callers you have incorporated into each workflow. If the majority of the callers (i.e., >50%) considers a variant high-confidence somatic mutation, the variant will be labeled PASS. If at least 1/3 of the callers (i.e., $\geq$ 33%), the variant will be labeled LowQual. Otherwise, the variant will be labeled REJECT. As an example, if there are five SNV callers used, then you need at least three callers to be labeled PASS and two callers to be labeled LowQual. If it is called by only one caller, it will be labeled REJECT.

In the INFO column of the VCF file, there is a string (e.g., MDUK = 1,1,0,1) that tells you each caller's binary classification on the variant (1 for positive classification and 0 otherwise). The string varies depending on the callers incorporated in the workflow, that is, M = MuTect/Indelocator/MuTect2, V = VarScan2, J = JointSNVMix2, S = SomaticSniper, D = VarDict, U = MuSE,

L = LoFreq, P = Scalpel, K = Strelka, T = TNscope, and Y = Platypus. NUM TOOLS tells you the number of callers where the variant is classified as a somatic mutation.

The following metrics are in the sample columns, for the tumor and normal samples separately:

- GT: genotyping. By default, it will be 1/1 if VAF $\geq$ 85%, 0/1 if between 1% and 85%, and 0/0 if VAF is under 1%.
- DP4: four numbers representing the number of forward reference-supporting reads, reverse reference-supporting reads, forward variant-supporting reads, and reverse variant-supporting reads.
- CD4: four numbers representing the number of concordant reference-supporting reads, discordant reference-supporting reads, concordant variant-supporting reads, and discordant variant-supporting reads.
- refMQ: average mapping quality score for reference-supporting reads.
- altMQ: average mapping quality score for variant-supporting reads.
- refBQ: average base-call quality score for reference-supporting bases.
- altBQ: average base-call quality score for variant-supporting bases.
- refNM: average edit-distance between reference-supporting reads and genome reference.
- altNM: average edit-distance between variant-supporting reads and genome reference.
- fetSB: Phred-scaled Fisher's Exact Test score for DP4 to measure strand bias in variant-supporting versus reference-supporting reads.
- fetCD: Phred-scaled Fisher's Exact Test score for CD4 to measure bias in concordant versus discordant reads in variant-supporting versus reference-supporting reads.
- zMQ: z-score for the mapping qualities between variant-supporting versus reference-supporting reads. The value will be positive if variant-supporting reads have higher MQs.
- zBQ: z-score for the base-call qualities between variant-supporting versus reference-supporting bases. The value will be positive if variant-supporting reads have higher BQs.
- MQ0: number of mapping quality 0 reads (multiply mapped reads) covering the variant position.
- VAF: variant allele frequency.

In the default consensus mode, the following files will be generated:

- *Ensemble.sSNV.tsv* and *Ensemble.sINDEL.tsv.*
- *Consensus.sSNV.vcf* and *Consensus.sINDEL.vcf.*

In training mode, the same files will also be generated, and each variant will be annotated as a true positive or false positive (i.e., TruePositive or FalsePositive in VCF's ID column, and 0 or 1 in the TSV's TrueVariant or False column). In addition, two classifiers will be created as well: *Ensemble.sSNV.tsv.ntChange.Classifier.RData* and *Ensemble.sINDEL.tsv.ntChange.Classifier.RData*.

If classifiers are supplied, the following files will be generated in prediction mode:

- *Ensemble.sSNV.tsv, Ensemble.sINDEL.tsv, SSeq.Classified.sSNV.tsv,* and *SSeq.Classified.sINDEL.tsv.*
- *SSeq.Classified.sSNV.vcf* and *SSeq.Classified.sINDEL.vcf.*

The difference between *SSeq.Classified* files and their consensus counterparts is that the former are scored by SomaticSeq classifiers.

3.3 Running Compatible Somatic Mutation Callers

To make it easy for new users to get things started, we have dockerized a number of commonly used somatic mutation callers that you may use before running SomaticSeq. SomaticSeq includes the *makeSomaticScripts.py* module that creates run scripts for those dockerized callers. Both tumor-normal paired runs and tumor-only jobs are supported.

To see the full options, you may run either of the following commands, one for tumor-normal (*paired*) mode and the other for tumor-only (*single*) mode.

```
makeSomaticScripts.py paired -h
makeSomaticScripts.py single -h
```

Here is an example to generate run scripts for the individual callers and SomaticSeq that combines the results of these callers. Do keep in mind that this module is not a core SomaticSeq algorithm. It simply calls for a number of third-party software tools that we have dockerized. The run scripts for the tools we generate are not extensively optimized. They may not run the latest version, and we cannot guarantee that they are fully compatible with your compute environment. If that is the case, you need to consult with the authors for those tools.

```
makeSomaticScripts.py paired \
--output-directory   /ABSOLUTE/PATH/TO/SomaticOutput \
--tumor-bam                /ABSOLUTE/PATH/TO/tumor.bam \
--normal-bam          /ABSOLUTE/PATH/TO/normal.bam \
```

```
--genome-reference    /ABSOLUTE/PATH/TO/GRCh38.fa \
--inclusion-region    /ABSOLUTE/PATH/TO/inclusion_region.bed \
--exclusion-region    /ABSOLUTE/PATH/TO/blacklist.bed \
--dbsnp-vcf                  /ABSOLUTE/PATH/TO/dbSNP.vcf \
--cosmic-vcf           /ABSOLUTE/PATH/TO/COSMIC.vcf \
--threads              12 \
--run-mutect2 --run-vardict --run-muse --run-lofreq --run-
strelka2
      --run-somaticseq
```

The *--threads 12* input invokes the program to create 12 subdirectories named 1, 2, ..., 12 in the output directory. In each subdirectory, a BED file is created to represent 1/12 of the total base pairs in the inclusion region BED file. If a BED file is not supplied, those sub-BED files will be based on the index file for the genome reference, that is, the .fa.fai file, although we recommend supplying a BED file even for whole genome sequencing (*see* **Note 2**).

In each of the subdirectory, a run script (ending in .cmd) for each somatic mutation caller invoked by *--run-XXX* flag is created in "logs." You will need to execute these scripts. If you invoke *--action qsub* in the command, then these scripts will be submitted to the compute management via the *qsub* command. You may also include extra arguments there, for example, *--action 'qsub -l h = node01'*. In addition, in each of the 12 subdirectories, a SomaticSeq directory will be created as well. The run scripts SomaticSeq/logs/somaticSeq.timestamp.cmd will need to be executed or qsubed manually *after* all the callers (in this thread) are completely successfully.

Furthermore, when more than one thread is invoked, there will be another script named SomaticOutput/logs/mergeResults.timestamp.cmd, which you may qsub or execute to merge the result file from all the threads together. SomaticSeq training will also be executed at this step if more than one thread is specified.

3.3.1 Inputs and Parameters

- *--output-directory*: absolute path to output all the results. Default is the current directory.
- *--somaticseq-directory*: name of the SomaticSeq directory inside the output directory. Default is SomaticSeq.
- *--tumor-bam*: absolute path to the indexed tumor bam file. This along with its .bai index file is required.
- *--normal-bam*: absolute path to the indexed normal bam file. This along with its .bai index file is required.
- *--tumor-sample-name*: sample name to place in the VCF file as the tumor. Default is TUMOR.

- --*normal-sample-name*: sample name to place in the VCF file as the normal. Default is NORMAL.
- --*genome-reference*: genome reference file in fasta format (typically .fa or .fasta extension). Always required, and also required are the existence of the index file (.fa.fai) and the dict file (.dict).
- --*inclusion-region*: if supplied, only calls within these regions will be considered.
- --*exclusion-region*: if supplied, calls outside these regions will be excluded.
- --*dbsnp-vcf*: dbSNP VCF file. This is required because some tools ask for it. Also required are the .vcf.gz file and the .vcf.gz.idx file if MuSE is invoked.
- --*cosmic-vcf*: COSMIC VCF file, purely for annotation purposes. Optional.
- --*run-mutect2*: a flag to create script to run GATK4's MuTect2.
- --*run-varscan2*: a flag to create script to run VarScan2.
- --*run-jointsnvmix2*: a flag to create script to run JointSNVMix2 (cannot be parallelized).
- --*run-somaticsniper*: a flag to create script to run SomaticSniper (cannot be parallelized).
- --*run-vardict*: a flag to create script to run VarDictJava.
- --*run-muse*: a flag to create script to run MuSE.
- --*run-lofreq*: a flag to create script to run LoFreq.
- --*run-scalpel*: a flag to create script to run Scalpel.
- --*run-strelka2*: a flag to create script to run Strelka2.
- --*run-somaticseq*: a flag to create script to run SomaticSeq.
- --*action*: the command for the caller scripts generator. Default is "echo," such that the paths of the scripts will be printed onto the command line terminal, but nothing will be done for them. A common choice would be "qsub" if you want to submit those scripts into your compute queue system. You may also include arguments for the scripts by using single quotes such as --*action 'qsub -l h = "node01|node02"'*.
- --*somaticseq-action*: same as above, but for the SomaticSeq script. Keep in mind the SomaticSeq cannot be executed until all the individual caller jobs have completed. Default is echo.
- --*snv-classifier*: absolute path to SomaticSeq SNV classifier, which will invoke prediction mode.
- --*indel-classifier*: absolute path to SomaticSeq SNV classifier, which will invoke prediction mode.
- --*truth-snv*: a VCF file containing true positive SNVs. When included, every SNV call in this VCF file will be labeled a true

positive, and everything a false positive. This is required in training mode. (If an exclusion region BED file is included, the calls inside the exclusion regions will be ignored.)

- *--truth-indel*: same as above, but for indels.
- *--train-somaticseq*: a flag to invoke training mode in SomaticSeq script if truth SNV and/or indel VCF files are supplied. Default is False.
- *--minimum-VAF*: minimum variant allele frequencies to be passed onto VarScan2 and VarDict callers. If not supplied, it will be the default 0.10 for VarScan2 and the recommended 0.05 for VarDict.
- *--threads*: number of threads. It will split the job into equal sized regions (based on inclusion BED file or the .fa.fai file) to execute each thread. Default = 1.
- *--exome-setting*: a flag to invoke exome setting in MuSE and Strelka2.
- *--mutect2-arguments*: extra argument to pass onto GATK4's Mutect2 command. Use single quotes to include multiple words, for example, *--mutect2-arguments '--min-base-quality-score 20 --tumor-lod-to-emit 10'*.
- *--mutect2-filter-arguments*: extra argument to pass onto GATK4's Filter-MutectCalls command. Use single quotes to include multiple words, see *--mutect2-arguments* for example.
- *--varscan-pileup-arguments*: extra argument to pass onto samtools mpileup prior to VarScan2. Use single quotes to include multiple words, see *--mutect2-arguments* for example.
- *--varscan-arguments*: extra argument to pass onto VarScan2. Use single quotes to include multiple words, see *--mutect2-arguments* for example.
- *--jsm-train-arguments*: extra argument to pass onto JointSNVMix2's train step. Use single quotes to include multiple words, see *--mutect2-arguments* for example.
- *--jsm-classify-arguments*: extra argument to pass onto JointSNVMix2's classify step. Use single quotes to include multiple words, see *--mutect2-arguments* for example.
- *--somaticsniper-arguments*: extra argument to pass onto SomaticSniper. Use single quotes to include multiple words, see *--mutect2-arguments* for example.
- *--vardict-arguments*: extra argument to pass onto vardict command. Use single quotes to include multiple words, see *--mutect2-arguments* for example.
- *--muse-arguments*: extra argument to pass onto MuSE. Use single quotes to include multiple words, see *--mutect2-arguments* for example.

- *--lofreq-arguments*: extra argument to pass onto LoFreq. Use single quotes to include multiple words, see *--mutect2-arguments* for example.
- *--scalpel-discovery-arguments*: extra argument to pass onto Scalpel's discovery step. Use single quotes to include multiple words, see *--mutect2-arguments* for example.
- *--scalpel-export-arguments*: extra argument to pass onto Scalpel's export step. Use single quotes to include multiple words, see *--mutect2-arguments* for example.
- *--scalpel-two-pass*: a flag to invoke the two-pass option for Scalpel.
- *--strelka-config-arguments*: extra argument to pass onto Strelka2's config step. Use single quotes to include multiple words, see *--mutect2-arguments* for example.
- *--strelka-run-arguments*: extra argument to pass onto Strelka2's run step. Use single quotes to include multiple words, see *--mutect2-arguments* for example.
- *--somaticseq-arguments*: extra argument to pass onto SomaticSeq. Use single quotes to include multiple words, see *--mutect2-arguments* for example.

Keep in mind that *makeSomaticScripts.py* is not a core SomaticSeq algorithm. It is to help get things started. The run scripts generated by the *makeSomaticScripts* module pull the docker images we have created. The docker containers access the system files by mounting the root directory to /mnt inside the container, and then access the files through /mnt/PATH/TO/files. Thus, when pointing to paths and files in the *makeSomaticScripts.py* command, it is imperative to use *absolute physical paths.*

3.4 Create Training Data Sets

SomaticSeq employs supervised machine learning that relies on good training data sets to create accurate classifiers. An ideal training data for SomaticSeq would be a pair of real tumor-normal NGS data sets, with every true somatic mutation and false positive accurately labeled. However, no such data set exists right now. Nevertheless, an excellent approach is to use two germline sequencing replicates as background, and then create in silico mutations into one of them as the designated tumor [15]. This way, only the in silico mutations we have created are true mutations, and everything else represents replicate-to-replicate noises that make up the false positives. This approach describes the SomaticSeq training mode depicted in Fig. 1c. There are number of well-characterized germline samples that have been repeatedly sequenced at multiple sequencing centers that you may try out [16, 17]. To make it easy to get things started, we have included a number of scripts in

SomaticSeq to run dockerized BAMSurgeon workflows to create training data on which SomaticSeq classifiers can be built.

3.4.1 Have Sequencing Replicates for the Normal

Here, we present an example for the entire workflow of creating SomaticSeq classifiers using two replicates of the NA12878 genome made public by Garvan Institute [17].

First, you may download the .fastq.gz files for both replicates, NA12878D and NA12878J, and then align them into BAM files. For inexperienced users, you may run the following command to generate a run script based on GATK's beat practices to create the BAM files [18]. Run the following command for both replicate D and J, assuming you have bwa indexed reference files [19].

```
somaticseq/utilities/dockered_pipelines/alignments/ fas-
tq2bam_pipeline.sh \
--output -dir        /ABSOLUTE/PATH/TO/replicateD \
--tumor-fq1/ABSOLUTE/PATH/TO/NA12878D_HiSeqX_R1.fastq.gz \
--tumor-fq2/ABSOLUTE/PATH/TO/NA12878D_HiSeqX_R2.fastq.gz \
--tumor-bam-header'@RG\tID:XTenD\tPL:illumina\tLB:X10\tSM:
NA12878D' \
--tumor-out-bam            NA12878D.bam \
--genome-reference   /ABSOLUTE/PATH/TO/GRCh38.fa \
--threads                   12
--bwa --pre-realign-markdup
```

In addition to the genome reference GRCh38.fa, there must also be GRCh38.fa.bwt, GRCh38.fa.sa, GRCh38.fa.pac, GRCh38.fa.ann, and GRCh38.fa.amb, which can be created with "*bwa index GRCh38.fa*" command [19].

Once you run the command above, a script /ABSOLUTE/PATH/TO/replicateD/logs/fastq2bam.timestampe.cmd will be created, which you may execute to use BAM MEM to align the reads into BAM files. Make sure to use different SM and ID tags in the BAM header for the two BAM files (designated tumor and normal) because MuTect2 requires it.

Once NA12878D.bam and NA12878J.bam are created, the command below will designate NA12878D.bam as the tumor and create up to 20,000 SNVs and 8000 indels into NA12878D.bam. BAMSurgeon creates in silico mutations by changing the base(s) in a subset of the reads covering the genomic position, and then realigns the reads after they are synthetically mutated. A common question is about the size of the training data, for example, how many samples or how many variants. A rule of thumb is >500 true positive variants plus a larger number of false positives (*see* **Note 3**). The somatic mutation rate in training data should not be vastly different from reality. If your data is small targeted panels, you may need more than one tumor-normal pair to create a training set large enough (*see* **Note 4**). The resulting semisynthetic tumor-normal

BAM files will be named syntheticTumor.bam and syntheticNormal.bam.

```
somaticseq/utilities/dockered_pipelines/bamSimulator/ BamSimulator_multiThreads.sh \
--output-dir          /ABSOLUTE/PATH/TO/trainingSet \
--genome-reference    /ABSOLUTE/PATH/TO/GRCh38.fa \
--tumor-bam-in        /ABSOLUTE/PATH/TO/NA12878D.bam \
--normal-bam-in            /ABSOLUTE/PATH/TO/NA12878J.bam \
--tumor-bam-out            syntheticTumor.bam \
--normal-bam-out           syntheticNormal.bam \
--num-snvs                  20000 \
--num-indels          8000 \
--min-vaf              0.0 \
--max-vaf              1.0 \
--left-beta                 2 \
--right -beta         5 \
--min-variant-reads  2 \
--threads              12 \
--action               qsub \
--merge -output -bams
```

Again, in addition to GRCh38.fa, bwa index files GRCh38.fa.bwt, GRCh38.fa.sa, GRCh38.fa.pac, GRCh38.fa.ann, and GRCh38.fa.amb are also required.

The four parameters in the command above, that is, *--min-vaf*, *--max-vaf*, *--left-beta*, and *--right-beta*, determine the VAF distribution of the in silico mutations. The following python script will display the VAF distribution of the settings in the previous command:

Once all the threads are completed successfully, you may execute the /ABSOLUTE/PATH/TO/trainingSet/logs/mergeFiles.timestamp.cmd to merge all the BAM files and ground truth VCF files (in silico mutations) into the output directory. These files may be used to train for SomaticSeq classifiers, for example,

```
makeSomaticScripts.py paired \
--normal-bam  /ABSOLUTE/PATH/TO/trainingSet/syntheticNormal.bam \
--tumor-bam/ABSOLUTE/PATH/TO/trainingSet/syntheticTumor.bam \
--truth-snv   /ABSOLUTE/PATH/TO/trainingSet/synthetic_snvs.vcf \
--truth-inde   /ABSOLUTE/PATH/TO/trainingSet/synthetic_indels.leftAlign.vcf \
--genome-reference   /ABSOLUTE/PATH/TO/GRCh38.fa \
--output-directory/ABSOLUTE/PATH/TO/trainingSet/somaticMutations \
--dbsnp-vcf           /ABSOLUTE/PATH/TO/dbSNP.hg38.vcf \
```

```
--inclusion-region    /ABSOLUTE/PATH/TO/genome.bed \
--action               qsub \
--threads                   12 \
--run-mutect2 --run-vardict --run-muse --run-strelka2 --run-
somaticseq
      --train -somaticseq
```

The *--action qsub* will submit the somatic mutation caller jobs via that command. Otherwise you may execute them yourself. Once all those jobs are complete, you may submit or execute all the SomaticSeq scripts created in /ABSOLUTE/PATH/TO/trainingSet/somaticMutations/1,2,3,.../SomaticSeq/logs/somaticSeq.timestamp.cmd.

After all the SomaticSeq threads are complete, you can finally submit or execute the script /ABSOLUTE/PATH/TO/trainingSet/somaticMutations/logs/mergeResults.timestamp.cmd to combine the results from the different threads. It will also train the labeled data sets into SomaticSeq classifiers *Ensemble.sSNV.tsv.ntChange.Classifier.RData* and *Ensemble.sINDEL.tsv.ntChange.Classifier.RData* in the output directory. These classifiers may then be used to classify your own mutation calls.

3.4.2 Split a Normal Data Set Into Designated Tumor and Normal

Another way to create training data set is to split a (relatively high coverage) germline sequencing data into two halves, with one designated as normal and the other designated as tumor (Fig. 2a). You may not get run-to-run biases from this method, but the data will still have plenty of false positives deriving from sequencing errors, sampling error of germline variants, and so on. An example command would be as follows, where the *--split-proportion* directs the fraction of reads to the designated normal. The following is an example command to split HighCoverageGenome.bam 50–50 into designated tumor and normal.

```
somaticseq/utilities/dockered_pipelines/bamSimulator/BamSimu-
lator_multiThreads.sh \
--output-dir           /ABSOLUTE/PATH/TO/trainingSet \
--genome-reference    /ABSOLUTE/PATH/TO/GRCh38.fa \
--tumor-bam-in   /ABSOLUTE/PATH/TO/HighCoverageGenome.bam \
--tumor-bam-out              syntheticTumor.bam \
--normal-bam-out             syntheticNormal.bam \
--split-proportion   0.5 \
--min-variant-reads  2 \
--threads              12 \
--action               qsub \
--num-snvs 10000 --num-indels 8000 --num-svs 1500 \
--min-vaf 0.0 --max-vaf 1.0 --left-beta 2 --right-beta 5 \
--split -bam --merge -output -bams
```

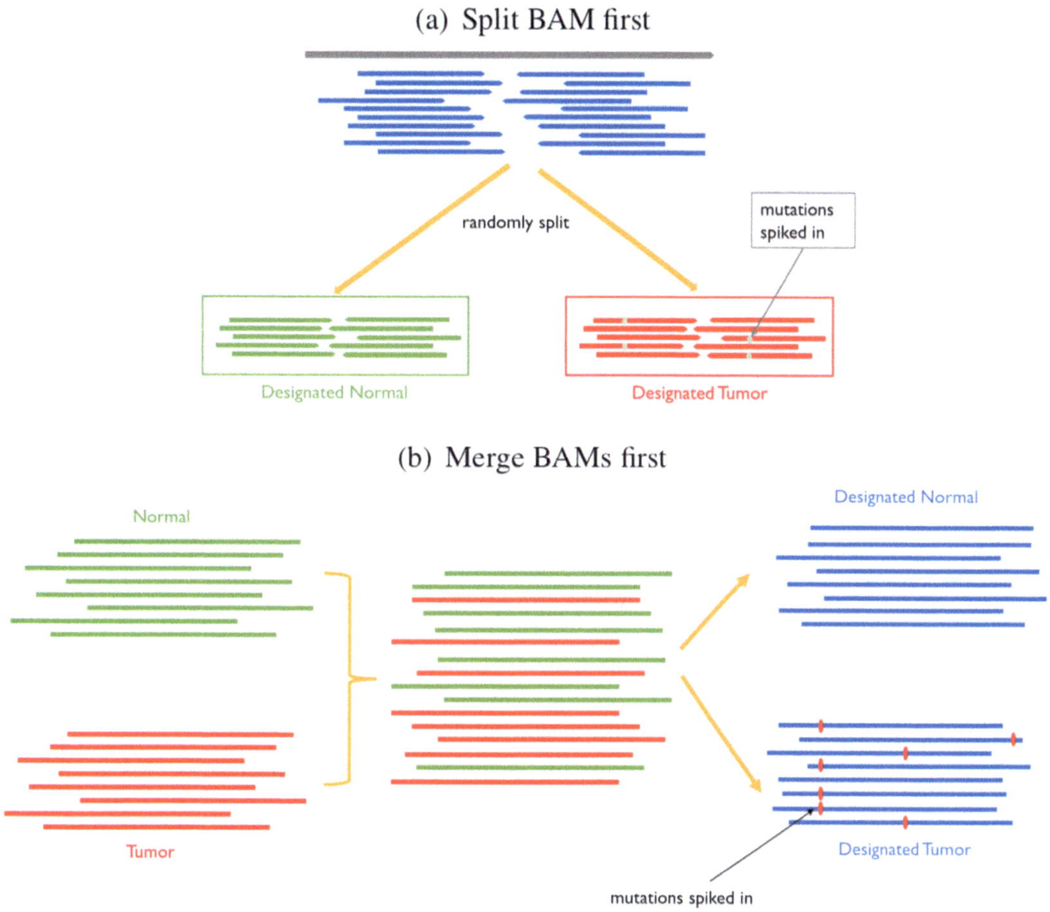

Fig. 2 Two additional scenarios to create synthetic tumor-normal pairs

3.4.3 Merge Tumor-Normal and Then Random Split Them

Another approach create training data set is to first merge the tumor and normal BAM files into a single BAM file, and then randomly split it into the designated tumor and normal (Fig. 2b). This way, the real somatic mutations in the original tumor BAM will be equally split into both designated tumor and normal (on average) and would effectively be germline variants that will be labeled as false positive if called by a caller. In silico mutations are then created in the designated tumor. In the absence of normal sequencing replicates or germline sequencing data with high enough coverage, this may be the next best option to create classifiers. Use *--merge-bam* flag to invoke this option:

```
somaticseq/utilities/dockered_pipelines/bamSimulator/BamSimu-
lator_multiThreads.sh \
--output-dir          /ABSOLUTE/PATH/TO/trainingSet \
--genome-reference    /ABSOLUTE/PATH/TO/GRCh38.fa \
--tumor-bam-in        /ABSOLUTE/PATH/TO/Tumor_Sample.bam \
```

```
--normal-bam-in      /ABSOLUTE/PATH/TO/Normal_Sample.bam \
--tumor-bam-out             syntheticTumor.bam \
--normal-bam-out            syntheticNormal.bam \
--split-proportion   0.5 \
--min-variant-reads  2 \
--threads             12 \
--num-snvs 30000 --num-indels 10000 --num-svs 1500 \
--min-vaf 0.0 --max-vaf 1.0 --left-beta 2 --right-beta 5 \
--merge -bam --split -bam --merge -output -bams
```

3.4.4 Inputs and Parameters

The following are all the options available for the BamSimulator multiThreads.sh script:

- *--output-dir*: absolute path to the output directory. Required.
- *--genome-reference*: absolute path to the genome reference, assuming the index files for the aligner (i.e., BWA) are available as well. Required.
- *--selector*: if provided, will create in silico mutations only in these regions.
- *--tumor-bam-out*: file name for the synthetic tumor BAM output. Default is syntheticTumor.bam.
- *--tumor-bam-in*: absolute path to input tumor BAM file. In scenario with two germline sequencing replicates, this is the sample where in silico mutations will be created. In scenario where a (relatively) high-coverage germline sample will be split, this is the path to that germline BAM file. In the final scenario where two BAM files are to be merged, this can point to any of the two BAM files. Required.
- *--normal-bam-out*: file name for the synthetic normal BAM output. Default is syntheticNormal.bam.
- *--normal-bam-in*: absolute path to input BAM file to be the designated normal. Required.
- *--split-proportion*: the fraction of total reads to be split into the designated normal. Not needed in the first scenario with two germline sequencing replicates. Otherwise the default is 0.5.
- *--down-sample*: downsamples the BAM files when creating synthetic tumor and normal BAM files. Default is 1, that is, no downsampling.
- *--num-snvs*: number of in silico SNVs to attempt. The actual SNVs will usually be lower because when certain conditions are not met (e.g., depth too low, etc.), an attempt will be skipped.
- *--num-indels*: number of in silico indels to attempt.
- *--num-svs*: number of in silico SVs to attempt. Default is 0.
- *--min-vaf*: minimum variant allele frequency create.

- --*max-vaf*: maximum variant allele frequency create.
- --*left-beta*: left beta for beta distribution for VAF.
- --*right-beta*: right beta for beta distribution for VAF.
- --*min-depth*: mimimum depth to attempt mutation.
- --*max-depth*: maximum depth to attempt mutation.
- --*min-variant-reads*: minimum number of variant reads to create for each in silico mutations.
- --*aligner*: what aligner to use to remap a read after the read is mutated in in silico. Default is bwa mem.
- --*seed*: choose a random number generator seed for reproducibility purposes.
- --*action*: what to do with the workflow scripts generated. Default is echo.
- --*threads*: number of threads. You would have to merge the results from each thread after all the threads are completed successfully.
- --*merge-bam*: flag to merge the input tumor and normal bam.
- --*split-bam*: flag to split the BAM files into designated tumor and normal.
- --*clean-bam*: flag to clean up the input BAM files if there are more than two reads of the same read names, by simply removing them.
- --*indel-realign*: flag to perform GATK's joint indel realignment for the designated tumor and normal BAM files.
- --*merge-output-bams*: flag to merge the output BAM and VCF files from different thread. You should use this for multithreaded jobs.
- --*keep-intermediates*: keep all the intermediate files for debugging purposes.

4 Notes

1. In SomaticSeq algorithms, each variant is defined by its genomic start position, reference base(s), and variant base(s), that is, the following four fields in a VCF file: CHROM, POS, REF, and ALT. Different mutations in the same genomic position are considered different variant calls and will have different features extracted.
2. SomaticSeq allows the input of an inclusion region. Without it, SomaticSeq will assume whole genome with the index file of the genome reference (typically .fa.fai). However, even for

whole genome data, we recommend that you create a BED file for the whole genome that only includes the major chromosomes (i.e., chr1, chr2, ..., chrX, chrY) because often the human reference files include alternate contigs, viral contigs, decoy contigs, and even chrM, and so on. Reads aligned to those contigs tend to be poorly aligned but often with very high apparent coverage, thus some mutations callers will then waste a disproportionate amount of time attempting to do local assembly on them to resolve variant calls in these contigs, when they are mostly wasted compute time. A BED file can be created to simply exclude those regions.

3. Generally speaking, machine learning works increasingly better with larger data sets. In our original publication, we measured the accuracy of SomaticSeq cross validation with increasing data size and have found that generally, the accuracy plateaus when there are >500 true mutations in the training data (assuming there are more false positives than true positives; Fig. 3).
4. There are cases where users need to combine samples to create a larger training set, for example, mutation call sets from targeted panel or even a whole exome sequencing may not be large enough to create a reliable classifier. In order to do so, you first need to merge the Ensemble.sSNV.tsv files and Ensemble.sINDEL.tsv files, but keeping only one header. Make sure those files were created with the same versions and parameters of SomaticSeq, so that each column means the same thing from different files. The files can be combined like this:

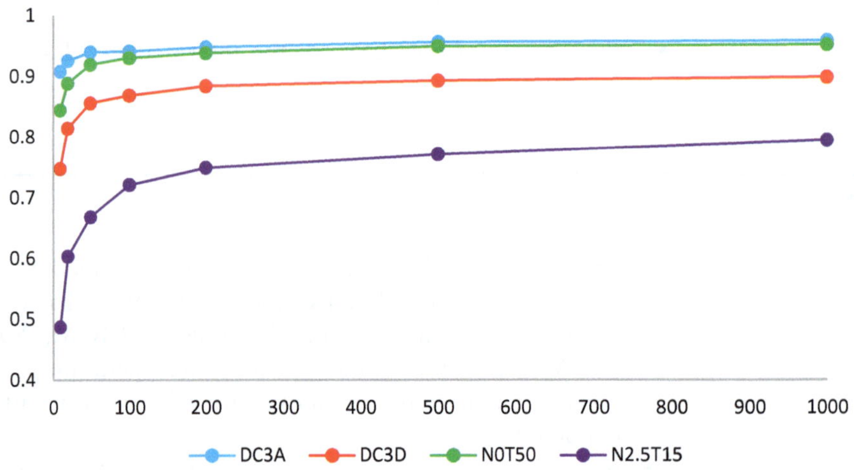

Fig. 3 Figure adapted from [12]. Y-axis represents the F_1 score of cross validation, and X-axis represents the number of true somatic mutations in a training data set. The four different plots represent four different data sets

```
cat /PATH/sampleA/SomaticSeq/Ensemble.sSNV.tsv /PATH/sampleB/
SomaticSeq/Ensemble.sSNV.tsv | awk 'NR==1 | $1 !~ /^CHROM/' >
Combined.sSNV.tsv
cat /PATH/sampleA/SomaticSeq/Ensemble.sINDEL.tsv /PATH/sam-
pleB/SomaticSeq/Ensemble.sINDEL.tsv | awk 'NR==1 | $1 !~ /
^CHROM/' > Combined.sINDEL.tsv
```

Then, you can invoke the machine learning training scripts in R directly:

```
r_scripts/ada_model_builder_ntChange.R Combined.sINDEL.tsv
Consistent_Mates Inconsistent_Mates Strelka_QSS Strelka_TQSS
r_scripts/ada_model_builder_ntChange.R Combined.sSNV.tsv Con-
sistent_Mates Inconsistent_Mates
```

The values after the TSV files are features to be excluded in training, and those features have not yet shown they improve accuracy, so by default, they were excluded.

References

1. Cibulskis K, Lawrence MS, Carter SL et al (2013) Sensitive detection of somatic point mutations in impure and heterogeneous cancer samples. Nat Biotechnol 31(3):213–219
2. Koboldt DC, Zhang Q, Larson DE et al (2012) VarScan 2: somatic mutation and copy numberalteration discovery in cancer by exome sequencing. Genome Res 22(3):568–576
3. Roth A, Ding J, Morin R et al (2012) JointSNVMix: a probabilistic model for accurate detection of somatic mutations in normal/tumour paired next-generation sequencing data. Bioinformatics 28(7):907–913
4. Larson DE, Harris CC, Chen K et al (2012) SomaticSniper: identification of somatic point mutations in whole genome sequencing data. Bioinformatics 28(3):311–317
5. Lai Z, Markovets A, Ahdesmaki M et al (2016) VarDict: a novel and versatile variant caller for next-generation sequencing in cancer research. Nucleic Acids Res 44(11):e108
6. Fan Y, Xi L, Hughes DST et al (2016) MuSE: accounting for tumor het- erogeneity using a sample-specific error model improves sensitivity and specificity in mutation calling from sequencing data. Genome Biol 17(1):178
7. Wilm A, Aw PPK, Bertrand D et al (2012) LoFreq: a sequence-quality aware, ultra-sensitive variant caller for uncovering cell-population heterogeneity from high-throughput sequencing datasets. NucleicAcidsRes 40(22):11189–11201
8. Narzisi G, O'Rawe JA, Iossifov I et al (2014) Accurate de novo and transmitted indel detection in exome-capture data using microassembly. Nat Methods 11(10):1033–1036
9. Kim S, Scheffler K, Halpern AL et al (2018) Strelka2: fast and accurate calling of germline and somatic variants. Nat Methods 15 (8):591–594
10. Freed D, Pan R, Aldana R (2018) Tnscope: accurate detection of somatic mutations with haplotype-based variant candidate detection and machine learning filtering. bioRxiv
11. Thorvaldsdottir H, Robinson JT, Mesirov JP (2013) Integrative genomics viewer (IGV): high-performance genomics data visualization and exploration. Brief Bioinform 14 (2):178–192
12. Fang LT, Afshar PT, Chhibber A et al (2015) An ensemble approach to accurately detect somatic mutations using somaticseq. Genome Biol 16(1):197
13. Johnson K, Culp M, Michailides G (2006) ada: an R package for stochastic boosting. J Stat Softw 17(2)
14. Quinlan AR, Hall IM (2010) BEDTools: a flexible suite of utilities for comparing genomic features. Bioinformatics 26(6):841–842
15. Ewing AD, Houlahan KE, Hu Y et al (2015) Combining tumor genome simulation with crowdsourcing to benchmark somatic single-nucleotide-variant detection. Nat Methods 12 (7):623–630

16. Genome in a bottle. https://www.nist.gov/programs-projects/genome-bottle
17. First publicly available XTen genome. http://allseq.com/knowledge-bank/1000-genome/get-your-1000-genome-test-data-set/
18. Roberts ND, Daniel Kortschak R, Parker WT et al (2013) A comparative analysis of algorithms for somatic snv detection in cancer. Bioinformatics 29(18):2223–2230
19. Li H (2013) Aligning sequence reads, clone sequences and assembly contigs with bwa-mem

Chapter 5

HLA Typing from RNA Sequencing and Applications to Cancer

Rose Orenbuch, Ioan Filip, and Raul Rabadan

Abstract

The human leukocyte antigen (HLA) complex is necessary for antigen presentation and regulates both innate and adaptive immune responses. In the context of cancer and treatment therapies, the HLA locus plays a critical role in tumor recognition and tolerance mechanisms. In silico HLA class I and class II typing, as well as expression quantification from next-generation RNA sequencing, can therefore have great potential clinical applications. However, HLA typing from short-read data is a challenging task given the high polymorphism and homology at the HLA locus. In this chapter, we present our highly accurate HLA typing solution, arcasHLA. We provide a detailed outline for practitioners using our protocol to perform HLA typing and demonstrate the applicability of arcasHLA in several clinical samples from tumors.

Key words Human leukocyte antigens, Major histocompatibility complex, Genotyping, Expression, RNA sequencing, Cancer

1 Introduction

The major histocompatibility complex (MHC) plays an integral role in immunity through antigen presentation. In all nucleated somatic cells, peptides derived from intracellular proteins are presented on the cell surface via MHC class I molecules to a variety of T lymphocytes, whose role, in turn, is to identify infected cells, tumor cells, or cells that are "nonself" (Fig. 1A). In doing so, CD8+ cytotoxic T cells (CTLs) bind to foreign antigens by means of T-cell receptors (TCRs) expressed on the cell surface, triggering apoptosis when the antigens are recognized as nonself. Natural killer cells (NKs) also interact with MHC class I molecules, albeit through a different mechanism: if NK receptors detect the presence of class I molecules and thereby recognize a target cell as being "self," they inhibit an NK cytotoxic response [1].

MHC class II is constitutively expressed in specialized immune cells, including dendritic cells, monocytes, B cells, as well as certain epithelial cells. Binding peptides derived from extracellular

Sebastian Boegel (ed.), *Bioinformatics for Cancer Immunotherapy: Methods and Protocols*, Methods in Molecular Biology, vol. 2120, https://doi.org/10.1007/978-1-0716-0327-7_5, © Springer Science+Business Media, LLC, part of Springer Nature 2020

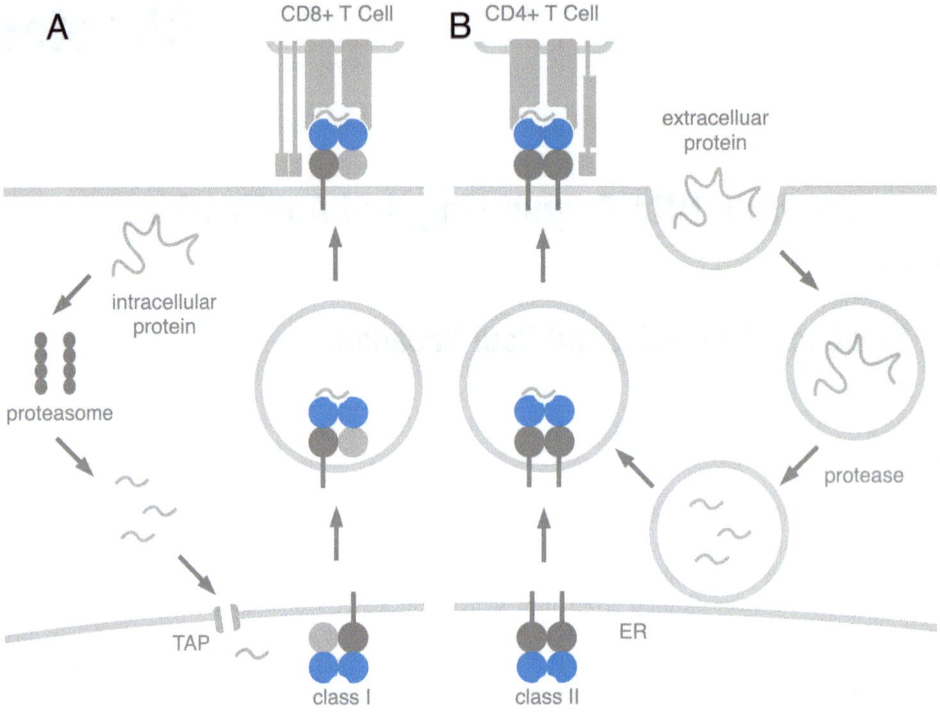

Fig. 1 (**A**) Intracellular proteins are broken down by proteasomes, transported into the endoplasmic reticulum via transporter associated with antigen processing (TAP) protein where they bind to MHC class I molecules. These complexes are then transported from the endoplasmic reticulum (ER) to the surface of the cell, where they are presented to cytotoxic T lymphocytes. (**B**) Extracellular proteins are endocytosed into lysosomes where they are broken down into smaller peptides. Lysosomes fuse with other vesicles that contain MHC class II dimers, and MHC-peptide binding occurs. The MHC-peptide complex is finally transported to the cell surface for antigen presentation to helper T cells

proteins, MHC class II presents antigens to CD4+ helper T cells which mediate adaptive immunity (Fig. 1B). When a CD4+ T cell recognizes a nonself antigen, it releases cytokines to recruit and activate other lymphocytes, thereby generating an immune response [2].

In humans, the MHC is known as the human leukocyte antigen (HLA) complex, and it is situated on the short arm of chromosome 6. MHC class I molecules are composed of a heavy chain and a nonvariant β2-microglobulin protein. Classical heavy chains are encoded by HLA-A, HLA-B, and HLA-C genes, while the non-classical genes include HLA-E, HLA-F, and HLA-G. MHC class II molecules (HLA-DP, HLA-DQ, HLA-DR, and several others) are heterodimers composed of two heavy chains denoted by α and β (Fig. 2). For example, HLA-DQA1 and HLA-DQB1 are encoded by two different genes whose protein products together comprise a single MHC class II molecule.

The HLA locus contains some of the most polymorphic sites in the human genome. The ImMunoGeneTics HLA (IMGT/HLA)

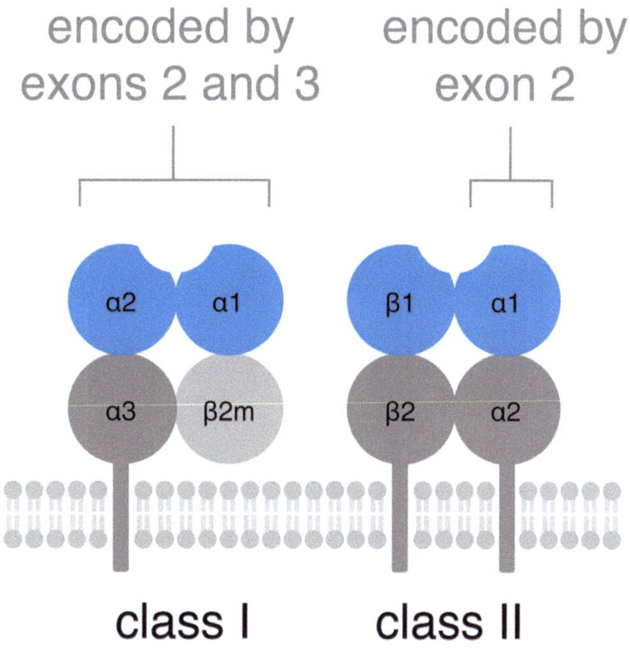

Fig. 2 The antigen-binding region (shown in blue), encoded by exons 2 and 3 in MHC class I and by exon 2 in MHC class II, determines which antigens can be bound and presented

database catalogues all known HLA alleles, the majority of which are not completely sequenced (Fig. 3A, B), many alleles missing information outside of the antigen-binding region—namely, exons 2 and 3 for class I, and exon 2 for class II (Fig. 2) [4]. The binding region exhibits the highest level of polymorphism, allowing different HLAs to bind preferentially to different antigens [5]. In addition to the functional HLA genes, there exist a multitude of class I pseudogenes, some of which are expressed and may interfere with typing of other HLA loci (Fig. 3C). The clinical impact of these pseudogenes is currently not well characterized.

Current HLA nomenclature consists of four fields separated by colons and a suffix (Fig. 4). High-resolution typing consisting of the first two fields (e.g., A*02:01) is the lowest resolution required to determine tissue compatibility. Lower resolution groupings include G-groups and P-groups. In the former, a group (e.g., A*02:01:01G) includes all alleles with the same nucleotide sequence in the antigen-binding domain. In the latter, a group (e.g., A*02:01P) includes all alleles that encode the same amino acid sequence for the antigen-binding region. Because the gold-standard benchmark sets available for in silico HLA typing from next-generation sequencing—including whole genome, whole exome, and RNA sequencing—is limited to G-group alleles, most HLA typing tools are only tested for this level of resolution even if they return three to four fields in resolution.

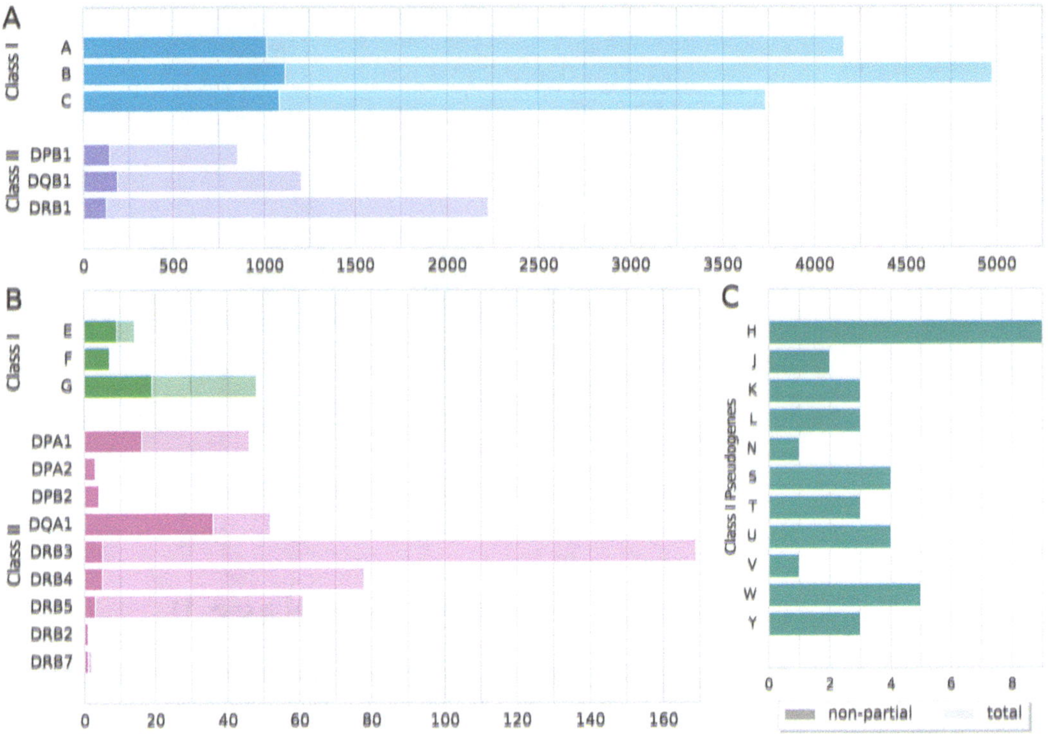

Fig. 3 The majority of alleles for MHC class I and class II loci, displaying (**A**) high polymorphism and (**B**) moderate polymorphism, are only partially sequenced. However, all known alleles of relatively invariant loci (e.g., HLA-F and HLA-DPA2) are fully characterized. (**C**) There are a multitude of HLA class I pseudogenes brought about by the duplication of the class I loci [3]

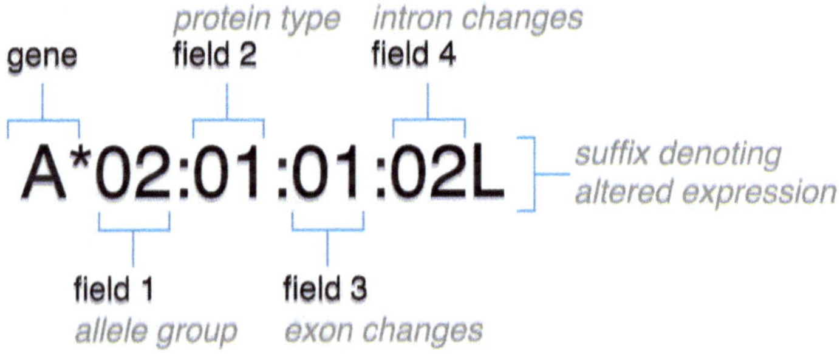

Fig. 4 HLA nomenclature: the first field denotes the allele group, and the second denotes the protein type (all alleles sharing these two fields produce identical proteins). The final two fields represent changes in the sequence that do not affect the protein product: the third field identifies synonymous changes in exons while the fourth indicates changes in noncoding regions. The suffix, present on select alleles, denotes altered expression, as follows: *L* for low expression, *N* for null expression, *S* for secreted, and *Q* for questionable

Here, we present the protocol necessary to run arcasHLA, an in silico HLA typing and quantification tool [6]. arcasHLA infers HLA class I and class II genotypes at three-field resolution from aligned RNA-sequencing BAM file inputs, either in the single-end or paired-end format. Additionally, with sufficient coverage which is often tissue dependent, arcasHLA is capable of typing nonclassical HLA genes (such as HLA-E, HLA-F, and HLA-G, known to be less polymorphic than the classical class I genes), as well as HLA pseudogenes (like HLA-H)—as long as the sequences in question have references that are included in the IMGT/HLA database. However, typing for these genes has not been validated due to a lack of benchmark datasets for nonclassical and HLA pseudogenes. As an optional step, arcasHLA also allows for typing including partial alleles, recommended only in paired-end samples with relatively high coverage of the HLA region.

Given the critical role of HLA genes in mediating the innate and adaptive immune responses, the HLA locus is also paramount in cancer development and response to therapy. First of all, HLA allele types may influence the immune response to tumor cells because antitumor activity of CTLs depends on HLA class I's ability to present tumor-derived antigens. Somatic mutations, along with chromosomal instability, drive carcinogenic activity, leading to the production of mutant proteins. The novel peptides, or neoantigens, derived from these proteins, sufficiently different from those in the normal genome, can thus provoke an immune response from CTLs which then eliminate the cancerous cells [7]. Homozygosity at one or more HLA class I loci reduces the variety of neoantigens that a cell is capable of presenting, and it has been recently reported in the literature that class I homozygosity can also influence cancer response to checkpoint blockade immunotherapy [8]. Indeed, certain HLA alleles preferentially bind known neoantigens characteristic of cancerous mutations, while others have structural features that interfere with antigen presentation [8]. As such, a tool that accurately types HLA loci from RNA sequencing is greatly useful, particularly in studies in which only RNA sequencing is performed.

Loss of HLA alleles, already reported in several tumor types including lung cancers [9], may give rise to false positives for homozygosity in addition to interfering with HLA typing overall. Hence, HLA typing from RNA sequencing is best performed with normal samples as input rather than tumor samples, although the latter can still be used to infer the correct HLA genotype when the tumor sample does not exhibit copy number loss in chromosome 6 and has sufficient coverage of the HLA loci. We end this chapter by illustrating several use cases for arcasHLA with matched normal and tumor samples, and we provide a detailed discussion of practical considerations and caveats in that setting.

2 Materials

2.1 Hardware

A Linux or Mac OS system is required to run arcasHLA. Our software was tested on an Amazon Web Services EC2 virtual instance with 16 vCPUs and 64 GiB of memory (x86_64).

2.2 Software

2.2.1 arcasHLA Installation

The source code for arcasHLA can be cloned from the Github repository with the following command:

```
git clone https://github.com/RabadanLab/arcasHLA.git
```

This will create the directory arcasHLA and populate it with the necessary python scripts and necessary arcasHLA resources. Either install to /usr/local/bin/ or install to your preferred location and add it to the path by adding the following line to ~/.bashrc:

```
PATH=/path/to/arcasHLA/:$PATH
```

2.2.2 Dependencies

The following tools and Python packages are needed to run arcasHLA:

1. arcasHLA was written in Python 3.6. In addition, arcasHLA requires the Python packages Biopython version 1.71 or newer [10], NumPy version 1.14.3 or newer [11], SciPy version 1.1.0 or newer [12], and Pandas version 0.23.0 or newer [13].
2. To use IMGT/HLA database versions ≥3.35.0, Git Large File Storage (https://git-lfs.github.com/) is necessary to clone the required files.
3. arcasHLA uses Samtools ≥v1.19 [14] to extract reads aligned to chromosome 6 and alternate contigs from BAM input before genotyping. In addition, arcasHLA uses bedtools v2.27.1 [15] to convert extracted reads to FASTQ files and pigz v2.3.1 (https://zlib.net/pigz/) to quickly compress these files.
4. arcasHLA uses Kallisto v0.44.0 to "pseudoalign" reads to an HLA reference, the results of which are used to genotype HLA alleles using arcasHLA's built-in transcript quantification and allele scoring. This RNAseq aligner and quantifier is also used for the quantification step when detecting allele imbalance.

2.3 Databases and Datasets

2.3.1 1000 Genomes Project

arcasHLA was developed using RNA sequencing of lymphoblastoid cell lines, derived from individuals included in the 1000 Genomes Project and provided by the Geuvadis project [16]. HLA G-groups for 358 of these samples were previously typed using Sanger sequencing [17].

2.3.2 AlleleFrequencies Net Database

At multiple points, arcasHLA uses HLA allele frequencies to break ties between alleles with the same read count. These allele frequencies were obtained from AlleleFrequencies Net Database (AFND) [18], selecting only submissions from "gold-standard" studies. The allele frequencies from AFND were grouped into overarching populations in a fashion similar to those catalogued by the National Marrow Donor Program [19]: Asian and Pacific Islander, Black, Caucasian, Hispanic, and Native American. The allele frequencies are smoothed using a Dirichlet prior, constructed from all population's allele frequencies as well as single counts for alleles in the IMGT/HLA database but not observed in the studies uploaded to AFND. arcasHLA requires no specialized enrichment for the HLA locus and can type HLA alleles from standard RNA-sequencing reads, either paired or single end.

2.3.3 Sequence Read Archive

We present a walkthrough of the arcasHLA protocol using two pairs of matched normal/tumor paired-end RNA-sequencing samples from nonsmall cell lung cancer originating from two separate studies: LC_S3 [20] and L511 [21]. Reads for both subjects are stored in the Sequence Read Archive (SRA): LC_S3 normal SRA ERX140379 and tumor SRA ERX040278; L511 normal SRA SRX1742067 and tumor SRA SRX1741906. Reads were extracted into split FASTQs using fasterq-dump (https://github.com/ncbi/sratools), then aligned to GRCh37 using STAR [22].

3 Methods

3.1 Selecting Reference

arcasHLA reference command builds a reference using the IMGT/HLA database file which provides the entirety of the known sequence as well as exon junctions. In building the database, the algorithm concatenates together the exons for each allele as well as adding 3′ and 5′ untranslated regions to the reference in order to draw away reads that may map to those regions. In addition, it cuts short alleles with stop losses, which continue into what is normally the untranslated region, preventing reads from erroneously being attributed to these alleles.

arcasHLA then uses Kallisto to build an index of the reference FASTA. This index is a De Bruijn graph, a graph in which nodes are k-mers (k-length sequences) and edges between them represent the next base in the sequence. For example, the 4-mer AAAA has edges that lead to AAAA, AAAT, AAAG, and AAAC. The maximum k-mer length allowed by many transcript quantifiers, including Kallisto, is 31. This is our default value too ($k = 31$). Unlike de-novo de Bruijn graphs, the one constructed from the FASTA reference only contains k-mers that are observed within that reference. In addition, each node has its own "equivalence class," the class of transcripts in which that k-mer is observed.

arcasHLA allows users to select which version of the IMGT/HLA reference they prefer to use. This is particularly useful when comparing results between tools with fixed versions (e.g., Polysolver [23] version 3.10.0, Optitype [24] version 3.14.0, HLAProfiler [25] version 3.24.0, seq2HLA version 3.6.0 class I, 3.7.0 class II [26]). With each iteration of the IMGT/HLA database, new alleles and sequences are added along with updates to the names and sequences of alleles in the previous version. In some rare cases, with corrections to old sequences, some alleles are found to be the same and are merged together, retiring one of the names.

```
arcasHLA reference \

 [--version VERSION] OR [--commit COMMITHASH] \
 [--rebuild] \
 [--version_list]
```

1. Versions 3.36.0 and older can be selected by their version name using --version. To see a list of the available versions, run arcasHLA reference with the --version_list flag.
2. Specific versions can also be fetched by their commithash using --commithash.
3. To force arcasHLA to rebuild the reference, use the --rebuild flag.
4. For the purposes of this protocol, we ran arcasHLA using the IMGT/HLA version it was validated on for the gold-standard benchmark, version 3.24.0:

```
arcasHLA reference --version 3.24.0 -v
```

3.2 Selecting a Population

arcasHLA uses population frequencies in order to distribute slightly more reads to more common alleles without overriding evidence of rarer alleles in the observed reads. In addition, when scoring pairs of alleles for both regular and partial typing, allele frequencies are used to break ties between equally likely pairs. With quality, high read count samples, omitting allele frequencies does not harm arcasHLA's performance. However, with low read count or single-end samples, using the individual's population or the Dirichlet prior significantly improves results, with the former outperforming the latter.

3.2.1 (Optional) Using Variants to Infer an Individual's Population

In this section, we outline an optional step that may be performed in order to infer an individual's population (when it is missing from the clinical data) for the purpose of supplying this population origin to the arcasHLA genotyping algorithm and increase the typing accuracy. Indeed, variants called from next-generation sequencing can be used to infer the population of an individual of unknown origin using a database such as the 1000 Genomes Project (*see* Subheading 2.3.1), which includes the following super populations: HG02879 African (AFR), HG02147 ad-mixed American (AMR), HG02136 East Asian (EAS), HG01700 European (EUR), and HG03928 South Asian (SAS). Using VCF tools [27], one can extract the exon region-enriched single-nucleotide polymorphisms (SNPs) identified by ExAC [28] as being indicative of population of origin from the Phase 3 genotype VCFs as follows:

```
vcftools --snps EXAC_rsids.txt --out snps.chr1 --gzvcf ALL.
chr1.phase3_shapeit2_mvncall_integrated_v5a.20130502.geno-
types.vcf.gz
```

Next, principal component analysis on the genotype matrix indexed by the ExAC population SNPs and individuals from the 1000 Genomes Project shows that the top three principal components are sufficient to separate super populations (although there still exists some blending between these populations, seen particularly between the ad-mixed American, European, and African populations, as can be expected). For "unknown" individuals whose population origin we have to identify, one can represent their population SNPs called from RNAseq [29] as an array and then perform clustering of these unknown samples in the principal component space together with the 1000 Genomes cohort in order to classify their unknown super populations (Fig. 5). This way, after calling variants from the BAM file of an individual with unknown population origin, it may be possible to infer the correct super population and use the latter as additional input for HLA typing. This step may be omitted altogether (the default value for arcasHLA is a population Prior composed of the weighted mean from all the populations in the Frequency database (*see* Subheading 2.3.2). This prior performs well in most cases.

3.3 Read Extraction

arcasHLA performs best when extracting reads from input BAM files (*see* **Note 1**). Unlike many tools which extract reads from HLA regions along chromosome 6, arcasHLA merely extracts all reads mapped to chromosome 6. This ensures that paired reads in the specified HLA region whose mate maps outside of it but to a similar gene along the length of chromosome 6 (e.g., an HLA pseudogene) are both extracted. arcasHLA automatically determines whether the reference to which the reads have been aligned contains the "chr" prefix or any of the HLA decoys or alternate chromosome 6 haplotypes.

```
arcasHLA extract <BAM> \

 [--paired] \
 [--unmapped] \
 [-o, --outdir /path/to/outdir/] \
```

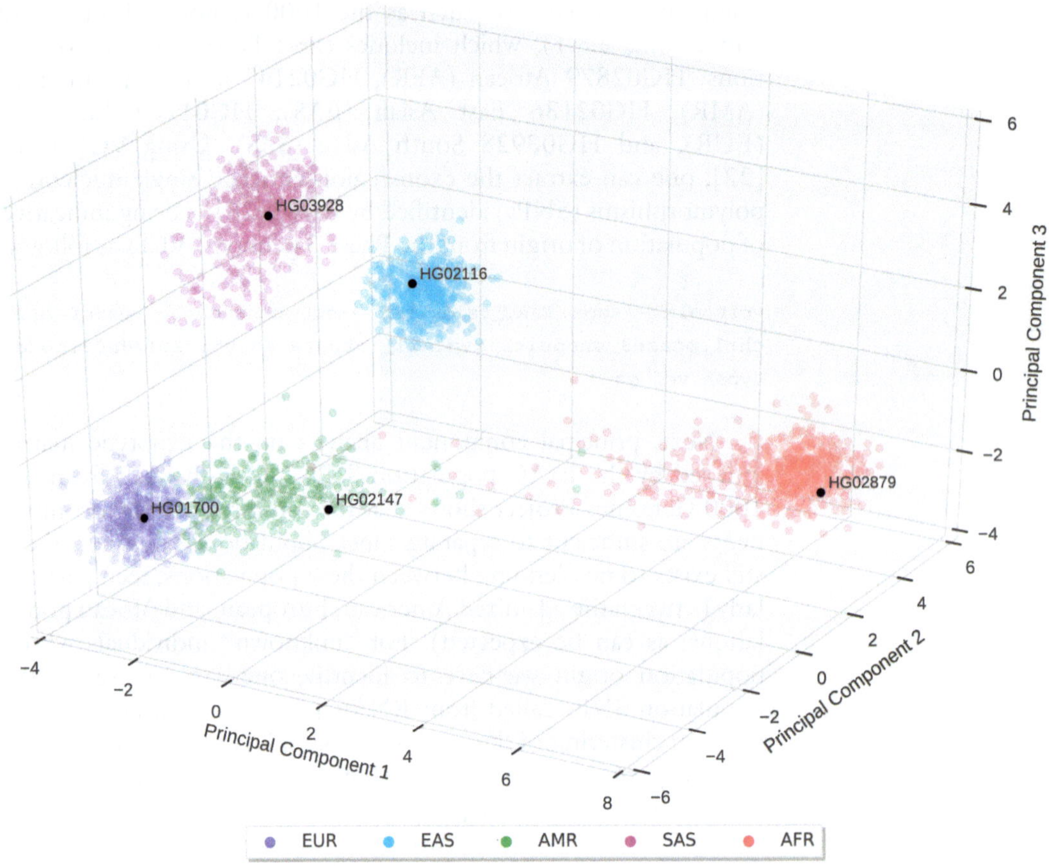

Fig. 5 5770 Phase 3 genotypes of 2504 individuals from 1000 Genomes (*see* Subheading 2.3.1) dprojected onto the top three principal components. Hypothetical test of "unknown" individuals that lie within the cluster of their true population of origin

```
[--temp /tmp/] \
[-t, --threads INT] \
[--keep_files] \
[-v, --verbose]
```

1. <BAM> denotes the input BAM file.
2. If your input BAM was aligned from paired-end reads, you must include the flag paired.
3. If your bam was aligned with a mapper that flags multimapping reads as unmapped (STAR for example) and has a relatively low read count, it is useful to extract unmapped reads along with chromosome 6 using the flag --unmapped. This is particularly helpful in samples with relatively low read counts and low uniquely mapped reads.
4. Both the target ouptut directory and the temporary directory can be set using --outdir /path/to/outdir/ and --temp /path/to/temp/, respectively.

5. The number of threads used by Samtools when extracting and sorting and Bedtools when converting to FASTQ is set by --threads INT.
6. To keep intermediate files for debugging issues, include --keep_files.
7. To view the information that is written to the log file in the terminal during the run, include the --verbose flag.

   For example, we extracted reads from each BAM file (LC_S3_normal.bam, LC_S3_tumor.bam, L511_normal.bam, L511_tumor.bam), using variations on the following command:

```
arcasHLA extract ~/bam/LC_S3_normal.bam --paired -o ~/ex-
tracted/ -v -t 8
```

3.4 Genotyping

3.4.1 Running arcasHLA Genotype

Next, to infer the most likely genotype (no partial alleles) from the observed reads, input the FASTQs produced by the extraction step to arcasHLA genotype. Using Kallisto's pseudo feature, reads are broken into k-length sequences and hashed into the index to find their compatible HLA allele transcripts. After the first few k-mers in the read, if it lies in a conserved region (nodes lie in a single path), Kallisto will skip these nodes until it either reaches a point of divergence or the end of the read. In the latter case, it will check the final k-mer to ensure that the read is properly aligned. It then takes the intersection of the equivalence classes of these nodes to find the read's equivalence class. A counter for this particular intersection is incremented each time it is observed. In the end, pseudo outputs a list of equivalence classes and the number of compatible reads.

After using Kallisto's pseudoalign command, arcasHLA performs transcript quantification for each individual gene using an expectation-maximization (EM) algorithm. Given the individual's population, it first distributes reads between alleles from the same equivalence class proportional to their allele frequency. This nudges the algorithm in the correct direction without overriding the influence of the observed reads. Next, it iteratively calculates overall abundance for each allele and redistributes reads between alleles in the same class based on their updated abundances. After a specified number of iterations (recommended is 20 for paired, 4 for single and low read count), alleles with low support are dropped (abundance less than a tenth that of the most abundant allele). Once transcript quantification terminates, at least one candidate allele will remain under consideration for the final genotyping output.

In an ideal case (usually when HLA expression is sufficiently high), a homozygote will have a single allele remaining, while a heterozygote will have two alleles. However, this is not always the case, and sometimes, more than two alleles are returned (seen far more often in very low read count or RIN samples). When this occurs, genotype compares the percent observed reads explained by each possible pair of alleles, selecting the one that explains the highest proportion. At this point, if there is a tie between pairs, allele frequencies are used to break the tie. Next, all genes that have two remaining alleles are checked for homozygosity. If the minor allele count to major allele count ratio is below 0.15, the gene is flagged as homozygous for the major allele.

```
arcasHLA genotype <FASTQ1> <FASTQ2> \

 [--genes GENES] \
 [--population POPULATION]
```

1. <FASTQ1> and <FASTQ2> denote the FASTQ files produced by the extract module.
2. By default, all HLA genes in the IMGT/HLA database are genotyped as long as enough reads are uniquely mapped to the locus. To specify a list of genes, include a comma-separated list of the genes behind --genes.
3. Include the sample's population (either asian_pacific_islander, black, caucasian, hispanic, native_american or prior) behind --population. If you want to genotype the sample without allele frequencies, set this flag to none.
4. Like extract, genotype takes the following flags: --threads, --temp, --outdir, --keep_files, and --verbose.

For example, we ran arcasHLA genotype on the extracted fastqs for LC_S3 and L511 as follows:

```
 arcasHLA genotype --population asian_pacific_islander -o ~/
genotypes/ -v -t 8
 ~/extracted/LC_S3_normal.extracted.1.fq.gz ~/extracted/
LC_S3_normal.extracted.2.fq.gz
```

The resulting genotype.json file contains the final genotype for the sample. If a gene is called as homozygous, only a single allele will be listed under that gene; if not enough reads are aligned to a gene, it will be omitted from the results. Below is the contents of LC_S3_normal.genotype.json.


```
{"A": ["A*24:02:01", "A*02:06:01"],
"B": ["B*48:01:01", "B*40:02:01"],
"C": ["C*08:01:01", "C*03:04:01"],
"DMA": ["DMA*01:01:01", "DMA*01:02"],
"DMB": ["DMB*01:01:01"],
```

```
"DOA": ["DOA*01:01:01", "DOA*01:01:02"],
"DOB": ["DOB*01:04:01", "DOB*01:01:01"],
"DPA1": ["DPA1*02:02:02"],
"DPB1": ["DPB1*05:01:01"],
"DQA1": ["DQA1*01:03:01", "DQA1*06:01:01"],
"DQB1": ["DQB1*06:01:01", "DQB1*03:01:01"],
"DRA": ["DRA*01:01:01", "DRA*01:02:02"],
"DRB1": ["DRB1*08:03:02", "DRB1*12:02:01"],
"DRB3": ["DRB3*03:01:01"],
"E": ["E*01:03:01", "E*01:03:02"],
"F": ["F*01:01:01"],
"H": ["H*01:01:01"],
"J": ["J*01:01:01"],
"K": ["K*01:02"],
"L": ["L*01:02", "L*01:01:01"]}
```


In addition to the options above, arcasHLA genotype's EM transcript quantification and scoring algorithms can be customized. To view the default parameters, run arcasHLA genotype with the --help flag.

1. --min_count INT denotes the minimum number of reads uniquely mapped to an HLA locus needed to continue genotyping. Default is 75.
2. --tolerance FLOAT denotes the tolerance (the difference in allele abundances between iterations) at which convergence is declared during EM.
3. --max_iterations INT denotes the maximum number of iterations of EM. By default, this is 1000 iterations; however, all samples used to build and test arcasHLA terminate well under this limit.
4. --drop_iterations INT denotes the number of EM iterations before arcasHLA begins to drop low support alleles. 20 is recommended for paired-end, while 4 is recommended for single-end or low read count samples.
5. --drop_threshold FLOAT denotes the proportion of the maximum abundance an allele needs to not be dropped.
6. --zygosity_threshold denotes the threshold of the minor to major allele read count ratio used to determine zygosity.

arcasHLA genotype outputs an intermediate alignment file before EM and scoring. This file can be used to rerun genotyping with different parameters, such as different population for HLA frequencies or a smaller minimum read count:

```
arcasHLA genotype [options] <sample.alignment.p>
```

3.4.2 Running arcasHLA Partial

arcasHLA partial takes the genotypes produced in arcasHLA genotype and looks for possible partial alleles that can explain the observed reads better than the complete genotype (*see* **Note 2**). The partial typing reference includes all possible exon combinations (that include the binding region for that particular allele). For example, if a class I partial allele only has exons 2, 3, and 4 sequenced, the reference includes a contig containing exons 2 and 3 and a contig containing exons 2, 3, and 4. In addition, all alleles with fully-sequenced coding sequences are included with all exon combinations. This allows for direct comparison of equivalence classes after pseudoalignment. All nonpartial alleles not called in genotyping are omitted from transcript quantification and the subsequent scoring of allele pairs. It is not recommended to call partial alleles from single-end reads as it is far more difficult to resolve ambiguities, even more so than with nonpartial genotyping due to the sheer number of possible alleles and the different lengths of transcripts. Some partial alleles are a part of the same P-group or G-group as a nonpartial allele, so it is not always necessary to call partial alleles, particularly, if one is interested in lower resolution calls. For the purposes of this tutorial, we have omitted this step:

```
arcasHLA partial [options] -G <genotype.json> <FASTQ1>
<FASTQ2>
```

All other options for arcasHLA partial are the same as arcasHLA genotype. This will output an intermediate partial alignment file, partial_alignment.p, and partial genotyping results, partial_genotype.json.

3.4.3 Merging Genotyping Results

arcasHLA provides a merge command that takes all genotype.json and partial_genotype.json files in a specified directory and outputs a json file and a tab-separated table containing all calls for all samples (genotypes and partial_gentoypes) for easy usage. This command will also output a table containing the gene-aligned read counts for each sample:

```
arcasHLA merge [--indir INDIR] \
[--genes GENES] \
[--outdir OUTDIR] \
[--run RUNNAME]
```

The structure of the merged json is a dictionary in which the key is the sample name (prefix of the original BAM file) and the value is the dictionary from the sample's genotype.json. Using the genes parameter with a comma-separated list of genes filters the output. For example, we merged the results from the four samples with the following command (Tables 1 and 2):

Table 1
3-field HLA class I genotypes for normal and tumor samples for LC_S3 and L511 in the merged file, classI.genotypes.tsv

Subject	A1	A2	B1	B2	C1	C2
LC_S3_normal	A*24:02:01	A*02:06:01	B*48:01:01	B*40:02:01	C*08:01:01	C*03:04:01
LC_S3_tumor	A*24:02:01	A*24:02:01	B*48:01:01	B*40:02:01	C*08:01:01	C*03:04:01
L511_normal	A*68:01:01	A*68:01:02	B*15:01:01	B*44:02:01	C*07:04:01	C*03:04:01
L511_tumor	A*68:01:01	A*68:01:02	B*44:02:01	B*15:01:01	C*07:04:01	C*03:04:01

Table 2
3-field HLA class II genotypes for normal and tumor samples for LC_S3 and L511 in the merged file, classII.genotypes.tsv

Subject	DPB11	DPB12	DQB11	DQB12	DRB11	DRB12
LC_S3_normal	DPB1*05:01:01	DPB1*05:01:01	DQB1*06:01:01	DQB1*03:01:01	DRB1*08:03:02	DRB1*12:02:01
LC_S3_tumor	DPB1*05:01:01	DPB1*05:01:01	DQB1*06:01:01	DQB1*03:01:01	DRB1*08:03:02	DRB1*12:02:01
L511_normal	DPB1*03:01:01	DPB1*04:01:01	DQB1*04:02:01	DQB1*03:03:02	DRB1*07:01:01	DRB1*08:01:01
L511_tumor	DPB1*03:01:01	DPB1*04:01:01	DQB1*04:02:01	DQB1*03:03:02	DRB1*07:01:01	DRB1*08:01:01

```
arcasHLA merge --indir ~/genotypes/ --run classI --genes A,B,C
arcasHLA merge --indir ~/genotypes/ --run classII --genes
DPB1,DQB1,DRB1
```

3.4.4 Converting HLA Genotypes

arcasHLA convert provides an easy way to decrease HLA genotyping resolution or convert to G-grouping or P-grouping, taking in either an arcasHLA-generated genotypes.json or genotypes.tsv file or any file with a similar architecture. Valid values for –resolution include the number of fields desired: 1, 2, or 3. In addition, P-group or G-group will convert the input HLA alleles to their respective groups (*see* **Note 3**):

```
arcasHLA convert <file> [--outfile output.tsv] [--resolution
RESOLUTION]
```

For the purposes of this protocol, we converted the three-field HLA genotypes to P-groups (Tables 3 and 4):

```
arcasHLA convert classI.genotypes.tsv -o classI.pgroup.geno-
types.tsv -r p-group
arcasHLA convert classII.genotypes.tsv -o classII.pgroup.gen-
otypes.tsv -r p-group
```

3.4.5 Interpreting Genotyping Results

Lowering the resolution of an HLA allele is useful in multiple circumstances. Take for example, the HLA-A genotype produced for L511 in both the normal and tumor samples. Although there is sufficient support for the presence of both HLA-A∗68:01:01 and HLA-A∗68:01:02, this individual is likely homozygous at the protein-coding level; the third field denotes synonymous changes in coding regions, any allele sharing the first two fields will produce the same protein product. In many studies, particularly those involving HLA alleles and their preference for specific neoantigens, changes outside of the antigen-binding region are of less importance (with the exception of null variants).

Table 3
P-group HLA class I genotypes for normal and tumor samples for LC_S3 and L511 in the merged file, classI.pgroup.genotypes.tsv

Subject	A1	A2	B1	B2	C1	C2
LC_S3_normal	A∗24:02P	A∗02:06P	B∗48:01P	B∗40:02P	C∗08:01P	C∗03:04P
LC_S3_tumor	A∗24:02P	A∗24:02P	B∗48:01P	B∗40:02P	C∗08:01P	C∗03:04P
L511_normal	A∗68:01P	A∗68:01P	B∗15:01P	B∗44:02P	C∗07:04P	C∗03:04P
L511_tumor	A∗68:01P	A∗68:01P	B∗44:02P	B∗15:01P	C∗07:04P	C∗03:04P

Table 4
P-group HLA class II genotypes for normal and tumor samples for LC_S3 and L511 in the merged file, classII.pgroup.genotypes.tsv

Subject	DPB11	DPB12	DQB11	DQB12	DRB11	DRB12
LC_S3_normal	DPB1*05:01P	DPB1*05:01P	DQB1*06:01P	DQB1*03:01P	DRB1*08:03P	DRB1*12:02P
LC_S3_tumor	DPB1*05:01P	DPB1*05:01P	DQB1*06:01P	DQB1*03:01P	DRB1*08:03P	DRB1*12:02P
L511_normal	DPB1*03:01P	DPB1*04:01P	DQB1*04:02P	DQB1*03:03P	DRB1*07:01P	DRB1*08:01P
L511_tumor	DPB1*03:01P	DPB1*04:01P	DQB1*04:02P	DQB1*03:03P	DRB1*07:01P	DRB1*08:01P

In some cases, changes denoted in the third or fourth field influence the expression type of the allele (e.g., L for low expression or Q for questionable expression). However, without proper ground truth to test HLA typing tools, calls beyond the second field are not well validated. When working with small sample sizes and diverse populations, lowering the resolution of HLA alleles facilitates association tests by increasing the size of groups sharing the same allele.

In these tutorial runs (*see* Subheading 2.3.2), arcasHLA genotyping calls do not differ between the normal and tumor samples (Tables 1 and 2). However, for a few cases, genotyping will fail or return different alleles, possibly due to mutations or changes in expression. For example, although there are sufficient reads in the L511 normal sample to genotype HLA-G, there are insufficient reads aligned to HLA-G alleles in the tumor sample, causing genotyping to fail. For the patient labelled as LC_S3, arcasHLA calls two HLA-A alleles (namely A*24:02:01 and A*02:06:01) in the normal sample, but it returns a homozygous call for HLA-A in the tumor sample (since only the presence of allele A*24:02:01 is inferred by the algorithm). The summaries of the genotyping results in the log files LC_S3_normal.genotype.log and LC_S3_tumor.genotype.log show that all reads aligned to the HLA-A locus are attributed to A*24:02:01 in the tumor sample but that there is a relatively even split between the two alleles A*24:02:01 and A*02:06:01 in the normal sample:

```
>LC_S3_normal.genotype.log
 [genotype] Genotyping HLA-A
 [genotype] 55580 reads aligned to HLA-A in 2860 classes ...
 [genotype] Top alleles by abundance:
 allele abundance A*24:02:01 58.66%
  A*02:06:01 41.34%
 [genotype] Pairs by % explained reads:
  allele pair explained
  A*24:02:01, A*02:06:01 98.62%
 [genotype] Checking zygosity
 [genotype] Likely heterozygous: minor/major nonshared count
0.70

>LC_S3_tumor.genotype.log
 [genotype] Genotyping HLA-A
 [genotype] 39669 reads aligned to HLA-A in 2418 classes ...
 [genotype] Top alleles by abundance:
 allele abundance A*24:02:01 100.00%
```

This inconsistency highlights the need to type HLAs from normal samples rather than tumor. Genotypes produced by arcasHLA can then be used to check for associations between either homozygosity at any locus or P-groups, G-groups, and higher resolution alleles with differences in survival or other phenotypes of interest.

After alignment, arcasHLA genotype outputs a file, genes.json, containing the uniquely mapped read count, number of equivalence classes and relative abundance for each HLA and HLA-associated gene (*see* **Note 4**). In our example, the normal samples for subjects LC_S3 and L511 follow similar trends for relative gene abundance which is expected as the samples are derived from the same tissue type (Fig. 6). Relative abundance levels are expected to vary between individuals as well as at different time-points for a single individual. The higher proportion of class II abundance in L511 may be indicative of greater immune cell infiltration of the lung tissue.

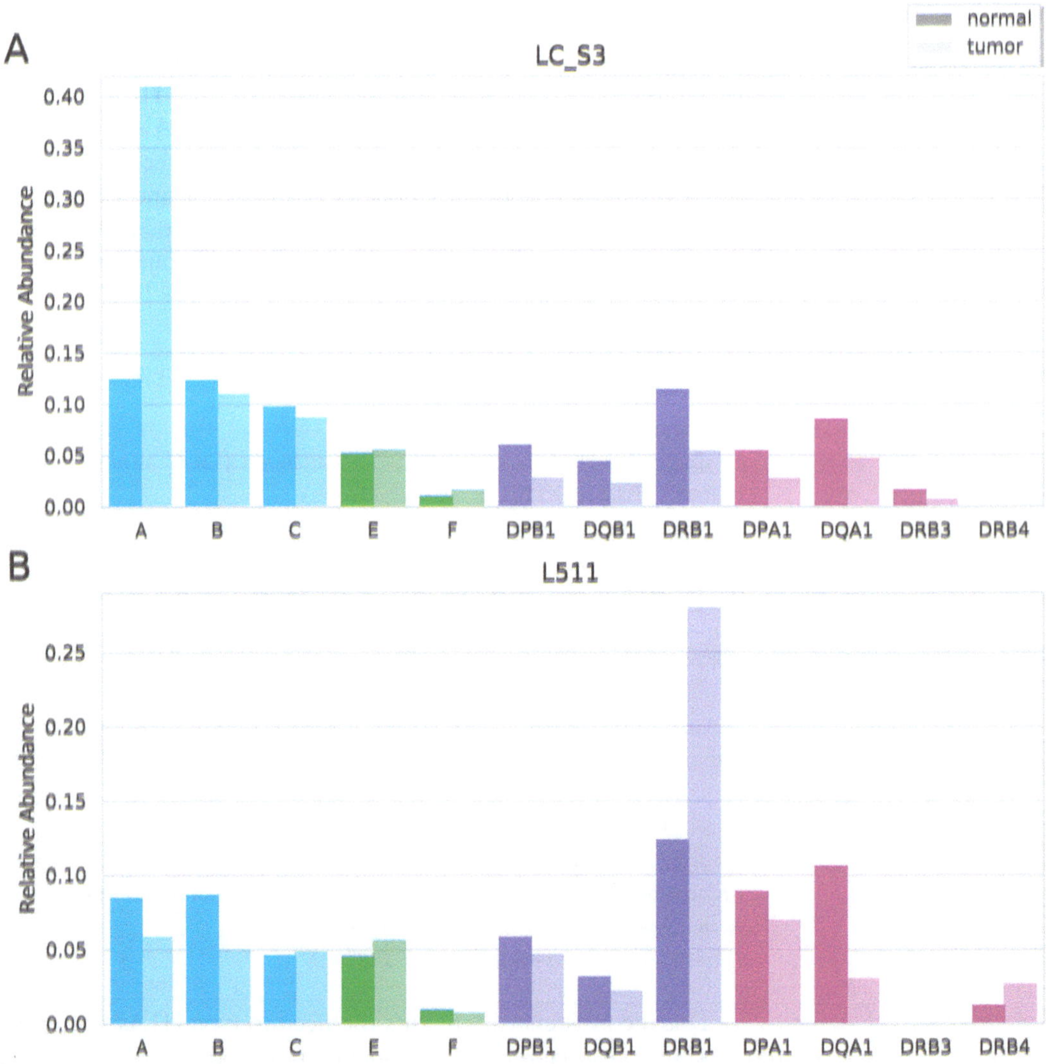

Fig. 6 Relative HLA gene abundances for (**A**) LC_S3_normal and LC_S3_tumor and (**B**) L511_normal and L511_tumor. Only genes with ≥0.01 abundance are shown

Strikingly, HLA-A abundance in LC_S3_tumor and HLA-DRB1 in L511_tumor are highly elevated relative to the abundance in the respective normal samples. Warranting further analysis, changes in relative abundance, especially in tumor samples, may be due to multiple factors, including normal variation of mRNA expression levels [30], loss of heterozygosity, or epigenetic modifications of the HLA loci. Abundances may also differ based on the level of immune infiltration into the sample. In short, arcasHLA provides a fast HLA genotyping solution, and it furthermore produces relative gene expression levels that are particularly useful for cancer studies with matched normal/tumor RNA sequencing.

4 Notes

1. It is possible to run arcasHLA genotyping and partial typing with unfiltered FASTQ files; however, this usage has not been validated to produce accurate results and should be used with caution.
2. Due to the small number of partial alleles in the 1000 Genomes benchmark set, arcasHLA partial typing is not as well validated as typing for nonpartial alleles. In addition, because arcasHLA typing is based on complete sequences (i.e., including exons outside the antigen-binding region), there exists uncertainty in confidently calling a partial allele over an allele that is completely sequenced.
3. Alleles with the same first two fields may fall into different G-groups (e.g., A*01:01:01 belongs to A*01:01:01G, while A*01:01:89 does not due to a synonymous change in the antigen-binding region); however, using resolution of 2 and G-group at the same time will place them under the same 2-field allele.
4. Relative gene abundance is determined prior to genotyping and allele selection. As such, the read count from which it was derived may include reads aligned to that particular locus but not the final genotype. In samples where the final genotype explains a relatively low percentage of observed reads, this relative abundance may be artificially inflated.

References

1. Elliott JM, Yokoyama WM (2011) Unifying concepts of MHC-dependent natural killer cell education. Trends Immunol 32:364–372. https://doi.org/10.1016/j.it.2011.06.001
2. Owen JA, Punt J, Stranford SA et al (2018) Kuby immunology, 8th edn. W.H. Freeman, Macmillan Learning, New York
3. Hughes AL (1995) Origin and evolution of HLA class I pseudogenes. Mol Biol Evol 12:247–258. https://doi.org/10.1093/oxfordjournals.molbev.a040201
4. Robinson J, Halliwell JA, McWilliam H et al (2013) IPD—The immuno polymorphism

database. Nucleic Acids Res 41:1234–1240. https://doi.org/10.1093/nar/gks1140

5. Sidney J, Peters B, Frahm N et al (2008) HLA class I supertypes: a revised and updated classification. BMC Immunol 9:1. https://doi.org/10.1186/1471-2172-9-1
6. Orenbuch R, Filip I, Comito D et al (2019) arcasHLA: high resolution HLA typing from RNA seq. Bioinformatics. https://doi.org/10.1093/bioinformatics/btz474
7. Schumacher TN, Schreiber RD (2015) Neoantigens in cancer immunotherapy. doi:https://doi.org/10.1126/science.aaa4971
8. Chowell D, Morris LG, Grigg CM et al (2018) Patient HLA class I genotype influences cancer response to checkpoint blockade immunotherapy. Science 359:582–587. https://doi.org/10.1126/science.aao4572
9. McGranahan N, Rosenthal R, Hiley CT et al (2017) Allele-specific HLA loss and immune escape in lung cancer evolution. Cell 171:1259–1271.e11. https://doi.org/10.1016/j.cell.2017.10.001
10. Cock PJ, Antao T, Chang JT et al (2009) Biopython: freely available Python tools for computational molecular biology and bioinformatics. Bioinformatics 25:1422–1423. https://doi.org/10.1093/bioinformatics/btp163
11. Oliphant TE (2006) Guide to NumPy
12. Oliphant TE (2007) SciPy: open source scientific tools for Python. Comput Sci Eng. https://doi.org/10.1109/MCSE.2007.58
13. Mckinney W et al (2019) pandas: powerful Python data analysis toolkit
14. Li H, Handsaker B, Wysoker A et al (2009) The sequence alignment/map format and SAMtools. Bioinformatics 25:2078–2079. https://doi.org/10.1093/bioinformatics/btp352
15. Quinlan AR, Hall IM (2010) BEDTools: a flexible suite of utilities for comparing genomic features. Bioinformatics 26:841–842. https://doi.org/10.1093/bioinformatics/btq033
16. Lappalainen T, Sammeth M, Friedländer M et al (2017) Transcriptome and genome sequencing uncovers functional variation in humans. Cancer Epidemiol Biomarkers 4:1264–1272. https://doi.org/10.1016/S2214-109X(16)30265-0
17. Gourraud PA, Khankhanian P, Cereb N et al (2014) HLA diversity in the 1000 genomes dataset. PLoS One 9:e97282. https://doi.org/10.1371/journal.pone.0097282
18. González-Galarza FF, Takeshita LY, Santos EJ et al (2015) Allele frequency net 2015 update: New features for HLA epitopes, KIR and disease and HLA adverse drug reaction associations. Nucleic Acids Res 43:D784–D788. https://doi.org/10.1093/nar/gku1166
19. Gragert L, Madbouly A, Freeman J et al (2013) Six-locus high resolution HLA haplotype frequencies derived from mixed-resolution DNA typing for the entire US donor registry. Hum Immunol 74:1313–1320. https://doi.org/10.1016/j.humimm.2013.06.025
20. Seo JS, Ju YS, Lee WC et al (2012) The transcriptional landscape and mutational profile of lung adenocarcinoma. Genome Res 22:2109–2119. https://doi.org/10.1101/gr.145144.112
21. Mezheyeuski A, Bergsland CH, Backman M et al (2018) Multispectral imaging for quantitative and compartment-specific immune infiltrates reveals distinct immune profiles that classify lung cancer patients. J Pathol 244:421–431. https://doi.org/10.1002/path.5026
22. Dobin A, Davis CA, Schlesinger F et al (2013) STAR: ultrafast universal RNA-seq aligner. Bioinformatics 29:15–21. https://doi.org/10.1093/bioinformatics/bts635
23. Shukla SA, Rooney MS, Rajasagi M et al (2015) Comprehensive analysis of cancer-associated somatic mutations in class I HLA genes. Nat Biotechnol 33:1152–1158. https://doi.org/10.1038/nbt.3344
24. Szolek A, Schubert B, Mohr C et al (2014) OptiType: precision HLA typing from next-generation sequencing data. Bioinformatics 30:3310–3316. https://doi.org/10.1093/bioinformatics/btu548
25. Buchkovich ML, Brown CC, Robasky K (2017) HLAProfiler utilizes k-mer profiles to improve HLA calling accuracy for rare and common alleles in RNA-seq data. Genome Med 9:86. https://doi.org/10.1186/s13073-017-0473-6
26. Boegel S, Löwer M, Schäfer M et al (2012) HLA typing from RNA-seq sequence reads. Genome Med 4:102. https://doi.org/10.1186/gm403
27. Danecek P, Auton A, Abecasis G et al (2011) The variant call format and VCFtools. Bioinformatics 27:2156–2158. https://doi.org/10.1093/bioinformatics/btr330
28. Lek M, Karczewski KJ, Minikel EV et al (2016) Analysis of protein-coding genetic variation in 60,706 humans. Nature 536:285–291. https://doi.org/10.1038/nature19057
29. GATK pipeline (2019) https://gatkforums.broadinstitute.org/gatk/discussion/3892/the-gatk-best-practices-for-variant-calling-on-rnaseq-in-full-detail. Accessed 31 May 2019
30. Johansson T et al (2018) HLA RNAseq reveals high allele-specific variability in mRNA expression. bioRxiv 413534. doi: https://doi.org/10.1101/413534

Chapter 6

Rapid High-Resolution Typing of Class I HLA Genes by Nanopore Sequencing

Chang Liu and Rick Berry

Abstract

Nanopore sequencing, enabled initially by the MinION device from Oxford Nanopore Technologies (ONT), is the only technology that offers portable, single-molecule sequencing and ultralong reads. The technology is ideal for the typing of human leukocyte antigen (HLA) genes for transplantation and cancer immunotherapy. However, such applications have been hindered by the high error rate of nanopore sequencing reads. We developed the workflow and bioinformatic pipeline, Athlon (accurate typing of human leukocyte antigen by Oxford Nanopore), to perform high-resolution typing of Class I HLA genes by nanopore sequencing. The method features a novel algorithm for candidate allele selection, followed by error correction through consensus building. Here, we describe the protocol of using Athlon packaged in a VirtualBox image for the above application.

Key words Human leukocyte antigen (HLA), HLA typing, Targeted sequencing, Nanopore sequencing

1 Introduction

High-resolution typing of human leukocyte antigen (HLA) genes provides critical information for the evaluation of the histocompatibility between transplant recipient and donor pairs [1]. Also, the HLA typings of cancer patients impact the presentation of tumor-specific neoantigens to T cells [2] and the treatment response to checkpoint blockade immunotherapy [3]. HLA typing in clinical laboratories are increasingly performed by next-generation squencing (NGS) on the Illumina [4], Ion Torrent [5], and PacBio [6–8] platforms. These technologies have substantially reduced the ambiguities encountered by Sanger sequencing and demonstrated excellent accuracy and throughput. However, the library preparation for most NGS platforms remains a lengthy process and may be prone to bias in regions with high GC content [9]. The length of sequencing reads are limited to hundreds of bases for Illumina and Ion Torrent, affecting their ability to phase variants that are far apart. Batching of

Sebastian Boegel (ed.), *Bioinformatics for Cancer Immunotherapy: Methods and Protocols*, Methods in Molecular Biology, vol. 2120, https://doi.org/10.1007/978-1-0716-0327-7_6,

samples is necessary for most HLA typing products to achieve cost-effectiveness.

Oxford Nanopore Technologies' (ONT) MinION is a portable sequencer with the size of a regular mobile phone. It performs parallel, single-molecule sequencing at hundreds of nanopores on a flow cell [10]. Single-stranded DNA or RNA molecules are ratcheted through these nanopores by attached motor proteins, and the dynamic changes in electric current across the pores are recorded and coverted to sequencing reads in real time. The superior length of the nanopore sequencing reads, theoretically as long as the fragments being sequenced, and the extreme portability of the MinION device make the technology an attractive option for delivering point-of-care molecular testing at the forefront of precision medicine.

Nanopore sequencing is especially suitable for applications that require a rapid turnaround time and haplotype phasing of variants across thousands of bases such as high-resolution HLA typing. However, the high error rate of the nanopore sequencing reads, ranging between 5 and 15% [11], and the lack of a reliable variant caller for diploid genomes have made such applications challenging. Recently, we developed Athlon (accurate typing of human leukocyte antigen by Oxford Nanopore), a bioinformatic pipeline that analyzes nanopore sequencing data using amplicons from class I HLA genes (*HLA-A*, *HLA-B*, and *HLA-C*) [12]. Athlon first maps gene-specific reads to reference alleles with unique exons 2 and 3 sequences in the IMGT database. These three-field reference alleles are treated as leaves (e.g., A*68:01:02G), with two tiers of parent nodes representing the two-field typing (e.g., A*68:01) and one-field typing (e.g., A*68) that comprise subgroups of reference alleles. Read coverage of each reference allele is summed at each two-field node and then at each one-field node. Up to two candidate alleles are selected from the top-ranked one-field nodes, two-field nodes, and three-field leaves, sequentially based on the read coverage. If the ratio of the coverage of the minor allele to that of the major allele is lower than the statistical thresholds (0.23 at the one-field level and 0.71 at the two-field level), only the major allele is called for a presumed homozygous genotype. Once the candidate alleles are selected, all gene-specific reads are realigned to the candidate alleles to build consensus sequences, which are subsequently used to query against all reference alleles to determine the final typing result. Our initial validation of Athlon using two independent data sets showed a 100% accuracy for the key exons of class I HLA genes.

Here, we describe the procedures for using Athlon to analyze whole-gene amplicon sequencing data generated on the MinION device (R9.4 chemistry) to determine the typing of class I HLA genes at the G-group resolution.

2 Materials

2.1 Software

1. We created a VirtualBox image to run Athlon and its dependencies on Ubuntu 16.04 LTS. The image includes the script for the Athlon pipeline, reference files, sample data set, and the following dependencies: BLASR version 2.0.0 [13], Bedtools version 2.25.0 [14], FreeBayes version 1.1.0 [15], vcf2fasta, a tool in vcflib version 1.0.0, and Blast version 2.2.3 [16]. The image can be downloaded at https://www.platformstl.com/post/hla-typing.
2. The current version of Oracle VM VirtualBox for Windows, OS X, or Linux host can be downloaded at https://www.virtualbox.org/wiki/Downloads. This chapter is based on VirtualBox Version 5.2.26.

2.2 Hardware

The procedure described in this chapter was tested on a desktop computer operating on Windows 10 with the following system specifications:

1. Processor: Intel® Core™ i7-8700 CPU@3.20GHz.
2. Installed memory (RAM): 32.0 GB.
3. System type: 64-bit Operating System, ×64-based processor.

2.3 Sample Data Set

Three fastq files, one for each of the class I genes (*HLA-A*, *HLA-B*, and *HLA-C*), are included in the /opt/Athlon/data directory. The methods to generate user data set and to separate the sample-level data into locus-specific fastq files are outlined in **Notes 1** and **2**, respectively.

3 Methods

3.1 Loading the VirtualBox Image

1. Install VirtualBox by double-clicking on the executable file downloaded from www.virtualbox.org and following the instructions in the Setup Wizard.
2. Import the downloaded image, HLA.ova, into VirtualBox by going to "File > Important Appliance." Select the image from the "Appliance to Import" screen, and select "Next." From the "Appliance Import" screen, select "Import."
3. Once the Virtual Machine (VM) is running, log in with "user" as the Username and "password" as the Password.

3.2 Run Athlon to Analyze the Sample Data Set

Go to the /opt/Athlon directory and run the Athlon script with this command:

```
./athlon.sh <number of reads>
```

Table 1
Typing result for the HLA-A locus of sample WU01 using 1000 reads

	Allele name	Identity	Alignment length	Mismatch	Number of edits made in candidate allele sequence
Allele 1	A*02:01:01G	100.00	546	0	1
Allele 2	A*80:01:01G	100.00	546	0	0

For example, "./athlon.sh 1000" will randomly sample 1000 reads from each of the locus-specific fastq files in the /opt/Athlon/ data directory and perform analysis to determine the HLA typing at the corresponding locus. *See* **Note 3** for additional information on the number of reads.

3.3 Interpreting the Output

The HLA typing results are listed in the <date&time>.data<read number>.rslt file, and up to two alleles are reported under each sample and locus. An example result is shown for the *HLA-A* locus of the sample data set in Table 1. Following the allele name are four additional columns that report the following parameters:

1. The sequence identity between the consensus sequence from the nanopore reads and the reference sequence of the reported allele.
2. The alignment length covering the key exons in the current version of Athlon (exons 2 and 3 for class I HLA).
3. The number of mismatches between the consensus sequence and the reference sequence.
4. The number of edits made in the sequence of the candidate allele to generate the consensus sequence during error correction.

3.4 Additional Analysis Files

In addition to the result file, Athlon generates a log file and a collection of intermediate analysis files in the rslt<read number> folder.

1. The log file is named <date&time>.data<read number>.log and located under the /opt/Athlon directory. The log file documents key steps of the analysis, including the initial read mapping; identification of candidate alleles based on the ranking of coverage at one-field, two-field, and three-field resolutions; the second read mapping; and consensus building. Information in the log file, especially the top-ranked alleles at different resolutions, may facilitate the troubleshooting when an incorrect genotype is encountered.

2. Intermediate files include (1) bam files from the initial read mapping, (2) csv files listing all one-field, two-field, and three-field typings sorted by coverage, (3) bam files from the second read mapping, (4) vcf files for error correction generated by freebayes, and (4) fasta files of candidate alleles and consensus sequences after error correction. Importantly, the <sample name>_<locus>candi_3f.fasta and <sample name>_<locus>_3fsm.sorted.bam files can be loaded onto the Integrative Genomics Viewer (igv) [17, 18] to visualize the coverage of candidate alleles after the second read mapping.

   See **Note 4** for limitations of the Athlon program.

3.5 Accessing User Data Set

Setting up the "shared folder" will allow access of user data set from within the VM. Select the HLA image in the VM and click "Settings." Go to "Shared Folders." For "Folder Path" and "Folder Name," specify the pathway to and the name of the directory on the host machine, where the user data set will be stored. Restart the VM, and the shared folder will be available under "/media" with the name starting with "sf_." User data files can be added to this folder and then copied to the /opt/Athlon/data directory for analysis in the VM (*see* **Note 2**).

4 Notes

1. Preparation of user data set: User data set can be generated by sequencing whole-gene amplicons from target genes (i.e., *HLA-A*, *HLA-B*, and *HLA-C*) using primers reported previously [19] and following the 1D library preparation protocol from ONT (Ligation Sequencing Kit, SQK-LSK109; https://community.nanoporetech.com/protocols/1d-lambda-control-sqk-lsk109/checklist_example.pdf). Barcodes can be added to amplicons of each sample using the Native Barcoding Kit (EXP-NBD104; https://community.nanoporetech.com/protocols/1d-native-barcoding-genomic-dna/checklist_example.pdf) to allow multiple samples to be sequenced in the same run. We recommend using Porechop (https://github.com/rrwick/Porechop) or Qcat (https://github.com/nanoporetech/qcat) to demultiplex nanopore reads into barcode-specific or sample-specific fastq files.
2. Before analysis by the Athlon program, sample-level sequencing data must be separated into locus-specific fastq files based on the primer sequences. Run the kmer_split.py script in the /opt/Athlon directory using the command below:

```
python3 kmer_split.py -k 7 -cutoff 0.2 -N 100 <fastq file name without ".fastq"> hla_primers_ABC
```

Importantly, the names for the sample-level fastq files must not contain "-" or "_" so that the Athlon program can extract the sample name and locus name during subsequent analysis. The sample-level fastq files should be copied from the shared folder into the opt/Athlon/data directory before running the command. Upon completion of splitting the data, remove the sample-level fastq files from the opt/Athlon/data directory before running the Athlon script.

3. Number of reads: Athlon requires 1000–2000 reads per locus to generate accurate typing results. If the read number specified is higher than the reads available, all available reads in a fastq file will be used for analysis. However, having less reads in the data than requested in the command line may negatively impact the error correction step because the accuracy of the consensus call is affected by the number of reads actually available. Therefore, for a batch of fastq files to be typed, we recommend setting the number of reads no higher than the lowest number of reads available in these fastq files.
4. Limitations of Athlon: The Athlon algorithm for candidate selection is supported by the presence of high sequence similarity within allele groups of class I HLA genes [20]. However, it may not be applicable to class II HLA genes, especially *HLA-DPB1*, where the nomenclature and sequence similarity are disconnected [21]. In addition, Athlon considers key exons only and limits the typing resolution to the G-group level. The algorithm must be expanded to include nonkey exons and introns to maximize the typing resolution. Athlon is also sensitive to allele imbalance, which could result in allele dropout when the coverage of the minor allele is below the threshold for calling heterozygous genotypes. Finally, the error rate of nanopore reads remains high and may not be fully overcome by the error correction step of Athlon, especially for indels. Considering these limitations, the typing results by Athlon need to be interpreted with caution.

References

1. Edgerly CH, Weimer ET (2018) The past, present, and future of HLA typing in transplantation. Methods Mol Biol 1802:1–10. https://doi.org/10.1007/978-1-4939-8546-3_1
2. Chen F, Zou Z, Du J, Su S, Shao J, Meng F, Yang J, Xu Q, Ding N, Yang Y, Liu Q, Wang Q, Sun Z, Zhou S, Du S, Wei J, Liu B (2019) Neoantigen identification strategies enable personalized immunotherapy in refractory solid tumors. J Clin Invest 130:2056–2070. https://doi.org/10.1172/jci99538
3. Chowell D, Morris LGT, Grigg CM, Weber JK, Samstein RM, Makarov V, Kuo F, Kendall SM, Requena D, Riaz N, Greenbaum B, Carroll J, Garon E, Hyman DM, Zehir A, Solit D, Berger M, Zhou R, Rizvi NA, Chan TA (2018) Patient HLA class I genotype influences cancer response to checkpoint blockade immunotherapy. Science 359(6375):582–587. https://doi.org/10.1126/science.aao4572
4. Allen ES, Yang B, Garrett J, Ball ED, Maiers M, Morris GP (2018) Improved accuracy of

clinical HLA genotyping by next-generation DNA sequencing affects unrelated donor search results for hematopoietic stem cell transplantation. Hum Immunol 79(12):848–854. https://doi.org/10.1016/j.humimm.2018.10.008

5. Barone JC, Saito K, Beutner K, Campo M, Dong W, Goswami CP, Johnson ES, Wang ZX, Hsu S (2015) HLA-genotyping of clinical specimens using ion torrent-based NGS. Hum Immunol 76(12):903–909. https://doi.org/10.1016/j.humimm.2015.09.014
6. Mayor NP, Robinson J, McWhinnie AJ, Ranade S, Eng K, Midwinter W, Bultitude WP, Chin CS, Bowman B, Marks P, Braund H, Madrigal JA, Latham K, Marsh SG (2015) HLA typing for the next generation. PLoS One 10(5):e0127153. https://doi.org/10.1371/journal.pone.0127153
7. Mayor NP, Hayhurst JD, Turner TR, Szydlo RM, Shaw BE, Bultitude WP, Sayno J, Tavarozzi F, Latham K, Anthias C, Braund H, Danby R, Perry JR, Wilson MC, Bloor AJ, Clark A, MacKinnon S, Marks DI, Pagliuca A, Potter MN, Russell NH, Thomason KJ, Madrigal JA, Marsh SGE (2018) Better HLA matching as revealed only by next generation sequencing technology results in superior overall survival post-allogeneic haematopoietic cell transplantation with unrelated donors. Biol Blood Marrow Transplant 24(3):S46
8. Turner TR, Hayhurst JD, Hayward DR, Bultitude WP, Barker DJ, Robinson J, Madrigal JA, Mayor NP, Marsh SGE (2018) Single molecule real-time DNA sequencing of HLA genes at ultra-high resolution from 126 international HLA and immunogenetics workshop cell lines. Hla 91(2):88–101. https://doi.org/10.1111/tan.13184
9. Lan JH, Yin Y, Reed EF, Moua K, Thomas K, Zhang Q (2015) Impact of three Illumina library construction methods on GC bias and HLA genotype calling. Hum Immunol 76(2-3):166–175. https://doi.org/10.1016/j.humimm.2014.12.016
10. Leggett RM, Clark MD (2017) A world of opportunities with nanopore sequencing. J Exp Bot 68(20):5419–5429. https://doi.org/10.1093/jxb/erx289
11. Rang FJ, Kloosterman WP, de Ridder J (2018) From squiggle to basepair: computational approaches for improving nanopore sequencing read accuracy. Genome Biol 19(1):90. https://doi.org/10.1186/s13059-018-1462-9
12. Liu C, Xiao F, Hoisington-Lopez J, Lang K, Quenzel P, Duffy B, Mitra RD (2018) Accurate typing of human leukocyte antigen class I genes by Oxford nanopore sequencing. J Mol Diagn 20(4):428–435. https://doi.org/10.1016/j.jmoldx.2018.02.006
13. Chaisson MJ, Tesler G (2012) Mapping single molecule sequencing reads using basic local alignment with successive refinement (BLASR): application and theory. BMC Bioinformatics 13:238. https://doi.org/10.1186/1471-2105-13-238
14. Quinlan AR, Hall IM (2010) BEDTools: a flexible suite of utilities for comparing genomic features. Bioinformatics 26(6):841–842. https://doi.org/10.1093/bioinformatics/btq033
15. Garrison E, Marth G (2012) Haplotype-based variant detection from short-read sequencing. arXiv. doi:arXiv:1207.3907
16. Camacho C, Coulouris G, Avagyan V, Ma N, Papadopoulos J, Bealer K, Madden TL (2009) BLAST+: architecture and applications. BMC Bioinformatics 10:421. https://doi.org/10.1186/1471-2105-10-421
17. Robinson JT, Thorvaldsdottir H, Winckler W, Guttman M, Lander ES, Getz G, Mesirov JP (2011) Integrative genomics viewer. Nat Biotechnol 29:24–26. https://doi.org/10.1038/nbt.1754
18. Thorvaldsdottir H, Robinson JT, Mesirov JP (2013) Integrative genomics viewer (IGV): high-performance genomics data visualization and exploration. Brief Bioinform 14(2):178–192. https://doi.org/10.1093/bib/bbs017
19. Hosomichi K, Jinam TA, Mitsunaga S, Nakaoka H, Inoue I (2013) Phase-defined complete sequencing of the HLA genes by next-generation sequencing. BMC Genomics 14:355. https://doi.org/10.1186/1471-2164-14-355
20. Robinson J, Guethlein LA, Cereb N, Yang SY, Norman PJ, Marsh SGE, Parham P (2017) Distinguishing functional polymorphism from random variation in the sequences of >10,000 HLA-A, -B and -C alleles. PLoS Genet 13(6):e1006862. https://doi.org/10.1371/journal.pgen.1006862
21. Klasberg S, Lang K, Gunther M, Schober G, Massalski C, Schmidt AH, Lange V, Schofl G (2019) Patterns of non-ARD variation in more than 300 full-length HLA-DPB1 alleles. Hum Immunol 80(1):44–52. https://doi.org/10.1016/j.humimm.2018.05.006

Chapter 7

HLApers: HLA Typing and Quantification of Expression with Personalized Index

Vitor R. C. Aguiar, Cibele Masotti, Anamaria A. Camargo, and Diogo Meyer

Abstract

The plethora of RNA-seq data which have been generated in the recent years constitutes an attractive resource to investigate HLA variation and its relationship with normal and disease phenotypes, such as cancer. However, next generation sequencing (NGS) brings new challenges to HLA analysis because of the mapping bias introduced by aligning short reads originated from polymorphic genes to a single reference genome. Here we describe HLApers, a pipeline which adapts widely used tools for analysis of standard RNA-seq data to infer HLA genotypes and estimate expression. By generating reliable expression estimates for each HLA allele that an individual carries, HLApers allows a better understanding of the relationship between HLA alleles and phenotypes manifested by an individual.

Key words HLA, HLA typing, HLA expression, RNA-seq, Immunogenetics

1 Introduction

If we were to draw a graph of nucleotide diversity across the genome, we would easily spot a small region on the short arm of chromosome 6 which harbors polymorphism at a level way beyond the genomic average. That region corresponds to the MHC (major histocompatibility complex) region, a gene-dense region which contains the HLA (human leukocyte antigens) genes that are involved in the presentation of self- or pathogen-derived antigens to the immune system [13, 18].

The MHC region—and in particular HLA genes—shows the highest density of disease associations in the entire genome [5, 13, 16, 18]. These associations involve a vast range of auto-immune diseases, phenotypes related to susceptibility and resistance to infectious diseases, psychiatric disorders, highlighting the central importance of HLA genes to the immune response, and also on the role of immune function in regulating cell and tissue-level phenotypes.

Sebastian Boegel (ed.), *Bioinformatics for Cancer Immunotherapy: Methods and Protocols*, Methods in Molecular Biology, vol. 2120, https://doi.org/10.1007/978-1-0716-0327-7_7,

Cancer also harbors characteristics of an immunological disease: in addition to the recognized tumor-promoting inflammation, a remarkable hallmark is the ability of cancer cells to evade immune destruction [12]. Tumor somatic mutations generate neoepitopes, which can be presented through MHC Class-I and recognized by CD8+ T-lymphocytes, resulting in a cytotoxic-specific response [27]. Therefore, the anti-tumor immunity is a barrier to cancer development and progression, and also a druggable target for disease treatment, with tumors harboring a high number of mutations potentially displaying a high number of neoepitopes, and then being more likely to elicit a cytotoxic T-cell response. In this context, HLA typing and HLA allele expression, along with predictions of epitope binding affinity, provide crucial information to derive immune profiles of tumors.

The omics revolution, bolstered by technological improvements which have led to next generation sequencing (NGS), has allowed valuable datasets to be sampled, comprising a rich catalog of genomic variation (e.g., The 1000 Genomes Project [25]), transcriptomic variation (e.g., the Geuvadis Consortium [14]), and joint catalogs of genomic and transcriptomic variation across healthy tissues (e.g., GTEx Consortium [11]) or of cancer types (e.g., The Cancer Genome Atlas (TCGA) [23] and the International Cancer Genome Consortium (ICGC) [24]).

However, understanding HLA variation and expression from genomic data brings unique challenges. Whole-genome data may contain incorrect genotype calls [3], and whole-transcriptome data may provide biased expression estimates for HLA genes [1]. This is a result of mapping bias, which can especially affect HLA genes because of two of their main characteristics: (1) the extreme polymorphism of HLA loci makes it difficult for sequencing reads originating from these genes to map to a reference genome, and (2) the presence of paralogues makes it difficult to map a read uniquely to a specific gene, potentially leading to the exclusion of many reads.

To overcome these challenges, multiple bioinformatic methods have been proposed to provide more reliable variant genotypes [6], HLA allele calls [7, 20, 22], and expression levels for HLA genes with omics data [1, 2, 15]. These methods share the principle of using references for read alignment which contemplate the known diversity at the HLA region instead of relying on a single and arbitrary reference genome.

With this challenge in mind we have developed *HLApers* [1], a computational pipeline which integrates currently available and highly used software in order to infer HLA genotypes and reliably estimate expression levels for HLA genes and alleles directly from whole-transcriptome RNA-seq data.

We have shown that these estimates provide improvements over alternative approaches which do not contemplate HLA diversity, and that with accurate estimates of expression levels we can obtain a better understanding of the regulation of HLA expression [1].

In addition to providing reliable expression estimates at the gene level, another important gain of using an HLA-tailored approach such as HLApers is the expression estimation at the level of HLA alleles, which is not a product of standard pipelines. This is relevant because many HLA alleles are associated with specific phenotypes of evolutionary and medical importance, making allele-level expression a natural unit of analysis.

2 Materials

2.1 Hardware

HLApers was developed on Ubuntu 16.04. The software is able to run on computers with a shell environment, such as Linux or MacOS.

2.2 Software

In order to run HLApers, the following software must be installed:

- R v3.4+ which can be downloaded from CRAN (cran.r-project.org)
- R packages from Bioconductor and GitHub. To this, the packages "BiocManager" and "devtools" are required. These can be installed with the following command in R:

  ```
  >install.packages(c("BiocManager","devtools"))
  ```

  Then, run the following commands in R:

  ```
  >BiocManager::install("Biostrings")
  >devtools::install_github("genevol-usp/
  hlaseqlib")
  ```

- Samtools 1.3+
- seqtk

As we will describe below, there are two pipelines built into HLApers: one based on STAR [8] and Salmon [21], and other based on kallisto [4]. Users can choose to install software below which are required for the pipeline of choice, or simply install all required tools.

For the STAR-Salmon pipeline, install:

- STAR v2.5.3a+
- Salmon v0.14.0+

Or, for the kallisto pipeline install:

- kallisto 0.45.0+

2.3 Download HLApers

Clone the HLApers repository from GitHub to your machine:

```
git clone https://github.com/genevol-usp/HLApers.git
```

2.4 Download IMGT and Gencode Datasets

Clone the IMGT repository from GitHub to your machine:

```
git clone https://github.com/ANHIG/IMGTHLA.git
```

Download the Gencode transcript sequences fasta file, and the corresponding annotations GTF file from the Gencode website (https://www.gencodegenes.org/human/).

3 Methods

3.1 Strategy

Our goal in developing HLApers was to use tools which are already available and widely adopted, adapting them to work under the strategy of an HLA-personalized reference. To do so, we developed the following pipeline. First, RNA-seq reads are aligned to the complete set of HLA allele coding sequences available from the IMGT database (https://www.ebi.ac.uk/ipd/imgt/hla/). The (up to 2) alleles which maximize the read counts are selected to compose the individual genotype at each HLA locus. Next, a personalized index is created, and quantification is performed (Fig. 1).

We have implemented HLApers with the goal of allowing the user to choose between different strategies for HLA genotyping and quantification. One approach uses a standard short-read aligner

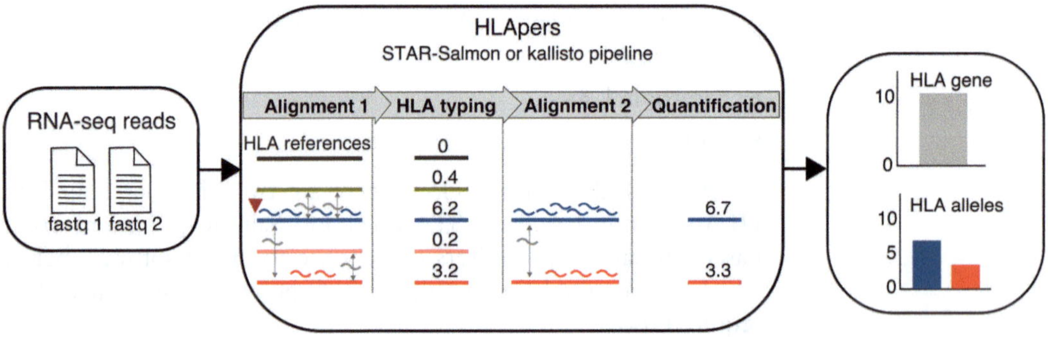

Fig. 1 Schematic representation of the HLApers pipeline. The pipeline uses as input RNA-seq paired-end reads in fastq format. These reads (represented by curved shapes) are aligned to an index containing references for all known HLA alleles (*Alignment 1*). The references to which most reads align are identified, so as to infer the HLA genotypes of the individual at that locus (in this example a heterozygote carrying a blue and a red allele, *HLA typing*). A second alignment is performed, using as an index only the alleles inferred to be present in that individual at that locus (*Alignment 2*). Expression is estimated based on the number of reads aligning to each reference, using a statistical model that accounts for instances where reads align to multiple HLA references (*Quantification*). HLApers outputs expression estimates for each HLA allele at each HLA locus, and gene-level estimates can be obtained by summing the allele-level estimates

Table 1
CPU times (in minutes) for the different pipelines implemented in HLApers, for the Geuvadis RNA-seq data for individual HG00096

Pipeline	Task	Whole fastq	Reduced fastq[a]
STAR-based	Genotyping	990	151
	Quantification	212	–
kallisto-based	Genotyping	26	6.8
	Quantification	28.5	–

[a]Fastq containing only MHC-mapped and unmapped reads extracted from a BAM file with "hlapers bam2fq -m"

(STAR [8]) and another uses a pseudoalignment approach (kallisto [4]). In our analysis of the GEUVADIS dataset, we found that the STAR-based pipeline was slightly more accurate to predict HLA genotypes (see S1 Table in Aguiar et al. [1]), and for that reason the STAR-based pipeline is the default one in HLApers (*see* **Note 1**). However this gain comes at a high cost of speed, with the kallisto-based pipeline being much faster (Table 1). We also found that the locus-level estimates obtained using either approach were extremely highly correlated ($r = 0.8$ for Class I and $r = 0.99$ for Class II genes).

Users can choose to use the kallisto pipeline by specifying the flag `--kallisto` for HLApers "index," "genotype," and "quant" modes.

3.2 Running HLApers

First, change to the HLApers directory:

```
cd HLApers
```

HLApers can be run in five different modes, called using an appropriate argument:

```
./hlapers --help
Usage: hlapers [modes]
```

`prepare-ref`	`Prepare transcript fasta files.`
`index`	`Create index for read alignment.`
`bam2fq`	`Convert BAM to fastq.`
`genotype`	`Infer HLA genotypes.`
`quant`	`Quantify HLA expression.`

Each module has its own help page, which can be accessed by executing the help command (e.g., for the genotype mode):

```
./hlapers genotype --help
```

Below, we use RNA-seq data for the individual HG00096 made available by the GEUVADIS Consortium (ENA ID ERR188040) to demonstrate how to run the pipeline and the results files it generates.

3.2.1 Preparing the Reference

The first step is to prepare a reference to which reads will be aligned. Here, we integrate Gencode data of annotated transcripts for the whole genome with IMGT data of HLA coding sequences to create a reference supplemented with the HLA diversity.

HLApers uses the entire set of HLA allele coding sequences, made available by IMGT, to populate the index to be used (*see* **Note 2**).

Because a subset of these alleles has missing information for some exons, we impute the most likely sequence in these cases. Our strategy consists in identifying the fully sequenced allele most similar to the one with missing data, and to assign the sequence of this closest sequence to the one with missing data.

The `hlapers prepare-ref` mode performs this processing on the HLA sequences, and supplements the Gencode reference with the HLA database. For example, we can run the prepare-ref mode with the following command:

```
./hlapers prepare-ref -t gencode.v25.transcripts.fa -a gencode.v25.annotation.gtf -i IMGTHLA -o hladb
```

Next, we call the "index" mode, which invokes the aligner of choice (STAR (default) or kallisto) to create the index for the read mapping. For example:

```
./hlapers index -t hladb/transcripts_MHC_HLAsupp.fa -p 4 -o index
```

3.2.2 In Silico HLA Genotyping

In order to build a personalized index (i.e., an index containing the reference sequences that correspond to the HLA alleles that the individual carries), we infer each individual's HLA genotype.

HLApers takes as input paired-end fastq files. Optionally, in order to speed up the process, these fastq can contain only reads mapped to the MHC region and unmapped reads from a previous alignment to the reference genome. The mode "bam2fq" can be used to extract those reads from a BAM file (see `hlapers bam2fq --help`).

The genotyping is done by executing the following command:

```
./hlapers genotype -i index/STARMHC -t
hladb/transcripts_MHC_HLAsupp.fa -1 HG00096_mhc_1.fq -2 HG00096_mhc_2.fq -p 8 -o results/HG00096
```

where

- `-i` denotes the index generated previously by "hlapers index";

- `-t` denotes the transcripts file generated previously with "hlapers prepare-ref";
- `-1` and `-2` denote the path to the fastq files;
- `-p` sets number of threads.

3.2.3 Quantification of HLA Expression

In order to quantify expression, we use the `quant` mode. If only a BAM file of a previous alignment to the genome is available, we first need to convert the BAM to fastq using the `bam2fq` mode. If the original fastq files are available, we can proceed directly to the quantification step.

We run the quantification step with the command:

```
./hlapers quant -t ./hladb -f ./results/
HG00096_index.fa -1 HG00096_1.fq.gz -2 HG00096_2.
fq.gz -o ./results/HG00096 -p 8
```

where

- `-t` denotes the path to the transcript database created by "hlapers prepare-ref" (i.e., the output directory given in the "prepare-ref" step);
- `-f` denotes the path to the personalized HLA index generated by "hlapers genotype."

3.3 Interpreting Output

The most important output file produced by HLApers, following the example above (Subheading 3.2.3), will be a file named "HG00096_hlaquant.tsv" written to the "results" directory.

Genotypic information is reported in this file by providing the alleles the individual carries at each surveyed locus (Fig. 2). Thus, for a locus detected as heterozygous, information is present in two lines, whereas it is present in one line if the genotype is hemi/homozygous. HLA genotypes are inferred at the 3-field resolution, and expression levels are given in number of reads, and also normalized to transcripts per million (*see* **Note 3** for a commentary on low expression values and loci with polymorphic presence/absence).

The HLA genotypes inferred by HLApers for individual HG00096 are concordant with those experimentally determined by Gourraud et al. [10]. In fact, we have shown that HLApers has a high genotyping accuracy ($\geq$97%) in a dataset with over 300 individuals (see S1 Table in Aguiar et al. [1]).

Having the HLA genotypes and expression levels, we can compare relative expression levels between HLA genes and alleles of interest (Fig. 3).

Other output files include:

- *_genotypes.tsv: contains the HLA alleles inferred by the "hlapers genotype" module;

Name	TPM	NumReads
IMGT_A*01:01:01	961.9	20568.7
IMGT_A*29:02:01	926.0	19620.3
IMGT_B*08:01:01	1368.6	29615.3
IMGT_B*44:03:01	1036.4	22273.5
IMGT_C*07:01:01	645.1	13643.1
IMGT_C*16:01:01	509.7	10998.1
IMGT_DMA*01:01:01	116.1	1385.8
IMGT_DMB*01:01:01	24.0	286.3
IMGT_DMB*01:04	26.8	317.5
IMGT_DOA*01:01:02	20.6	250.0
IMGT_DOB*01:01:01	6.0	73.2
IMGT_DOB*01:01:03	8.8	107.8
IMGT_DPA1*01:03:01	285.6	3543.5
IMGT_DPA1*02:01:02	273.5	3388.5
IMGT_DPB1*01:01:01	219.6	2934.2
IMGT_DPB1*02:01:02	120.4	1563.8
IMGT_DPB2*03:01:01:03	0.6	7.0
IMGT_DQA1*02:01:01	449.1	4740.6
IMGT_DQA1*05:01:01	370.1	3901.9
IMGT_DQB1*02:01:01	110.1	1463.9
IMGT_DQB1*02:02:01	150.8	2011.9
IMGT_DRA*01:02:02	771.0	9098.6
IMGT_DRA*01:01:01	905.6	10804.4
IMGT_DRB1*03:01:01	709.9	9084.8
IMGT_DRB1*07:01:01	483.5	6149.8
IMGT_DRB3*01:01:02	171.9	2250.1
IMGT_DRB4*01:01:01	83.1	1054.4
IMGT_DRB5*01:31	0.7	9.0
IMGT_E*01:03:02	206.2	4304.7
IMGT_E*01:01:01	225.1	4692.5
IMGT_F*01:01:01	211.8	4257.3
IMGT_G*01:01:09	0.6	11.9
IMGT_H*02:02	0.9	19.5
IMGT_H*02:01:01	1.4	29.8
IMGT_J*02:01	0.3	5.2
IMGT_K*01:03	0.8	17.4
IMGT_K*01:01:01:02	0.6	11.6
IMGT_L*01:02	0.7	15.8
IMGT_L*01:01:01	0.8	16.2
IMGT_Y*03:01	0.8	17.6
IMGT_Y*01:01	0.4	9.0

Fig. 2 HLApers output file "*_hlaquant.tsv"

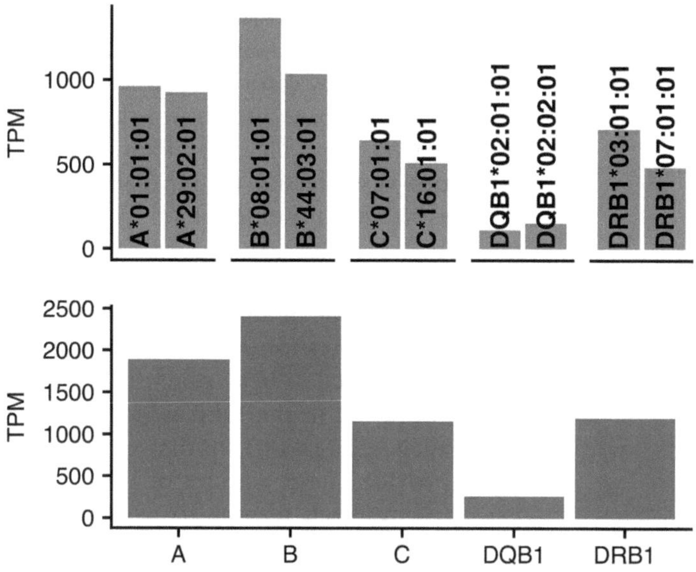

Fig. 3 Expression levels at the HLA allele level (top), and at the gene level (bottom) for the individual HG00096. Gene-level quantifications are simply the sum of the allelic estimates at each locus. TPM: transcripts per million

- *_logs: directory where log files generated by the aligners are saved to;
- *_quant: directory which contains expression levels for all other genes/transcripts in the genome, which can be used to compare with the expression of HLA genes, and also in downstream analyses such as eQTL mapping.

4 Example Applications

4.1 Estimation of HLA Expression at the Population Level and Downstream Analyses

The level at which a gene is expressed can have important influence on the phenotype of an organism, including its predisposition to develop diseases. The HLA genes are those most implicated in diseases in the human genome, and understanding how HLA expression levels vary across individuals and populations, and how this variation is connected to the genotypic and regulatory variation will be essential to understand the basis of HLA-related phenotypes.

As an example, we [1] applied HLApers to RNA-seq data for 358 European individuals from GEUVADIS [14], for which whole-genome variant genotype data are available from The 1000 Genomes Project [25]. We were able to reliably estimate gene-level and HLA allele-level expression estimates, which we used for downstream analyses such as eQTL mapping (eQTL, expression Quantitative Trait Locus). We then integrated the information on eQTLs mapped for HLA genes with the HLA allele expression data to show that, in some cases, variation in expression levels across HLA alleles can be explained by the regulatory variation for the gene. We have also shown that expression levels are relatively even between two alleles at a locus, and investigated coordinated regulation of expression for HLA alleles at different HLA loci lying on the same chromosomes [1].

4.2 Identification of Biomarkers for Cancer Susceptibility and Treatment Response

HLA typing provides crucial information for algorithms that predict epitope binding affinity (i.e., neoepitope prediction [19]), and HLA expression estimates allow immune profiles of tumors to be determined [26], potentially with information beyond that contained in the HLA genotypic data (e.g., when there is monoallelic expression). Initiatives such as TCGA [23] and ICGC [24] provide a large amount of DNA and RNA sequencing data, allowing research groups to explore tumor mutation signatures and molecular subtypes in a population scale.

Specific analyses that rely on HLA typing and expression include understanding how the patient's MHC-I genotype shapes the landscape of oncogenic mutations by restricting the neoantigen presentation during tumor formation (e.g., [17]). Furthermore, the questions surrounding whether HLA expression does [9] or does not [1] show evidence of allele-specific expression (i.e., a single allele is expressed) can likewise have a profound impact on the immunological response to a tumor.

HLApers can be employed for HLA typing and expression estimation directly from NGS data, thus contributing to understand how to increase success in predicting personal cancer susceptibility, tumor clonal evolution, disease progression, and drug response to immunotherapy centered on the individual's HLA.

5 Notes

1. The default pipeline built in HLApers uses the STAR aligner to map reads directly to the transcriptome. Reads longer than paired-end 75 bp are expected to produce less multimaps, then the user can choose to reduce the threshold for parameters "–outFilterMultimapNmax" and "–winAnchorMultimapNmax" in "./script/run_genotyping.sh" script.
2. By default, HLApers uses all HLA genes in the IMGT database, including pseudogenes. Users can choose to not include these pseudogenes by removing the corresponding files from the "IMGTHLA/alignments" directory.
3. The low expression (<15 TPM) levels of some HLA genes may be due to noise because of similarity to other highly expressed genes (it is possible that *HLA-DRB5* in Fig. 2 is an example).

On the other hand, for genes with polymorphic presence/absence status, high expression levels is a strong evidence that the locus is present (such as *HLA-DRB3* in Fig. 2).

Acknowledgements

Funding was provided by the National Institutes of Health, USA (GM 075091). VRCA was supported by a postdoc fellowship from the São Paulo Funding Agency (FAPESP, http://www.fapesp.br/en/) (2014/12123-2 and 2016/24734-1).

References

1. Aguiar VRC, Cesar J, Delaneau O et al (2019) Expression estimation and eQTL mapping for HLA genes with a personalized pipeline. PLoS Genet 15(4):e1008091
2. Boegel S, Löwer M, Schäfer M et al (2012) HLA typing from RNA-Seq sequence reads. Genome Med 4(102):1–12
3. Brandt DYC, Aguiar VRC, Bitarello BD et al (2015) Mapping bias overestimates reference allele frequencies at the HLA genes in the 1000 Genomes Project phase I data. G3 5 (5):931–941
4. Bray N, Pimentel H, Melsted P et al (2016) Near-optimal RNA-Seq quantification. Nat Biotechnol 34(5):525–527
5. Dendrou C, Petersen J, Rossjohn J et al (2018) HLA variation and disease. Nat Rev Immunol 18(5):325–339
6. Dilthey A, Cox C, Iqbal Z et al (2015) Improved genome inference in the MHC using a population reference graph. Nat Genet 47(6):682–688
7. Dilthey A, Gourraud P, Mentzer A et al (2016) High-accuracy HLA type inference from whole-genome sequencing data using population reference graphs. PLoS Comput Biol 12 (10):e1005151
8. Dobin A, Davis C, Schlesinger F et al (2013) STAR: ultrafast universal RNA-seq aligner. Bioinformatics 29(1):15–21
9. Gensterblum-Miller E, Weisheng W, Sawalha A (2018) Novel transcriptional activity and extensive allelic imbalance in the human MHC region. J Immunol 200(4):1496–1503
10. Gourraud PA, Khankhanian P, Cereb N et al (2014) HLA diversity in the 1000 genomes dataset. PLoS One 9(7):e97282
11. GTEx Consortium (2017) Genetic effects on gene expression across human tissues. Nature 550(7675):204–213
12. Hanahan D, Weinberg RA (2011) Hallmarks of cancer: the next generation. Cell 144 (5):646–674
13. Horton R, Wilming L, Rand V et al (2004) Gene map of the extended human MHC. Nat Rev Genet 5(12):889–899
14. Lappalainen T, Sammeth M, Friedländer M et al (2013) Transcriptome and genome sequencing uncovers functional variation in humans. Nature 501(7468):506–511
15. Lee W, Plant K, Humburg P et al (2018) AltHapAlignR: improved accuracy of RNA-seq analyses through the use of alternative haplotypes. Bioinformatics 34 (14):2401–2408
16. Lenz TL, Spirin V, Jordan DM et al (2016) Excess of deleterious mutations around HLA genes reveals evolutionary cost of balancing selection. Mol Biol Evol 33(10):2555–2564
17. Marty R, Kaabinejadian S, Rossell D et al (2017) MHC-I genotype restricts the oncogenic mutational landscape. Cell 171 (6):1272–1283
18. Meyer D, Aguiar VRC, Bitarello BD et al (2017) A genomic perspective on HLA evolution. Immunogenetics 70(1):5–27
19. Nielsen M, Andreatta M (2016) NetMHCpan-3.0; improved prediction of binding to MHC class I molecules integrating information from multiple receptor and peptide length datasets. Genome Med 8:33
20. Orenbuch R, Filip I, Comito D et al (2018) arcasHLA: high resolution HLA typing from RNA seq. BioRxiv 479824

21. Patro R, Duggal G, Love M et al (2017) Salmon provides fast and bias-aware quantification of transcript expression. Nat Methods 14 (4):417–419
22. Szolek A, Schubert B, Mohr C et al (2014) OptiType: precision HLA typing from next-generation sequencing data. Bioinformatics 30(23):1–7
23. The Cancer Genome Atlas Research Network, Weinstein JN, Collisson EA et al (2013) The cancer genome atlas pan-cancer analysis project. Nature 45:1113–1120
24. The International Cancer Genome Consortium (2010) International network of cancer genome projects. Nature 464:993–998
25. The 1000 Genomes Project Consortium (2015) A global reference for human genetic variation. Nature 526(7571):68–74
26. Thorsson V, Gibbs DL, Brown SD et al (2018) The immune landscape of cancer. Immunity 48 (4):821–830
27. Yarchoan M, Johnson III BA, Lutz ER et al (2017) Targeting neoantigens to augment antitumour immunity. Nat Rev Cancer 17 (569):209–222

Chapter 8

High-Throughput MHC I Ligand Prediction Using MHCflurry

Timothy O'Donnell and Alex Rubinsteyn

Abstract

MHCflurry is an open source package for peptide/MHC I binding affinity prediction. Its command-line and programmatic interfaces make it well-suited for integration into high-throughput bioinformatic pipelines. Users can download models fit to publicly available data or train predictors on their own affinity measurements or mass spec datasets. This chapter gives a tutorial on essential MHCflurry functionality, including generating predictions, training new models, and using the MHCflurry Python interface. MHCflurry is available at https://github.com/openvax/mhcflurry.

Key words Epitope prediction, MHC, HLA, Neoantigen, Immunoinformatics

1 Introduction

T cells recognize antigenic peptides in complex with major histocompatibility complex (MHC) proteins on cell surfaces. There are thousands of MHC variants (alleles) in the human population, and each MHC allele exhibits a strong bias for binding a distinct repertoire of peptides. MHCflurry is an open source package for modeling the peptide sequence characteristics that confer binding to a particular MHC allele [1]. While accurate prediction servers for this task have existed for several years [2], MHCflurry introduces a fast, documented, and modifiable implementation that is well suited for integration into high-throughput bioinformatic pipelines, such as those required for precision oncology. Users can download trained MHCflurry models fit to publicly available affinity measurements or train a MHCflurry predictor on their own data. MHCflurry is licensed under the Apache 2.0 license.

This chapter describes MHCflurry 1.2, which uses separate artificial neural networks (ANNs) trained for each MHC allele, an approach referred to as *allele-specific* prediction. Each ANN solves a regression problem, in which the nanomolar binding affinity is predicted for a specified peptide. The released models support a fixed set of common MHC class I alleles for which sufficient

Sebastian Boegel (ed.), *Bioinformatics for Cancer Immunotherapy: Methods and Protocols*, Methods in Molecular Biology, vol. 2120, https://doi.org/10.1007/978-1-0716-0327-7_8, © Springer Science+Business Media, LLC, part of Springer Nature 2020

published training data are available to train predictors. Pan-allele predictors, in which the input to the ANN includes a representation of the MHC allele (allowing prediction for any allele), are under development and not described here. For further technical details and API documentation, see the MHCflurry documentation at https://openvax.github.io/mhcflurry/. MHCflurry is developed at https://github.com/openvax/mhcflurry.

2 Materials

MHCflurry supports OS X and Linux using Python versions 2.7 and 3.4+. It uses the Keras neural network library [3] via either the TensorFlow [4] or Theano [5] backends. A graphics processing unit (GPU) may optionally be used, which can substantially accelerate model training.

We suggest installing MHCflurry using a conda environment to isolate its dependencies from other tools (*see* **Note 1**). Whether or not you are using a conda environment, you can install the latest version of MHCflurry from the Python package index by running:

```
$ pip install mhcflurry
```

Then download the trained models:

```
$ mhcflurry-downloads fetch models_class1
```

Here, `models_class1` is the name of the standard models recommended for MHC I ligand prediction. Files downloaded with the `mhcflurry-downloads` tool are stored in a platform-dependent directory. To get the path to downloaded data, you can use:

```
$ mhcflurry-downloads path models_class1
/Users/tim/Library/Application Support/mhcflurry/4/1.2.0/models_class1/
```

You can run `mhcflurry-downloads info` to get a list of available downloads and `mhcflurry-downloads -h` for details on using the `mhcflurry-downloads` command.

3 Methods

3.1 Generating Predictions

The `mhcflurry-predict` command generates predictions from the command-line. By default, it will use the trained models you downloaded above; other models can be used by specifying the `--models` argument. With the models downloaded, you can now generate predictions:

```
$ mhcflurry-predict \
    --alleles HLA-A0201 HLA-A0301 \
    --peptides SIINFEKL SIINFEKD SIINFEKQ \
    --out /tmp/predictions.csv
```

The above command will create a CSV file that looks like:

```
allele,peptide,mhcflurry_prediction,mhcflurry_prediction_
low,mhcflurry_prediction_high,mhcflurry_prediction_
percentile
HLA-A0201,SIINFEKL,4571.98373389,2093.54161999,
11331.8276571,6.381625
HLA-A0201,SIINFEKD,20649.7202275,14353.6110341,
32423.3688234,55.223
...
```

The predictions are given as nanomolar affinities (K_d) in the *mhcflurry_prediction* column. The *mhcflurry_prediction_ low* and *mhcflurry_prediction_high* fields indicate the 5th and 95th percentile predictions across the models in the ensemble, respectively. The last field, *mhcflurry_prediction_ percentile*, gives the quantile of the affinity prediction among a large number of random peptides tested on that allele (*see* **Note 2**).

It is usually more convenient to provide an input CSV file instead of passing peptides and alleles as command-line arguments. In this case, the prediction command would look like:

```
$ mhcflurry-predict --out /tmp/predictions.csv input.csv
```

The input CSV file is expected to have two columns, named `peptide` and `allele`. For example,

```
peptide,allele
AAAAAAAAA,HLA-A0201
```

Every row in the input will have a corresponding row in the output CSV file. See *mhcflurry-predict --help* for more options.

If you would prefer to supply full length protein sequences to scan for predicted MHC ligands, the *mhctools* package provides a convenient interface for doing so (*see* **Note 3**).

MHCflurry accepts MHC allele names in variety of formats, which are parsed into standard MHC nomenclature [6] using the *mhcnames* package (https://github.com/openvax/mhcnames). An allele name may omit any of the following elements of the MHC nomenclature:

- the ":" separating its two numeric fields (e.g., *HLA-A*0201*),

- the "∗" character separating the gene name from numeric fields (e.g., *HLA-A0201*),
- the species prefix, which is assumed to be "HLA" if omitted (e.g., *A∗02:01*).

This leniency in MHC nomenclature parsing was included since a large number of allele name variations are commonly found in publications and databases.

3.2 Training Your Own Predictors

MHCflurry enables users to train their own binding predictors, for example, to apply in-house training datasets or alternative neural network architectures. In this section, we walk through the approach used to train the production models released with MHCflurry. This is a more advanced topic that is not required to use MHCflurry to generate predictions.

The released MHCflurry predictors are generated in a two-step process. First, a large number of neural networks are trained for each allele using different neural network architectures. Second, model selection is performed to select 8–16 networks for each allele that perform well, based on data held-out from the training step. These selected networks will be used as an ensemble to generate predictions by taking the geometric mean of their affinity outputs. Alleles with poorly-performing ensembles are also eliminated in the model selection step.

To generate the released models, in the first step, we trained four replicates across 80 architectures (320 networks total) for each allele. The networks were trained on a random 80% sample of the data for the allele, with 10% held out for early stopping and 10% held out for model selection. In the model selection step, we selected 8–16 networks for each allele. We attempted to train predictors for 130 alleles and eliminated 18 alleles due to low performance in the model selection step, resulting in 112 supported alleles for the production models.

MHCflurry can make use of two kinds of training data: affinity measurements and MHC ligands identified by mass spec (reviewed in [7]). MHCflurry depends predominantly on affinity measurements, but the addition of mass spec data can give an accuracy improvement. MHCflurry 1.2 supports only "monoallelic" mass experiments, in which each mass spec hit is associated unambiguously with a single MHC allele. Mass spec can be used both to train MHCflurry models and in the model selection step. The currently recommended models use mass spec for model selection but not for model training (Fig. 1). This is a conservative choice that avoids the risk of biases in mass spec (such as a depletion of cysteines) from strongly affecting the predictors. We also release alternative predictors that do not use mass spec at all or use mass spec for both training and model selection (Fig. 2).

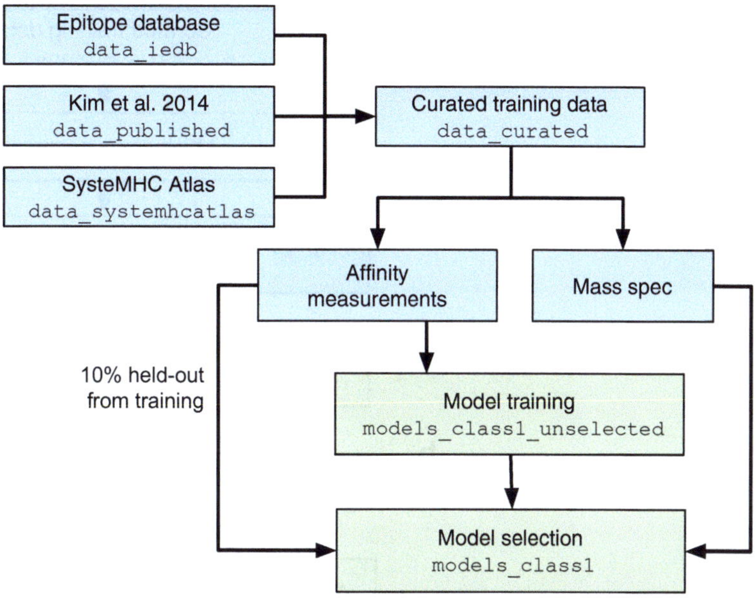

Fig. 1 Recommended trained models (green) and datasets (blue) available for download using the `mhcflurry-downloads` tool. Arrows indicate the data flow we used to generate these datasets and models. Names of datasets and models are given in `monospace` font. For example, the results of the final model selection step can be downloaded using the command `mhcflurry-download fetch models_class1`

The affinity measurements used to train MHCflurry come from two sources, the Immune Epitope Database (IEDB) [8] and a benchmarking study that combined several data sources [9]. The mass spec data come from IEDB, the SysteMHC Atlas project [10], and one additional study [11]. All data sources are combined and filtered using the curation script available in `downloads- generation/data_curated` in the MHCflurry repository.

3.2.1 Fitting Models

The combined training dataset can be downloaded using `mhcflurry-downloads`:

```
$ mhcflurry-downloads fetch data_curated
```

The local path to the downloaded training data can be found with.

```
$ mhcflurry-downloads path data_curated
/Users/tim/Library/Application Support/mhcflurry/4/1.2.0/data_curated/
```

The file named `curated_training_data.no_mass_spec.csv.bz2` in this directory contains the MHCflurry training data after filtering

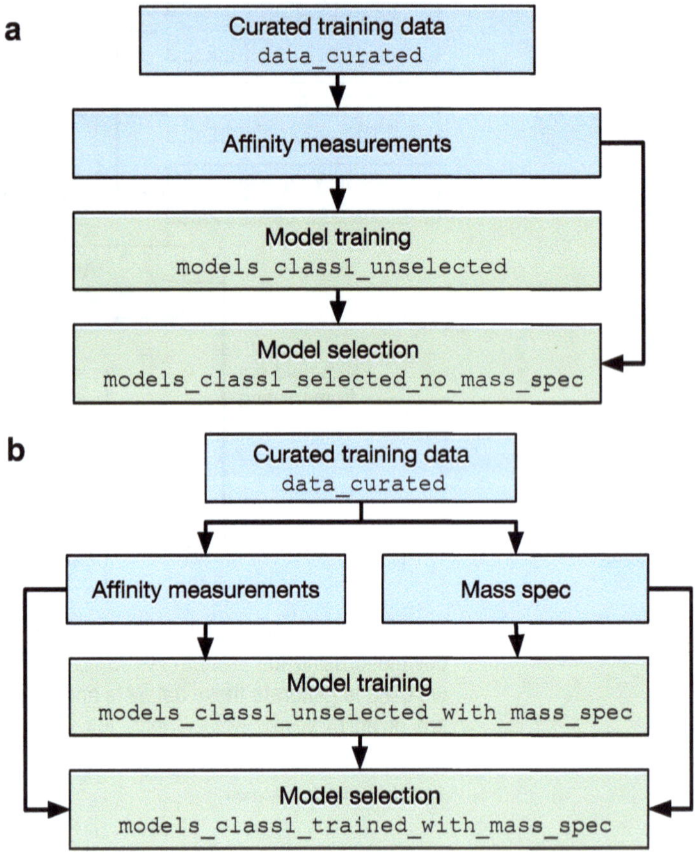

Fig. 2 Alternative models available for download. These consist of models trained and selected without the use of mass spec datasets (**a**) and models for which mass spec was used for both training and model selection (**b**). The standard models, in contrast, use mass spec datasets for model selection but not training

to affinity measurements only. The file named *curated_training_data.with_mass_spec.csv.bz2* contains combined affinity and mass spec data. The training data look like this:

```
allele,peptide,measurement_value,measurement_inequality,
measurement_type,measurement_source,original_allele
BoLA-1*21:01,AENDTLVVSV,7817.0,=,quantitative,Barlow -
purified MHC/competitive/fluorescence,BoLA-1*02101
BoLA-1*21:01,NQFNGGCLLV,1086.0,=,quantitative,Barlow -
purified MHC/direct/fluorescence,BoLA-1*02101
```

The *mhcflurry-class1-train-allele-specific-models* command is used to fit models. This tool trains models but does not perform model selection, which is done later using a different tool that takes the trained models from the first step as input. The models we release with MHCflurry 1.2 are trained using a command similar to the following:

```
$ mhcflurry-class1-train-allele-specific-models \
    --data TRAINING_DATA.csv \
    --hyperparameters hyperparameters.yaml \
    --min-measurements-per-allele 75 \
    --out-models-dir models
```

Here, `TRAINING_DATA.csv` can be set to the path to the training data downloaded above. The *hyperparameters.yaml* file is a yaml file (similar to JSON) that contains a list of hyperparameter settings describing the neural network architectures to train (*see* **Note 4**).

The command above will write the models to the output directory specified by the *--out-models-dir* argument.

See *downloads-generation/models_class1_unselected* in the MHCflurry repository for the precise commands and hyperparameters used to train the MHCflurry models.

3.2.2 Model Selection

Model selection is performed using the *mhcflurry-class1-select-allele-specific-models* command. This tool takes as input trained models and test data (i.e., data that were not used to train the models). It scores the trained models and selects those that perform well for retention. Here is an example invocation:

```
mhcflurry-class1-select-allele-specific-models \
   --data TEST_DATA.csv \
   --models-dir /PATH/TO/TRAINED_MODELS \
   --out-models-dir models \
   --scoring combined:mass-spec,mse,consensus \
   --consensus-num-peptides-per-length 10000 \
   --combined-min-models 8 \
   --combined-max-models 16 \
   --unselected-accuracy-scorer combined:mass-spec,mse \
   --unselected-accuracy-percentile-threshold 95 \
   --mass-spec-min-measurements 500
```

The *--scoring combined:mass-spec,mse,consensus* option indicates that individual models should be evaluated by averaging their performance on mass spec validation data (computed as AUC), mean square error (MSE) on held-out affinity data, and agreement with all other trained models for the allele on a large number of random peptides (consensus). The *--unselected-accuracy-scorer* option specifies the criteria for deciding which alleles should be dropped due to insufficient performance; here, we specify that mass spec and affinity measurements should be used to make this determination. In most cases, users will not need to change these arguments. See the usage information for the tool for further details.

The exact workflow used to perform model selection for the released models is available in *downloads-generation/models_class1* in the MHCflurry repository.

3.3 Using MHCflurry from Python

The MHCflurry Python interface is the most convenient way to use MHCflurry when developing analyses or bioinformatic workflows in Python. It also exposes additional options and features beyond those supported by the command-line tools. This tutorial gives a basic overview of the most important functionality. See the API Documentation (https://openvax.github.io/mhcflurry/api.html) for further details.

The *lass1AffinityPredictor* class is the primary user-facing interface. Use the *load* static method to load a trained predictor from disk. With no arguments this method will load the predictor released with MHCflurry. If you pass a path to a models directory, then it will load that predictor instead.

```
>>> from mhcflurry import Class1AffinityPredictor
>>> predictor = Class1AffinityPredictor.load()
>>> predictor.supported_alleles[:10]
['BoLA-6*13:01', 'Eqca-1*01:01', 'H-2-Db', 'H-2-Dd',
'H-2-Kb', 'H-2-Kd', 'H-2-Kk', 'H-2-Ld', 'HLA-A*01:01',
'HLA-A*02:01']
```

The `predict` method will return binding predictions as nanomolar affinities:

```
>>> predictor.predict(allele="HLA-A0201",
peptides=["SIINFEKL", "SIINFEQL"])
array([ 4571.98373389, 3583.33113994])
```

For more detailed results, we can use the `predict_to_dataframe` method:

```
>>> predictor.predict_to_dataframe(allele="HLA-A0201",
peptides=["SIINFEKL", "SIINFEQL"])
peptide      allele     prediction    prediction_low
prediction_high  \
0   SIINFEKL   HLA-A0201   4571.983734      2093.541620
11331.827657
1   SIINFEQL   HLA-A0201   3583.331140      1388.942403
10621.888497
         prediction_percentile
     0                6.381625
     1                5.284625
```

Instead of a single allele and multiple peptides, we may need predictions for allele/peptide pairs. We can predict across pairs by specifying the `alleles` argument instead of `allele`. The list of alleles must be the same length as the list of peptides (i.e., it is predicting over corresponding items).

```
>>> predictor.predict(alleles=["HLA-A0201", "HLA-B*57:01"], peptides=["SIINFEKL", "SIINFEQL"])
array([ 4571.98345747, 20648.92075372])
```

The *Class1AffinityPredictor* class can also be used to train predictors. If you have not already, run this shell command to download the MHCflurry training data:

```
$ mhcflurry-downloads fetch data_curated
```

We can get the path to this data from Python using *mhcflurry.downloads.get_path*:

```
>>> from mhcflurry.downloads import get_path
>>> data_path = get_path("data_curated", "curated_training_data.no_mass_spec.csv.bz2")
>>> data_path
'/Users/tim/Library/Application Support/mhcflurry/4/1.2.0/data_curated/curated_training_data.no_mass_spec.csv.bz2'
```

Now we load this dataset using the *pandas* library and select peptides with lengths between 8 and 15:

```
>>> import pandas
>>> df = pandas.read_csv(data_path)
>>> df = df.loc[(df.peptide.str.len() >= 8) & (df.peptide.str.len() <= 15)]
>>> df.head(5)
allele  peptide measurement_value measurement_inequality \
   0 BoLA-1*21:01 AENDTLVVSV      7817.0          =
   1 BoLA-1*21:01 NQFNGGCLLV      1086.0          =
  measurement_type           measurement_source \
  0      quantitative  Barlow - purified MHC/competitive/fluorescence
  1  quantitative   Barlow - purified MHC/direct/fluorescence
   original_allele
  0  BoLA-1*02101
  1  BoLA-1*02101
```

Here we instantiate an untrained `Class1AffinityPredictor` and use the `fit_allele_specific_predictors` method to train models:

```
>>> new_predictor = Class1AffinityPredictor()
>>> single_allele_train_data = df.loc[df.allele ==
"HLA-B*57:01"].sample(100)
>>> new_predictor.fit_allele_specific_predictors(
...     n_models=1,
...     architecture_hyperparameters_list=[{
...         "layer_sizes": [16],
...         "max_epochs": 5,
...         "random_negative_constant": 5,
...     }],
...     peptides=single_allele_train_data.peptide.values,
...       affinities=single_allele_train_data.measurement_
value.values,
...     allele="HLA-B*57:01")
[ 100 peptides ]  Epoch   0 /   5: loss=0.308307. Min val
loss (None) at epoch None
[<mhcflurry.class1_neural_network.Class1NeuralNetwork
object at 0x121bd9ad0>]
```

The `fit_allele_specific_predictors` method can be called any number of times on the same instance to build up ensembles of models across alleles. The architecture hyperparameters we specified are for demonstration purposes. To fit production models, you would usually train for more epochs.

Now, we can generate predictions:

```
>>> new_predictor.predict(["SYNPEPII"], allele="HLA-
B*57:01")
array([ 4481.70925883])
```

We can save our predictor to the specified directory on disk by running:

```
>>> new_predictor.save("/tmp/new-predictor")
```

And restore it:

```
>>> new_predictor2 = Class1AffinityPredictor.load("/tmp/
new-predictor")
>>> new_predictor2.supported_alleles
['HLA-B*57:01']
```

The high-level `Class1AffinityPredictor` delegates to low-level `Class1NeuralNetwork` objects, each of which represents a single neural network. The purpose of `Class1AffinityPredictor` is to implement several important features:

- *Ensembles*: More than one neural network can be used to generate each prediction. The predictions returned to the user are the geometric mean of the individual model predictions. This gives higher accuracy in most situations.
- *Multiple alleles*: A `Class1NeuralNetwork` generates predictions for only a single allele. The `Class1AffinityPredictor` maps alleles to the relevant `Class1NeuralNetwork` instances.
- *Serialization*: Loading and saving predictors is implemented in `Class1AffinityPredictor`.

Sometimes, however, it is easiest to work directly with `Class1NeuralNetwork`. Here is a simple example of doing so:

```
>>> from mhcflurry import Class1NeuralNetwork
>>> network = Class1NeuralNetwork()
>>> network.fit(
...     single_allele_train_data.peptide.values,
...     single_allele_train_data.measurement_value.values,
...     verbose=0)
>>> network.predict(["SIINFEKLL"])
array([ 27009.78916591])
```

4 Notes

1. Running MHCflurry in a conda virtual environment is a convenient way to isolate the dependencies installed by MHCflurry (such as Keras) from the libraries in use by other tools on your system. Some users have also reported that using conda environments can avoid problems installing TensorFlow:

```
$ conda create -q -n mhcflurry-env python=3.6 tensorflow
$ source activate mhcflurry-env
$ mhcflurry-downloads fetch
```

2. The predictions shown above were generated with MHCflurry 1.2.0. Different versions of MHCflurry can give different results. Even on the same version, exact predictions may vary (up to about 1 nM) depending on the Keras backend and other details.
3. A common application of MHC ligand prediction is to scan protein sequences (e.g., viral proteins) for peptides predicted to bind any of a set of alleles. This functionality is not implemented in MHCflurry but is provided by a simple, open source tool called *mhctools* (https://github.com/openvax/mhctools). The *mhctools* package provides support for scanning protein

sequences to find predicted epitopes using a number of binding predictors, including MHCflurry. Here is an example.

First, install *mhctools* if it is not already installed:

```
$ pip install mhctools
```

We will generate predictions across `example.fasta`, a FASTA file with two short sequences:

```
>protein1
MDSKGSSQKGSRLLLLLVVSNLLLCQGVVSTPVCPNGPGNCQV
EMFNEFDKRYAQGKGFITMALNSCHTSSLPTPEDKEQAQQTHH
>protein2
VTEVRGMKGAPDAILSRAIEIEEENKRLLEGMEMIFGQVIPGA
ARYSAFYNLLHCLRRDSSKIDTYLKLLNCRIIYNNNC
```

Here is an example *mhctools* invocation to scan these sequences for predicted ligands for two MHC alleles. See *mhctools -h* for more information:

```
$ mhctools \
--mhc-predictor mhcflurry \
--input-fasta-file example.fasta \
--mhc-alleles A02:01,A03:01 \
--mhc-peptide-lengths 8,9,10,11 \
--extract-subsequences \
--output-csv /tmp/subsequence_predictions.csv
```

This will write a file giving predictions for all subsequences of the specified lengths to the given file:

```
source_sequence_name,offset,peptide,allele,affinity,
percentile_rank,prediction_method_name,length
protein2,42,AARYSAFY,HLA-A*03:01,3270.87565337,4.056625,
mhcflurry,8
protein2,42,AARYSAFYN,HLA-A*03:01,9140.1869546,10.101125,
mhcflurry,9
...
```

In most cases, users will need to filter this result to only tightly-binding peptides. The most common threshold to use is 500 nM (i.e., keep peptides with affinities <500 nM), owing to an early observation that most T-cell epitopes have peptide/MHC affinities tighter than 500 nM [12]. Another common threshold is 2.0 percentile rank. More recent work has recommended different affinity thresholds for each allele [13]. In practice, the 500 nM threshold is usually a reasonable choice.

4. The neural network architectures supported by MHCflurry are described using options referred to as hyperparameters. Table 1

Table 1
Commonly-used MHCflurry 1.2 neural network hyperparameters

Hyperparameter	Default value	Meaning
`activation`	`tanh`	Activation applied to dense layers. Any Keras activation can be used
`batch_normalization`	`False`	Whether to apply batch normalization after dense layers. See [14]
`dense_layer_l1_regularization`	`0.001`	L1 and L2 regularization applied to dense layers
`dense_layer_l2_regularization`	`0`	
`dropout_probability`	`0`	Dropout rate applied after dense layers. *See* [15]
`early_stopping`	`True`	Whether to use early stopping
`init`	`glorot_uniform`	Weights initialization approach. Any Keras initializer can be used
`kmer_size`	`15`	Maximum peptide length
`layer_sizes`	`[32]`	List of dense layer sizes.
`learning_rate`	`None`	Learning rate. Defaults to optimizer's default value
`left_edge, right_edge`	`4`	Peptide encoding. The first `left_edge` characters in the input always map to the first `left_edge` characters in the output. Similarly for the last `right_edge` characters. The middle characters are filled in based on the length, with a special sentinel *X* character filling in the blanks. The minimum peptide length is `left_edge` + `right_edge`
`locally_connected_layers`	`[{kernel_size: 3, filters: 8, activation: tanh}]`	List of dicts describing locally connected layers applied to the peptide. See Keras locally connected layers documentation for details
`loss`	`custom:mse_with_inequalities`	Loss function. Can also be set to any Keras supported loss function Setting this to the special value `custom:mse_with_inequalities` allows support for training data with inequalities (e.g., mass spec hits can be used with a "< 500 nM" affinity value)

(continued)

Table 1
(continued)

Hyperparameter	Default value	Meaning
max_epochs	500	Maximum number of training epochs. This will rarely be hit if early stopping is enabled
minibatch_size	128	Training batch size. Affects both training speed and potentially model performance
optimizer	rmsprop	Any Keras supported optimizer can be used
output_activation	sigmoid	Activation function for output layer
patience	20	Number of epochs to wait after best validation epoch before early stopping
peptide_amino_acid_encoding	BLOSUM62	Vector encoding for peptide
random_negative_affinity_max	50,000	Random negative peptides are sampled with affinities between the indicated nanomolar affinity values
random_negative_affinity_min	20,000	
random_negative_constant	25	Number of random negative peptides sampled will be random_negative_constant + n * random_negative_rate, where n is the size of the training data for the allele
random_negative_rate	0	
random_negative_distribution_smoothing	0	Pseudocount to add to the occurrence count of each amino acid when computing the amino acid distribution of the training data
random_negative_match_distribution	True	If True, the random negative peptides will be sampled so as to match the amino acid distribution of the peptides in the training data
validation_split	0.1	Fraction of data to use to monitor validation score during training. Also useful as a way of generating diversity across the ensemble. For example, if set to 0.2, then each model is trained on a random **80%** sample of the data. If you are using early stopping (recommended), validation_split must be >0

describes the most commonly used hyperparameters. *See* [1] for details on the MHCflurry neural network architectures and peptide encoding.

References

1. O'Donnell TJ, Rubinsteyn A, Bonsack M et al (2018) MHCflurry: open-source class I MHC binding affinity prediction. Cell Syst 7:129–132.e4
2. Jurtz V, Paul S, Andreatta M et al (2017) NetMHCpan-4.0: improved peptide–MHC class I interaction predictions integrating eluted ligand and peptide binding affinity data. J Immunol 199:3360–3368
3. Chollet F, Others (2015), Keras. https://github.com/keras-team/keras
4. Abadi M, Agarwal A, Barham P, et al (2016) TensorFlow: large-scale machine learning on heterogeneous distributed systems. http://arxiv.org/abs/1603.04467
5. Al-Rfou R, Alain G, Almahairi A, et al (2016) Theano: a Python framework for fast computation of mathematical expressions
6. Marsh SGE, Albert ED, Bodmer WF et al (2010) Nomenclature for factors of the HLA system, 2010. Tissue Antigens 75:291–455
7. Creech AL, Ting YS, Goulding SP et al (2018) The role of mass spectrometry and proteogenomics in the advancement of HLA epitope prediction. Proteomics 1700259:1–10
8. Vita R, Mahajan S, Overton JA et al (2019) The immune epitope database (IEDB): 2018 update. Nucleic Acids Res 47:D339–D343
9. Kim Y, Sidney J, Buus S et al (2014) Dataset size and composition impact the reliability of performance benchmarks for peptide-MHC binding predictions. BMC Bioinformatics 15:241
10. Shao W, Pedrioli PGA, Wolski W et al (2018) The SysteMHC atlas project. Nucleic Acids Res 46:D1237–D1247
11. Abelin JG, Keskin DB, Sarkizova S et al (2017) Mass spectrometry profiling of HLA-associated peptidomes in mono-allelic cells enables more accurate epitope prediction. Immunity 46:315–326
12. Sette A, Vitiello A, Reherman B et al (1994) The relationship between class I binding affinity and immunogenicity of potential cytotoxic T cell epitopes. J Immunol 153:5586–5592
13. Paul S, Weiskopf D, Angelo MA et al (2013) HLA class I alleles are associated with peptide-binding repertoires of different size, affinity, and immunogenicity. J Immunol 191:5831–5839
14. Ioffe S, Szegedy C (2015) Batch normalization: accelerating deep network training by reducing internal covariate shift
15. Srivastava N, Hinton GE, Krizhevsky A et al (2014) Dropout: a simple way to prevent neural networks from overfitting. J Mach Learn Res 15:1929–1958

Chapter 9

In Silico Prediction of Tumor Neoantigens with TIminer

Alexander Kirchmair and Francesca Finotello

Abstract

Tumor neoantigens are at the core of immunological tumor control and response to immunotherapy. In silico prediction of tumor neoantigens from next-generation sequencing (NGS) data is possible but requires the assembly of complex, multistep computational pipelines and extensive data preprocessing. Using public data from two cancer cell lines, here we show how TIminer, a framework to perform immunogenomics analyses, can be easily used to assemble and run customized pipelines to predict cancer neoantigens from multisample NGS data.

Key words Neoantigen prediction, HLA typing, Gene expression, Cancer immunology, Immuno-oncology

1 Introduction

Tumor neoantigens are immunogenic peptides that are about 8-to-11 amino acids long, derived from the expression of mutated genes, and bound to the class-I human leukocyte antigen (HLA) molecules of tumor cells. As neoantigens are only displayed by tumor cells but not by normal cells, they can elicit strong, tumor-specific immune responses that are not limited by central tolerance. Recognition of class-I neoantigens by cytotoxic $CD8^+$ T cells is at the core of immunological tumor control and response to anticancer therapies [1–3]. Moreover, neoantigens are major determinants of response to immune checkpoint blockade and are the targets of personalized cancer vaccines and T-cell-based immunotherapies [4–6].

Neoantigens can be identified using mass spectrometry (MS) measurements of eluted HLA-binding peptides [7]. However, this approach suffers from some drawbacks, including (1) limited sensitivity; (2) large amount of starting material needed; (3) dependence on protein-sequence databases for data analysis, which do not represent mutated proteins. Alternatively, candidate neoantigens can be predicted in silico from RNA sequencing (RNA-seq),

Sebastian Boegel (ed.), *Bioinformatics for Cancer Immunotherapy: Methods and Protocols*, Methods in Molecular Biology, vol. 2120, https://doi.org/10.1007/978-1-0716-0327-7_9,

whole-exome (WES), and/or whole-genome sequencing (WGS) data using computational pipelines integrating different analytical tools. However, the assembly of multistep pipelines is complex and involves the integration of different tools with specific software dependencies, parameter settings, and data input requirements, requirements that preclude this kind of analyses to nonexpert users. TIminer (Tumor Immunology miner) [8] is a user-friendly framework that enables integrative immunogenomics analyses [9] of tumor RNA-seq and mutational data from single patients/samples, including the prediction of candidate neoantigens. All tools needed for the analysis are embedded in a Docker (https://www.docker.com) image to simplify installation and analysis. TIminer is available at: https://icbi.i-med.ac.at/software/timiner/timiner.shtml.

An overview of the TIminer framework is presented in Fig. 1. Starting from somatic mutations derived from WES or WGS data, TIminer predicts mutated proteins derived from nonsynonymous variants using the Ensembl Variant Effect Predictor (VEP) [10]. Class-I HLA alleles are determined with Optitype from RNA-seq data [11]. The binding affinities of mutated peptides (derived from the predicted proteins) for the reconstructed HLA alleles are predicted with NetMHCpan [12–14]. Peptides with high binding affinity (called *strong binders*) are then selected as candidate neoantigens. These predictions can be further refined by considering additional molecular information about the antigens, most importantly, whether they are actually expressed in the sample of interest. TIminer uses Kallisto [15] to quantify gene expression levels from RNA-seq data. Finally, candidate neoantigens can be prioritized by filtering only peptides generated from expressed genes. Each tool needed for data analysis, as well as the required software dependencies and code for data pre−/postprocessing can be easily accessed through the Python application programming interface (API) of TIminer. The TIminer API simplifies the assembly of complex analytical pipelines to the simple invocation and chaining of a few API modules (Fig. 1).

In this chapter, we show how the TIminer API can be used to easily assemble a pipeline for neoantigen prediction that can be run with no additional handling and processing of input files. As an example, we use public data from two breast cancer cell lines: MCF7 and MDA-MB-453. We provide guidance through the implementation of the single analytical steps and assembly of the final pipeline for neoantigen prediction shown in Fig. 1 (dark gray boxes). Further details on TIminer API modules, parameters, and output files can be found in the online documentation (*see* **Note 1**): https://icbi.i-med.ac.at/software/timiner/doc.

Besides neoantigen prediction, TIminer provides simplified access to tools for immunological analysis of bulk-tumor RNA-seq data [16]. We only briefly mention these analytical modules (shown

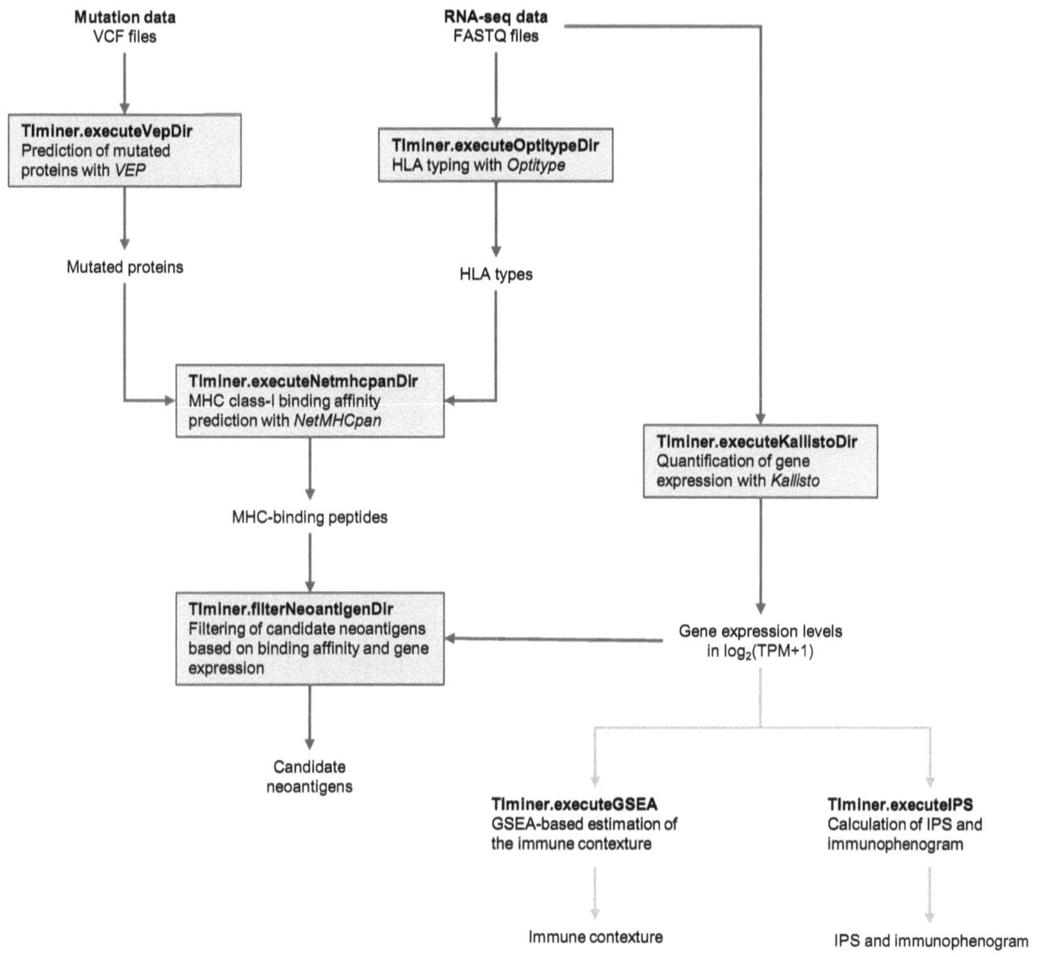

Fig. 1 Workflow for neoantigen prediction and other immunogenomics analyses with TIminer. Mutation data are used to determine mutated proteins with *Timiner.executeVepDir*. *Timiner.executeOptitypeDir* uses RNA-seq data for HLA typing. Mutated proteins and HLA types are used to predict peptides binding to the predicted HLA by *Timiner.executeNetmhcpanDir*. RNA-seq data are also used to compute normalized gene expression values as $\log_2(\text{TPM} + 1)$ by *Timiner.executeKallistoDir*. Finally, binding peptides are filtered with *Timiner.filterNeoantigenDir* to consider only expressed genes to derive a list of candidate neoantigens. In addition to neoantigen prediction, additional analyses (shown in light gray) can be performed starting from the normalized gene expression values: gene set enrichment analysis (GSEA) of the tumor immune contexture can be run with *Timiner.executeGSEA*, whereas the immunophenogram and immunophenoscore (IPS) can be computed with *Timiner.executeIPS*

in light gray in Fig. 1) and refer the interested reader to the original publication [8] and online documentation for additional details.

2 Materials

2.1 Hardware

The hardware requirements for TIminer are mainly related to the memory and disk space needed for analysis of large next-generation sequencing (NGS) datasets. Besides the disk space for input files, the TIminer installation requires the download and storage of genomics databases that need ~40 Gb of disk space. Additionally, sufficient memory is required for processing the input files, which is about the size of the FASTQ files for HLA typing with Optitype.

2.2 Software

The following software requirements need to be fulfilled for the installation of TIminer:

- Operating system: Linux (Fedora 23, Centos 7, or newer) or Mac OS X.
- Installation of Docker (https://www.docker.com) (*see* **Note 2**).
- Installation of Python 2.7 (*see* **Note 3**).

2.3 Dataset Availability

To compute neoantigens, TIminer requires somatic mutation data in VCF format and RNA-seq data in FASTQ format. Optionally, WES or WGS data (FASTQ) can be used instead of RNA-seq data for HLA typing, but only when the TIminer modules are used singularly. Small example files containing mutation and RNA-seq data are provided with the TIminer download files and can also be downloaded separately at https://icbi.i-med.ac.at/software/timiner/timiner.shtml ("Test data" in the "Download" window).

Here, we use data from two breast cancer cell lines (BCCL), MCF7 and MDA-MB-453. Mutation files can be downloaded from the Cell Model Passports database [17]:

- MCF7: https://cog.sanger.ac.uk/cmp/data/MCF7.cave.annot.vcf.gz.
- MDA-MB-453: https://cog.sanger.ac.uk/cmp/data/MDA-MB-453.cave.annot.vcf.gz.

Paired-end RNA-seq data (FASTQ files) can be downloaded from the Sequence Read Archive (SRA) using the ***fasterq-dump*** function of the SRA toolkit (https://www.ncbi.nlm.nih.gov/sra/docs/toolkitsoft).

In particular, the MCF7 RNA-seq data (accession number: SRR925723) can be downloaded as.

```
$ fasterq-dump -S -O outputDir/ -t tempDir/ SRR925723
```

where the option –S is used to split the paired-end data into two files.

The MDA-MB-453 RNA-seq data (accession number: SRR1283038) can be downloaded as.

```
$ fasterq-dump -S -O outputDir/ -t tempDir/ SRR1283038
```

When the download is finished, the files SRR925723_1.fastq and SRR925723_2.fastq for MCF7, and SRR1283038_1.fastq and SRR1283038_2.fastq for MDA-MB-453 should be available in the working directory.

3 Methods

3.1 TIminer Download and Installation

TIminer can be downloaded from the website: https://icbi.i-med.ac.at/software/timiner/timiner.shtml ("Download" window). The download can be performed only after agreement to the terms and license conditions and provision of a valid NetMHCpan license. After the files are downloaded and unzipped, installation can be performed by executing the installation script from the "install" directory.

This can be done in Linux using the following command:

```
$ bash install_red_hat.sh
```

Or, in Mac OS X, by executing:

```
$ bash install_mac.command
```

As part of the installation process, genomics databases of ~20 Gb size are downloaded into a new directory called "databases" (after extraction, the directory size increases to ~40 Gb). During the installation, the user is requested to indicate a location for storage of the database files.

3.2 Running TIminer

TIminer can be run as a full computational pipeline for integrative immunegenomics analyses (*see* **Note 4**). Alternatively, the individual analytical tools can be called separately through the TIminer API on multiple or single samples (*see* **Notes 5** and **6**), as done in the example described in this chapter. Python 2.7 is used to run the single modules from the TIminer python API. The best approach to call these modules is to write the Python commands into a script to be executed subsequently. To this end, a simple text editor can be used to save a script invoking one or more TIminer modules. An example script is shown in the following:

```
from TIminer import TIminerAPI
TIminerAPI.executeKallistoDir(inputInfofile="path/to/input
File.txt",
                 outputFile="path/to/outputFile.txt",
                 threadCount=8)
```

A python script for calling one or more TIminer functions must contain, on the first line, the instruction to import the TIminer

API, and then the actual call to the single function(s), together with the specified parameters. Parameters must be divided by commas and can be specified on separate lines to ensure readability. Once saved (e.g., with the name "my_pipeline.py"), the script can be run with python from a terminal by executing.

```
$ python my_pipeline.py
```

3.3 Pipeline for Neoantigen Prediction with TIminerAPI

To simplify the execution of the following example and easily retrieve the output files, we suggest to create a new directory for the analysis and three subdirectories, named,

- "data": where the mutation and RNA-seq data will be store.
- "analysis": working directory, where the script(s) needed for the analysis will be stored.
- "results": where all output files generated by the analysis will be stored.

Once downloaded, the example FASTQ and VCF files for the MCF7 and MDA-MB-453 cell lines have to be moved into the "data" directory.

To perform a multisample analysis, two input info files needs to be created and saved into the "analysis" directory (*see* also **Note 7**). A tab-delimited text file that we call "BCCL_FASTQ_info.txt" has to be created to specify the locations of the RNA-seq FASTQ input files. It should specify, for each cell line (on the rows), the cell line identifier, the path to the first FASTQ file, and the path to the second FASTQ file, as shown in the following:

MCF7	../data/SRR925723_1.fastq	../data/SRR925723_2.fastq
MDA-MB-453	../data/SRR1283038_1.fastq	../data/SRR1283038_2.fastq

Similarly, for the mutation data, a tab-delimited text file called "BCCL_VCF_info.txt" contains the cell line names, and the path to the corresponding VCF files has to be created:

MCF7	../data/MCF7.cave.annot.vcf
MDA-MB-453	../data/MDA-MB-453.cave.annot.vcf

3.3.1 Protein Mutation Prediction with TIminerAPI.executeVepDir

TIminer performs prediction of mutations in protein sequences from nonsynonymous mutations present in the input VCF files using VEP (*see* **Note 8**), which can be called with *TIminerAPI.executeVepDir*. Performing VEP analysis with the *TIminerAPI.executeVepDir* function requires the VCF sample information file ("BBCL_VCF_info.txt") as input, which specifies the path to the VCF files from each sample (in case, multiple files for each sample).

In addition, the output directory, the output summary file (needed for downstream analyses), the cache directory (i.e., the path to the VEP database directory set in the installation phase), and the correct genome version can be set.

```
TIminerAPI.executeVepDir(inputInfoFile="BCCL_VCF_info.txt",
             outputDir="../results",
             outputSummaryFile="../results/BCCL_VEP_info.txt",
             cacheDir="/home/user/databases/vep/",
             genomeVersion=37,
             threadCount=6)
```

The module generates a file in the output directory with the amino acid sequences of the mutated proteins and a file in VEP format with the info about the considered mutation for each sample, including the location, change of the codon(s) in the DNA sequence, and change of the amino acid(s) in the protein.

3.3.2 HLA-Type Determination with TIminerAPI.executeOptitypeDir

Class-I HLA types of each sample can predicted at four-digit resolution from RNA-seq, WES or WGS data with Optitype, using the TIminer function *TIminerAPI.executeOptitypeDir*. The predictions are based on the alignment of sequencing reads to HLA loci and subsequent determination of the HLA types based on read coverage. As the alignment is very memory intensive, it should be ensured that enough memory is available (*see* **Notes 2** and **9**). To run the HLA prediction, the type of sequencing data ("dna" or "rna") and the RNA-seq input info file "BCCL_FASTQ_info.txt" need to be specified. The name of the output file containing the predicted HLA types ("BCCL_HLA.txt") and the number of threads for computation can be also specified:

```
TIminerAPI.executeOptitypeDir(inputtype="rna",
             inputInfoFile="BCCL_FASTQ_info.txt",
             outputFile="../results/BCCL_HLA.txt",
             threadCount=2)
```

3.3.3 MHC I Binding Affinity Prediction with TIminerAPI.executeNetmhcpanDir

From the HLA types determined with *TIminerAPI.executeOptitypeDir* and the mutated protein sequences identified with *TIminerAPI.executeVepDir*, HLA-binding, mutated peptides can be predicted with the TIminer function *TIminerAPI.executeNetmhcpanDir* (Fig. 1). NetMHCpan 3.0 is used to estimate the binding affinity of mutated peptides to the class-I HLA alleles identified in the corresponding samples. The *TIminerAPI.executeNetmhcpanDir* function takes as inputs the files "BCCL_VEP_info.txt" containing the paths to VEP result file with the mutated proteins generated by *TIminerAPI.executeVepDir* and the file "BCCL_HLA.txt" containing the HLA types generated by *TIminerAPI.executeOptitypeDir* (*see* **Note 10**). The output directory, output summary file, and the parameters for the determination of HLA-binding peptides (minimum and maximum peptide length, threshold for maximum binding affinity in nM, or maximum rank) can be also specified. By default, all mutated peptides of lengths between 8 and 11 amino acids are considered, which correspond to the range of class-I HLA-binding peptides. These values can be changed with the *minPeptideLength* and *maxPeptideLength* parameters. In this example, we consider only peptides with the canonical length of nine amino acids to speed up the analysis. As minimum affinity threshold, 500 nM is used by default to select strong binders, but this can be changed using the *affinityThresh* parameter. Alternatively, a *rankThresh* parameter can be used to select only peptides with binding affinities in nM smaller than the top N% of a reference library of random peptides (with the *affinityThresh* parameter being set to "None"). Due to the intrinsic variability of binding affinities between different HLA alleles, the rank-based method is preferable over a fixed affinity threshold. Here, we use a rank-threshold of 0.5% to select strong binders:

```
TIminerAPI.executeNetmhcpanDir(hlaInputFile="../results/B
CCL_HLA.txt",
                inputInfoFile="../results/BCCL_VEP_info.txt",
                outputDir="../results/",
                outputSummaryFile="../results/BCCL_NetMHCpan_info.
                txt",
                minPeptideLength=9,
                maxPeptideLength=9,
                affinityThresh=None,
                rankThresh=0.5,
                threadCount=6)
```

Accordingly, all mutated peptides with affinity ranks lower than 0.5% are selected as strong binders and reported in the output file. The paths to the result files containing the predicted binding peptides for each sample can be found in the output summary file ("BCCL_NetMHCpan_info.txt").

3.3.4 Gene Expression Quantification with TIminerAPI.executeKallistoDir

To obtain information about the expression of potential neoantigen-producing genes, gene expression levels have to be determined. In addition, gene expression levels can also be used for other analyses, such as *TIminerAPI.executeGSEA*, *TIminerAPI.executeIPS,* or selection of differentially expressed genes. The quantification of gene expression from paired-end or single-end RNA-seq FASTQ files is performed with Kallisto and can be run with *TIminerAPI.executeKallistoDir.* The function *TIminerAPI.executeKallistoDir* requires the input info file "BCCL_FASTQ_info.txt" and the output file name as main arguments (*see* **Note 11**). The number of threads and the location of the index of the reference genome in the database directory can be also specified.

```
TIminerAPI.executeKallistoDir(inputInfoFile="BCCL_FASTQ_i
nfo.txt",
              outputFile="../results/BCCL_Kallisto.txt",
              index="/home/databases/kallisto/hg19_M_rCR
              S_kallisto.idx",
              threadCount=6)
```

The gene expression quantification results are reported in three separate files, containing,

- File of normalized gene expression levels as $\log_2(\text{TPM} + 1)$, whose name can be specified by the *outputFile* parameter.
- Raw gene counts ("raw_counts_BCCL_Kallisto.txt" file, saved in the same directory as the output file).
- Gene TPM ("tpm_BCCL_Kallisto.txt" file, saved in the same directory as the output file).

These files are tab-delimited files with gene names on the rows and sample identifiers on the columns.

3.3.5 Filtering of Candidate Neoantigens

Neoantigens cannot arise from mutations in genes that are not expressed. The *TIminerAPI.filterNeoantigenDir* function can be used to select only neoantigens predicted from expressed genes. This function takes as input the output summary file from *TIminerAPI.executeNetmhcpanDir* (*see* **Notes 12** and **13**), the file of normalized gene expression levels generated by *TIminerAPI.executeKallistoDir*, and, optionally, the output directory and threshold for identifying expressed genes:

```
TIminerAPI.filterNeoantigenDir(inputInfoFile="../results/
BCCL_NetMHCpan_info.txt",
              geneExpressionInputFile="../results/BCCL_K
              allisto.txt",
              filteredNeoantigenOutputDir="../results/",
              expressionThreshold=2)
```

This filtering step generates, for each sample, a result file in the specified output directory that contains information on the candidate neoantigens, including gene and protein of origin, binding HLA allele, binding affinity of the reference and mutant peptides, and normalized gene expression in $\log_2(\text{TPM} + 1)$.

3.4 Additional Analyses with TIminer

Beyond neoantigen prediction, the gene expression data generated by *TIminerAPI.executeKallistoDir* can be used for complementary immunogenomics analyses with TIminer. In particular, the composition of tumor immune infiltrates can be estimated with gene set enrichment analysis (GSEA) [18] (*see* **Note 14**). Moreover, an immunophenogram (IPG) and a composite immunophenoscore (IPS) can be generated, which can serve as indicators of the immunological state of the tumor [16] (*see* **Note 15**).

3.5 Full and Custom Pipelines

TIminer offers the option to run all TIminer modules of Fig. 1 in once, using a single command (*see* **Note 4**):

```
$ python my_TIminerPipeline.py
```

In addition, a graphical user interface is available to run the analysis on single samples (*see* **Note 16**).

For neoantigen prediction, we recommend building a custom pipeline as explained in the example analysis in this chapter, and then run it as:

```
$ python my_pipeline.py
```

This approach uses only the functions that are relevant for this purpose, thereby saving computational time and resources, allowing greater control over time and memory requirements, and facilitating troubleshooting. The functions can be either called through distinct scripts (each starting with the "import" command of Subheading 3.2), or assembled into a single script (requiring the "import" command only once, before the API function invocations). The ordering of the function calls is not strict, but it should ensure that all the input files required by a module have been generated in the previous steps.

3.6 Results Interpretation

All results from this example analysis can be found in the "results" directory. In the first step of the pipeline, mutated proteins were predicted from the original mutation data with VEP. For all samples, the paths to the results from *TIminerAPI.executeVepDir* can be found in the file "BCCL_VEP_info.txt". The files "MCF7_mutprotein_info.txt" and "MDA-MB-453_mutprotein_info.txt" contain the VEP results, and the files "MCF7_mutprotein_info.txt" and "MDA-MB-453_mutprotein_info.txt" contain the corresponding sequences of the mutated proteins for MCF7 and MDA-MB-453, respectively. Overall, VEP reported 6726 mutations in 4460 genes for MCF7 and 2587 mutations in 1765 genes for MDA-MB-453 (Fig. 2a, b). Most of the mutated genes were unique to a single cancer cell line, but 728 were shared (Fig. 2d).

The "BCCL_HLA.txt" file that was generated in the second step of the pipeline by *TIminerAPI.executeOptitypeDir* contains HLA types for all samples determined at four-digit resolution, as reported in the following:

MDA-MB-453	HLA-A01:01	HLA-A01:01	HLA-B08:01	HLA-B08:01	HLA-C07:01	HLA-C07:01
MCF7	HLA-A02:01	HLA-A02:01	HLA-B44:02	HLA-B44:02	HLA-C05:01	HLA-C05:01

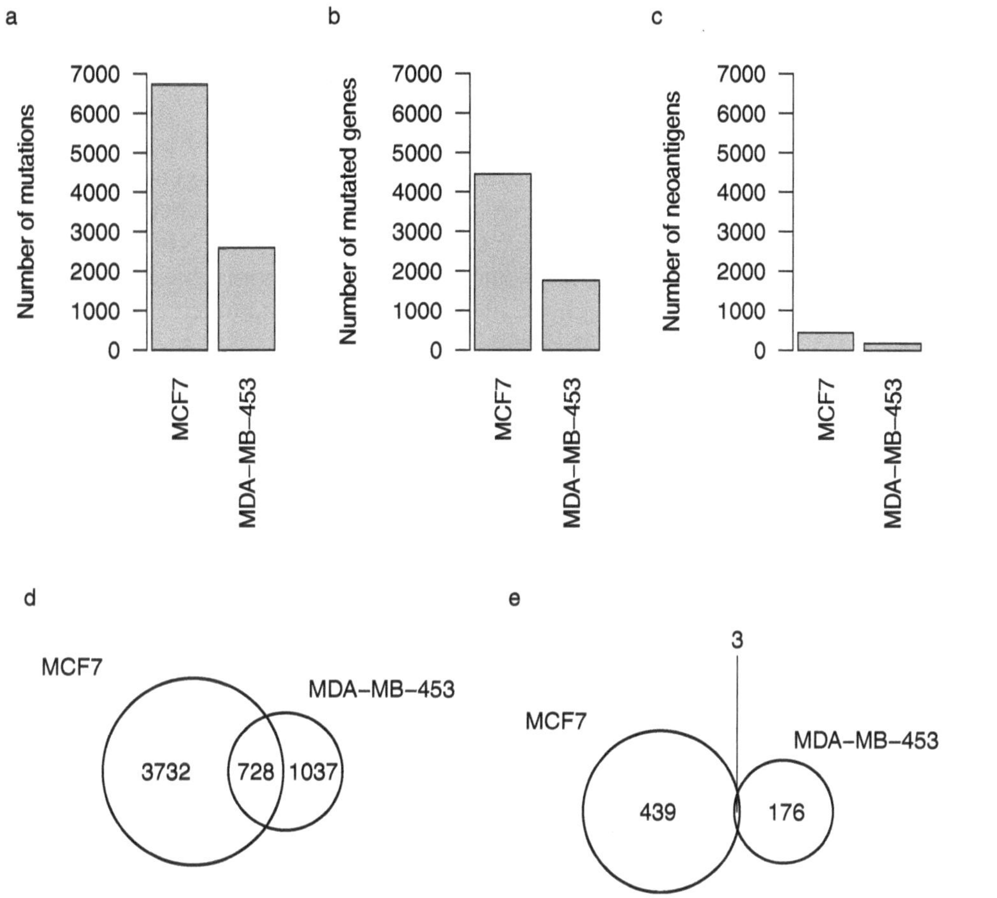

Fig. 2 Summary of TIminer results for the two breast cancer cell lines. (**a**) Number of nonsynonymous mutations after VEP preprocessing. (**b**) Number of mutated genes after VEP preprocessing. (**c**) Number of candidate neoantigens (unique peptides) selected after filtering. (**d**) Venn diagram of the mutated genes for the two cancer cell lines after VEP preprocessing. (**e**) Venn diagram of candidate neoantigens (unique peptides) for the two cancer cell lines selected after filtering

Both cell lines were predicted to be homozygous on all alleles. This could indicate the loss of HLA types presenting immunogenic peptides, a common immune evasion mechanism, but it might also be due to the impossibility to reconstruct the HLA loci with very low RNA-seq coverage [19]. Comparison with previous results on the same cell lines are hampered not only by slight differences in computational HLA typing methods but also by the true genetic variability of cancer cell lines used in the different studies and datasets [20].

In the third step of the pipeline, binding affinities of mutated peptides to the determined HLA proteins were predicted with NetMHCpan. The locations of the result files with the candidate neoantigens generated by *TIminerAPI.executeNetmhcpanDir* can be found in the "BCCL_NetMHCpan_info.txt" info file. The files report 762 strong binders arising from 652 genes for MCF7 and 313 strong binders arising from 232 genes for MDA-MB-453.

After filtering for a minimum gene expression of $\log_2(\text{TPM} + 1)$, 442 neoantigens (unique peptides) from 379 genes were selected for MCF7, and 179 neoantigens from 132 genes for MDA-MB-453, with three neoantigenic peptides shared between the two cell lines (Fig. 2c, e): YLDNRFFTL from gene NBPF9, FSSPKTSHI from gene ERI2, and FEMDKGTYI from gene FRG1.

Although the identification of immunogenic peptides is still an open issue in personalized oncology, features such as binding affinity (or rank) of the mutated peptide, gene expression levels (which has a major impact on the actual peptide abundance), and difference in binding affinity between the wild-type and mutated peptides have been associated with efficiency of peptide-HLA presentation and recognition by T cell [4, 5]. These parameters can be extracted from the final files of filtered peptides and analyzed in concert to identify candidate neoantigens which have low rank affinity, high RNA expression, and/or high ratio of binding affinity in nM between the wild-type and the mutated peptide (Fig. 3). Two of the most highly expressed candidate neoantigens in MCF7 originated from the BCAS3 (Breast Carcinoma-Amplified Sequence 3) and MRPS21 (Mitochondrial Ribosomal Protein S21) genes and presented also a low affinity rank of the mutated peptides (Fig. 3a). However, of the two, only the mutated peptide originated from MRPS21 (MEMAPKINF) had a stronger binding affinity for the HLA-B44:02 allele than its nonmutated counterpart (data not shown). When considering the binging affinity log-ratio, the mutated peptides derived from ZNF619 (Zinc Finger Protein 619) and NBPF14 (Neuroblastoma Breakpoint Family Member 14) obtained a predicted affinity much higher (i.e., lower in terms of nM) than their wild-type versions. The two most highly expressed candidate neoantigens in MDA-MB-453 cells, also characterized by low affinity rank, were derived from RPL4 (Ribosomal Protein L4) and SLC25A5 (Solute Carrier Family 25 Member 5) genes (Fig. 3c). Both mutated peptides were stronger binders than their wild-type versions (data not shown). When considering the binging affinity log-ratio of wild-type vs. mutated peptides, the candidate neoantigens derived from HSD17B7 (Hydroxysteroid 17-Beta Dehydrogenase 7) and from SRRM2 (Serine/Arginine Repetitive Matrix 2) bound ~100–200 times more strongly than their nonmutated counterparts (data not shown).

In summary, TIminer results can be considered to characterize the predicted peptides in terms of binding affinity of the mutated and wild-type peptides and corresponding gene expression to prioritize top candidates. However, we point out that the final prioritization strategy inevitably depends on the downstream application

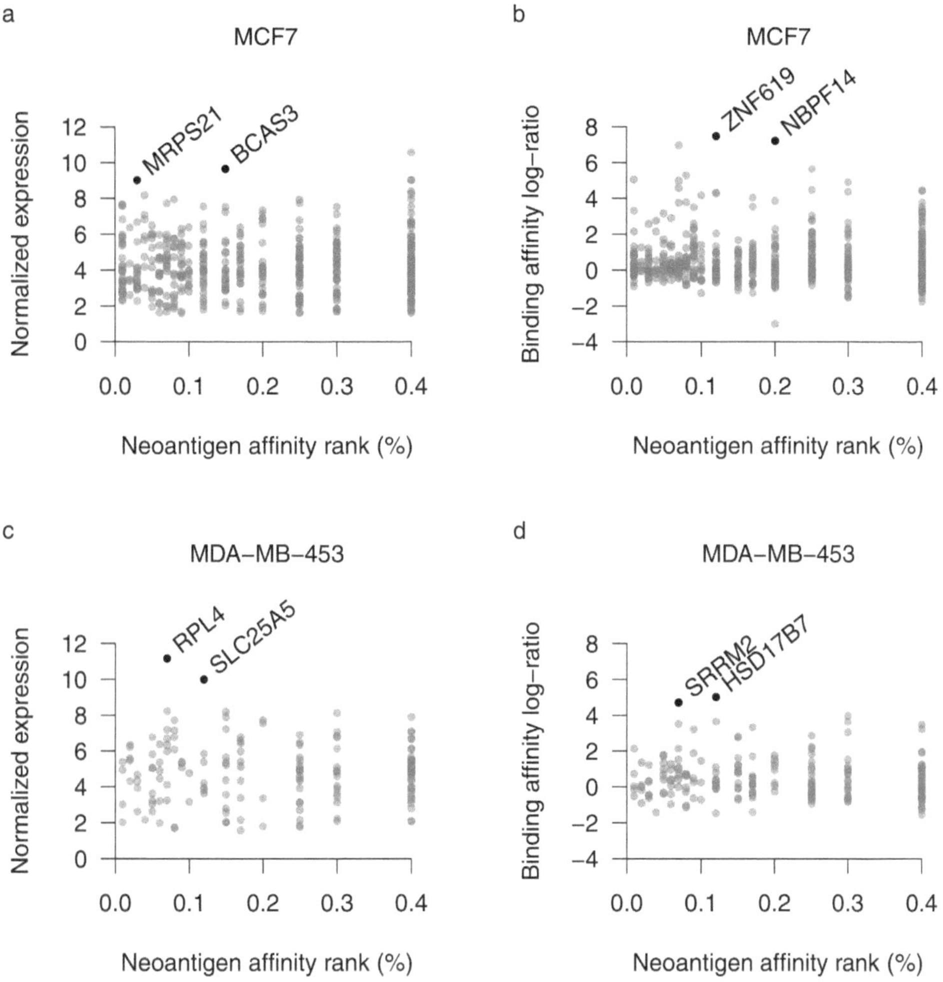

Fig. 3 Visualization of the features of the final candidate neoantigens. Scatterplot of the affinity rank % of mutated peptides versus the expression of the corresponding genes in $\log_2(\text{TPM} + 1)$ for the MCF7 (**a**) and MDA-MB-453 (**c**) cell line. Scatterplot of the affinity rank % of mutated peptides versus the binding affinity log-ratio of wild-type and mutant peptides for the MCF7 (**b**) and MDA-MB-453 (**d**) cell line

(e.g., simple computation of the overall neoantigen load or accurate design of personalized anticancer vaccines), and at the present, a consensus on the optimal approach for integrating and prioritizing these features is still lacking.

4 Notes

1. The full TIminer documentation is accessible online at https://icbi.i-med.ac.at/software/timiner/doc/index.html.
2. For Docker installation, see instructions here: https://docs.docker.com/install. Recent versions of Docker for Mac OS X

limit the available memory and CPUs. To increase CPUs, memory, and swap in Docker settings, go to "Settings" > "Preferences" > "Advanced".

3. Python 2.7 is required for TIminer and can be installed from: https://www.python.org/downloads.

   Other Python versions should not be used. To check which version(s) are currently installed, use.

```
$ python -version
```

4. The full pipeline including all the tools mentioned before can be run with python TIminerPipeline.py from the TIminer "scripts" directory. This requires a tab-separated input info file with the paths to the FASTQ and VCF files as argument to --input. The output directory has to be set with --out. Additional arguments can be set to specify the database location and the number of threads to be used.

```
$ python TIminerPipeline.py --input INPUT --out OUT \
                  --database DATABASE \
                  --threadcount THREADCOUNT
```

5. Most of the functions allow parallelization, which can be set with the *threadcount* or the *maxthreads* parameter and is particularly useful for larger numbers of samples.
6. Instead of running batch analyses of multiple sample files, individual files can be processed by calling the corresponding TIminer function without the "Dir" suffix. This may be useful for conducting test runs before doing the full analysis. Note that in some cases, the input arguments of the batch and the single-sample function versions differ as the single inputs do not require the use of input information files. The input info files can also be used for just a single sample as well.
7. The input information files are used to run multiple samples in one batch. For each sample, specify the location of the corresponding sample files that are used by a particular function. Both absolute and relative paths can be used. For each of the different input data types, separate input info files have to be made. The files are plain text files; each line corresponds to one sample and begins with the sample name, followed by the path(s) to the input file(s). The individual elements need to be tab-separated (blank characters cannot be used). In addition, a header starting with # can be included. An example is shown below:

Sample_A	Filepath_A1	Filepath_A2
Sample_B	Filepath_B1	Filepath_B2

In the FASTQ input info file, in case of unpaired RNA-seq data, only one FASTQ file has to be specified per sample and the second column has to be set to "None," whereas the two corresponding FASTQ files (for forward and reverse reads) have to be used for paired-end data. In the VCF info file, multiple VCF files can be provided for each sample.

8. The input VCF files for *TIminerAPI.executeVepDir* have to be in the correct format, including a header that specifies the VCF format and the genome version. If the input files are correct, the input info file or the path to the database directory may be misspecified.
9. The input files for *TIminerAPI.executeOptitypeDir* have to be in FASTQ format and data from paired-end sequencing reads need to be given as two separate input files. If the input info file and the input FASTQ files are correct, generation of the results may still fail when not enough memory is available. Monitoring memory usage while running the Optitype module can indicate whether more memory is required. If there is not enough memory available, it can be increased by expanding the swap space (e.g., by making a new swap file), or alternatively, the number of threads to be used can be lowered, which decreases memory usage.
10. Errors while running *TIminerAPI.executeNetmhcpanDir* may be related to problems with the input files generated by the previous steps in the pipeline.
11. To use Kallisto for the quantification of gene expression, the correct input FASTQ files have to be given.
12. If the number of candidate neoantigens is low or the count threshold too high, it might happen that no expressed antigens are retained. Otherwise, there might be a problem related to gene expression data. TIminer uses HGNC symbols for gene naming. Thus, if user-provided gene expression files are used, the gene names need to follow the HGNC nomenclature; other gene identifiers necessarily result in empty output files.
13. Highly mutated samples can result in underestimation of gene expression levels due to differences with respect to the human reference genome. To limit this problem, instead of considering sample-specific gene expression levels, TIminer also allows to compute median gene expression levels across the input samples belonging to the same cancer type. To this end, a tab-delimited text file that maps the sample identifiers (first

column) to the cancer type (second column) should be provided through the *subjectCancertypeMappingInputFile* option.

14. If the samples originate from bulk tumor tissue, they contain not only tumor cells but also other cell types (note that this analysis is not meaningful for cell line/in vitro samples). Of particular relevance for tumor immunology are the levels of different immune cell populations in the tumor. Gene set enrichment analysis can be used to determine the composition of immune infiltrates using immune cell gene expression signatures. *TIminerAPI.executeGSEA* uses GSEA on one of two preestablished signature datasets or a custom gene set to perform the estimation:

```
TIminerAPI.executeGSEA(inputFile="Kallisto.txt",
                       outputDir="../results"
```

    This function generates for each sample the GSEA outputs for all cell types and an overview HTML file. In addition, it reports two text files containing normalized enrichment scores ("NES.txt") and FDR-corrected p-values ("FDR_q-val.txt").

15. The immunophenogram gives an overview of immunological parameters of the tumor and the aggregate immunophenoscore (IPS) can be used to predict the reponse to immune checkpoint inhibitors. TIminer computes both the immunophenogram and the immunophenoscore with the function TIminerAPI.executeIPS. For details, we refer to the original publication [16]. The function can be used for both single and multiple samples by providing the gene expression output file from *TIminerAPI.executeKallistoDir* as input. Note that this, like *TIminerAPI.executeGSEA*, is only meaningful if the analysis is done for bulk tumor samples:

```
TIminerAPI.executeIPS(inputFile="Kallisto.txt",
                      outputDir="../results")
```

    Immunophenogram plots and a file containing the numerical results are generated in the output directory.

16. TIminer can be used with a graphical user interface for single-sample analysis by starting python TIminerUI.py from the "scripts" directory in a terminal. This requires *tk-inter* to be installed on Linux. A user-defined database location can be set with the optional database parameter by executing:

```
$ python TIminerUI.py [--database DATABASE]
```

All other options can be defined in the GUI, and the TIminer example files can be loaded with the "load examples" button.

References

1. Chen DS, Mellman I (2013) Oncology meets immunology: the cancer-immunity cycle. Immunity 39(1):1–10. https://doi.org/10.1016/j.immuni.2013.07.012
2. Galluzzi L, Chan TA, Kroemer G et al (2018) The hallmarks of successful anticancer immunotherapy. Sci Transl Med 10(459):eaat7807. https://doi.org/10.1126/scitranslmed.aat7807
3. Fridman WH, Pages F, Sautes-Fridman C et al (2012) The immune contexture in human tumours: impact on clinical outcome. Nat Rev Cancer 12(4):298–306. https://doi.org/10.1038/nrc3245
4. Havel JJ, Chowell D, Chan TA (2019) The evolving landscape of biomarkers for checkpoint inhibitor immunotherapy. Nat Rev Cancer 19(3):133–150. https://doi.org/10.1038/s41568-019-0116-x
5. Lee CH, Yelensky R, Jooss K et al (2018) Update on tumor neoantigens and their utility: why it is good to be different. Trends Immunol 39(7):536–548. https://doi.org/10.1016/j.it.2018.04.005
6. Schumacher TN, Scheper W, Kvistborg P (2018) Cancer neoantigens. Annu Rev Immunol. https://doi.org/10.1146/annurev-immunol-042617-053402
7. Gfeller D, Bassani-Sternberg M (2018) Predicting antigen presentation-what could we learn from a million peptides? Front Immunol 9:1716. https://doi.org/10.3389/fimmu.2018.01716
8. Tappeiner E, Finotello F, Charoentong P et al (2017) TIminer: NGS data mining pipeline for cancer immunology and immunotherapy. Bioinformatics 33(19):3140–3141. https://doi.org/10.1093/bioinformatics/btx377
9. Hackl H, Charoentong P, Finotello F et al (2016) Computational genomics tools for dissecting tumour-immune cell interactions. Nat Rev Genet 17(8):441–458. https://doi.org/10.1038/nrg.2016.67
10. McLaren W, Gil L, Hunt SE et al (2016) The ensembl variant effect predictor. Genome Biol 17(1):122. https://doi.org/10.1186/s13059-016-0974-4
11. Szolek A, Schubert B, Mohr C et al (2014) OptiType: precision HLA typing from next-generation sequencing data. Bioinformatics 30(23):3310–3316. https://doi.org/10.1093/bioinformatics/btu548
12. Nielsen M, Lundegaard C, Blicher T et al (2007) NetMHCpan, a method for quantitative predictions of peptide binding to any HLA-A and -B locus protein of known sequence. PLoS One 2(8):e796. https://doi.org/10.1371/journal.pone.0000796
13. Hoof I, Peters B, Sidney J et al (2009) NetMHCpan, a method for MHC class I binding prediction beyond humans. Immunogenetics 61(1):1–13. https://doi.org/10.1007/s00251-008-0341-z
14. Nielsen M, Andreatta M (2016) NetMHCpan-3.0; improved prediction of binding to MHC class I molecules integrating information from multiple receptor and peptide length datasets. Genome Med 8(1):33. https://doi.org/10.1186/s13073-016-0288-x
15. Bray NL, Pimentel H, Melsted P et al (2016) Near-optimal probabilistic RNA-seq quantification. Nat Biotechnol 34(5):525–527. https://doi.org/10.1038/nbt.3519
16. Charoentong P, Finotello F, Angelova M et al (2017) Pan-cancer immunogenomic analyses reveal genotype-immunophenotype relationships and predictors of response to checkpoint blockade. Cell Rep 18(1):248–262. https://doi.org/10.1016/j.celrep.2016.12.019
17. van der Meer D, Barthorpe S, Yang W et al (2019) Cell model passports-a hub for clinical, genetic and functional datasets of preclinical cancer models. Nucleic Acids Res 47(D1): D923–D929. https://doi.org/10.1093/nar/gky872
18. Subramanian A, Tamayo P, Mootha VK et al (2005) Gene set enrichment analysis: a knowledge-based approach for interpreting genome-wide expression profiles. Proc Natl Acad Sci U S A 102(43):15545–15550. https://doi.org/10.1073/pnas.0506580102
19. Boegel S, Lower M, Bukur T et al (2014) A catalog of HLA type, HLA expression, and neo-epitope candidates in human cancer cell lines. Oncoimmunology 3(8):e954893. https://doi.org/10.4161/21624011.2014.954893
20. Liu Y, Mi Y, Mueller T et al (2019) Multi-omic measurements of heterogeneity in HeLa cells across laboratories. Nat Biotechnol 37(3):314–322. https://doi.org/10.1038/s41587-019-0037-y

Chapter 10

OpenVax: An Open-Source Computational Pipeline for Cancer Neoantigen Prediction

Julia Kodysh and Alex Rubinsteyn

Abstract

OpenVax is a computational workflow for identifying somatic variants, predicting neoantigens, and selecting the contents of personalized cancer vaccines. It is a Dockerized end-to-end pipeline that takes as input raw tumor/normal sequencing data. It is currently used in three clinical trials (NCT02721043, NCT03223103, and NCT03359239). In this chapter, we describe how to install and use OpenVax, as well as how to interpret the generated results.

Key words Neoantigen, Cancer vaccine, Bioinformatics pipeline, NGS, Docker, Immunoinformatics

1 Introduction

Mutations acquired by cancer cells can be used to distinguish tumor cells from normal cells, and as such, they create actionable targets for the body's immune system [1]. Genetic mutations that result in a mutated protein sequence can be recognized as "foreign" when they are displayed to the immune system on MHC molecules, eliciting a T-cell response. Such mutations are known as neoantigens [2]. Personalized neoantigen vaccination is a therapeutic strategy which aims to direct the patient's immune system to mount a response against the specific mutated cancer proteins present in his or her tumor. The contents of a personalized vaccine can be chosen in silico by identifying expressed mutations from genomic sequencing of the patient's normal and tumor DNA as well as tumor RNA, and by further filtering those mutated protein subsequences by predicted MHC binding. Here, we present the bioinformatics pipeline for selecting patient-specific cancer neoantigen vaccines developed by the OpenVax group at Mount Sinai. This pipeline is currently the basis for three phase I clinical trials using synthetic long peptides (NCT02721043, NCT03223103, and NCT03359239).

Sebastian Boegel (ed.), *Bioinformatics for Cancer Immunotherapy: Methods and Protocols*, Methods in Molecular Biology, vol. 2120, https://doi.org/10.1007/978-1-0716-0327-7_10,

The OpenVax pipeline starts by aligning tumor and normal DNA sequencing FASTQ files to a user-specified reference genome using BWA [3]. These initial DNA alignments are then transformed and modified using a series of steps from GATK 3.7 [4]: MarkDuplicates, IndelRealigner, and BQSR. Tumor RNA sequencing reads are aligned by STAR [5] to the same reference genome. Aligned RNA reads are grouped into two sets: those spanning introns and those aligning entirely within an exon (determined based on the CIGAR string). The latter group is passed through IndelRealigner, and the two groups of reads are merged. Somatic variant calling is performed by running MuTect 1.1.7 [6], Mutect 2, and/or Strelka version 1 [7]. The user is expected to specify which of those three variant callers they would like to use. Once both somatic variants and aligned RNA are ready, a custom tool called Vaxrank coordinates variant effect annotation, expression estimation, and MHC binding prediction. Vaxrank prioritizes somatic variants based on expression and predicted MHC class I binding affinity. A diagram of these steps is shown in Fig. 1.

The OpenVax pipeline is implemented using the Snakemake workflow management system [8]. We recommend running the pipeline using our provided Docker image, which includes all needed software dependencies. In this protocol, we focus on instructions for installing and running the pipeline using Docker.

2 Materials

2.1 Hardware

This pipeline was developed and optimized for a machine with the following specifications:

- 16 or more cores.
- 32GB RAM in order to run the full pipeline to compute vaccine peptides, or to run the pipeline for the first time with a custom reference genome which has not yet been processed. Otherwise, 8GB of RAM is enough to run somatic variant calling.
- Suggested: 500GB free disk space, if running on human exome sequencing data.

2.2 Software

Since this pipeline is fully Dockerized, the only requirement is an installation of Docker. *See* **Note 5** for information about running the OpenVax pipeline outside of Docker.

2.3 Data

To run the OpenVax pipeline, the user must have gzip-compressed FASTQ files from the tumor DNA, normal DNA, and tumor RNA (*see* **Note 1** for suggested sequencing parameters). These genomic inputs are sufficient to get somatic variant calls and RNA alignments. Subsequent neoantigen prediction steps also require specifying a patient's MHC class I alleles.

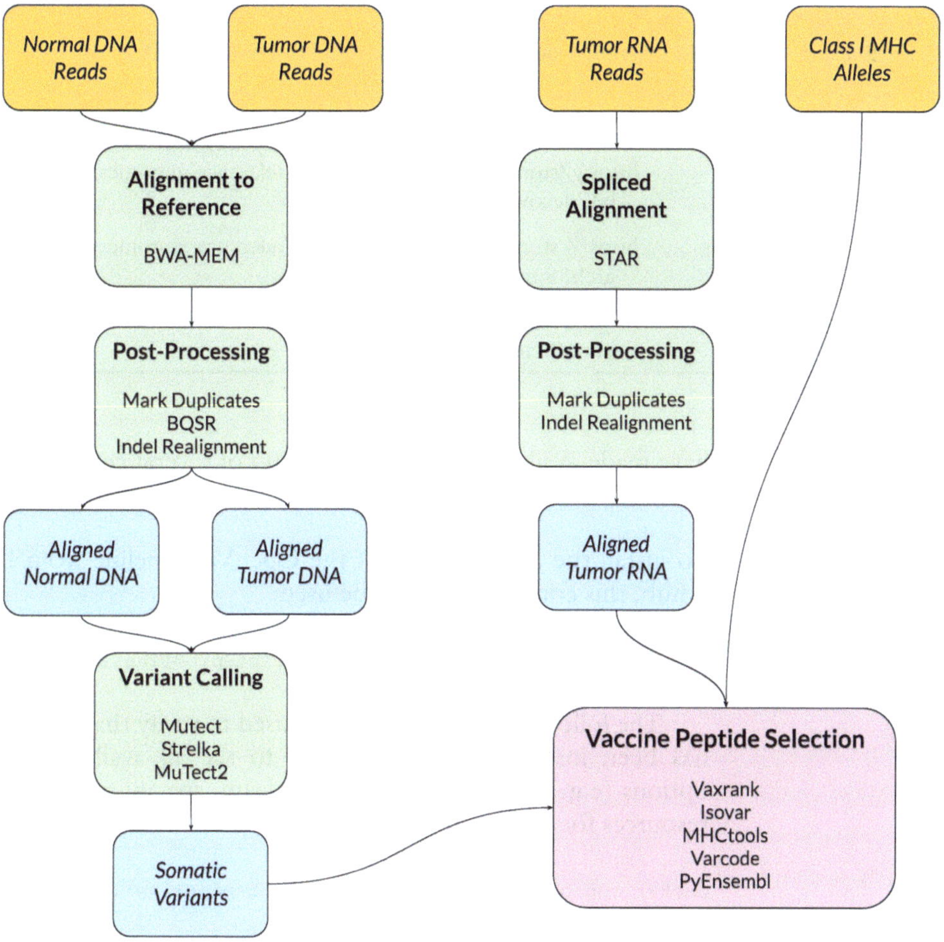

Fig. 1 Steps involved in the OpenVax computational pipeline

In addition to sample data, the pipeline requires the user to provide the following reference genome and associated data:

- Reference sequence file (FASTA).
- Known transcripts file (GTF).
- dbSNP [9] known mutation file for this reference genome (VCF).

Optionally, the user may also provide the following:

- COSMIC [10] cancer mutations file (VCF). If provided, will be used in somatic variant calling.
- Exome capture kit coverage file (BED). If provided, will be used to compute certain sequencing metrics (see the Broad's GATK documentation on HsMetrics).

While the pipeline supports preparing a custom reference genome for use by aligners and other tools, for a quick start, we

Table 1
List of available preprocessed reference genomes with associated download URLs

Reference genome	Download link	File size
b37 with decoys	https://storage.cloud.google.com/reference-genomes/b37decoy.tar.gz	30GB
GRCh38 with decoys, virus sequences	https://storage.cloud.google.com/reference-genomes/grch38.tar.gz	31GB
mm10	https://storage.cloud.google.com/reference-genomes/mm10.tar.gz	28GB

have made available a processed version of several commonly used reference genomes in Google Cloud (Table 1).

2.4 Installation

To install the latest version of the OpenVax pipeline from Dockerhub, this command should be used:

```
docker pull openvax/neoantigen-vaccine-pipeline:latest
```

The following command can be used to verify that the pipeline has been installed successfully and to see all available pipeline options (e.g., ability to execute a dry run, specify memory/CPU resources for the pipeline):

```
docker run openvax/neoantigen-vaccine-pipeline:latest -h
```

3 Methods

3.1 Setup Instructions

The pipeline is run by invoking a Docker entrypoint in the image while providing three directories as mounted Docker volumes:

- `/inputs`: FASTQ files and a configuration YAML.
- `/outputs`: directory to write results to.
- `/reference-genome`: reference genome data, shared across patients.

In the example above and for the remainder of this protocol, the three local directories corresponding to these volumes will be referred to as follows:

- `/your/path/to/fastq/inputs`
- `/your/path/to/pipeline/outputs`
- `/your/path/to/pipeline/outputs`
- `/your/path/to/reference/genome`

For the duration of the pipeline execution, all three directories and their contents must be world-writable. This is necessary because the Docker pipeline runs as an unprivileged user, and the pipeline will write data to one or more of these directories.

It is easiest to use one of the provided processed reference genomes (*see* Table 1). To do that, the user must save the relevant link to `/your/path/to/reference/genome`. The archived file can also be downloaded using `gsutil` from the command-line, here showing an example for GRCh38:

```
gsutil -m cp gs://reference-genomes/grch38.tar.gz /your/path/
to/reference/genome/
```

Once the compressed reference genome data are downloaded to the right location, the file must be uncompressed using `tar`:

```
cd /your/path/to/reference/genome/ && tar -zxvf grch38.tar.gz
```

These provided reference genomes include DNA/RNA aligner and GATK index files, which must exist before the pipeline can proceed. If a custom user-provided reference genome is used, the pipeline will first generate the required index files, which may take several hours. The Docker image also supports a run of the pipeline that performs standalone reference processing (*see* Subheading 3.3.1).

3.2 Configuration

A sample configuration file (Fig. 2) contains sample-specific settings, as well as those that may be common across samples and shared in pipeline runs. This configuration file must exist in the directory `/your/path/to/fastq/inputs`, the same directory as the input FASTQ files. Various aspects of this configuration file are explained in the following section.

3.2.1 Sample Configuration

We note several instructions for defining the sample-specific portion of the configuration file:

- In specifying locations of the input FASTQ files, the `/inputs` part of the filename should remain unchanged. The user needs to modify only the basename of the input file paths.
- The configuration sample ID (here and for the remainder of this chapter, we will use test-sample as the sample ID) will determine the outputs subdirectory to which all pipeline result files will be written (`/your/path/to/pipeline/outputs/test-sample`). This will include the final vaccine peptide results as well as many intermediate files (*see* **Note 2**).

```
# Sample-specific configuration
input:
  id: test-sample
  mhc_alleles:
    - HLA-A*02:01
    - HLA-A*24:01
    - HLA-B*03:01
    - HLA-B*40:01
    - HLA-C*07:01
    - HLA-C*07:02
  normal:
    - fragment_id: L001
      type: single-end
      r: /inputs/normal.fastq.gz
  tumor:
    - fragment_id: L001
      type: single-end
      r: /inputs/tumor.fastq.gz
  rna:
    - fragment_id: L001
      type: single-end
      r: /inputs/rna.fastq.gz

# Pipeline parameter configuration
workdir: /outputs  # should remain unchanged
reference:
  genome: /reference-genome/b37decoy/b37decoy.fasta
  dbsnp: /reference-genome/b37decoy/dbsnp.vcf
  cosmic: /reference-genome/b37decoy/cosmic.vcf
  transcripts: /reference-genome/b37decoy/transcripts.gtf
  capture_kit_coverage_file: /reference-genome/b37decoy/S04380110_Covered_grch37.bed
mhc_predictor: netmhcpan-iedb
variant_callers:
  - mutect
  - strelka
```

Fig. 2 Example OpenVax pipeline configuration file

- If the input data are paired-end FASTQ files, the two files must be specified as `r1` and `r2` entries instead of the singular entry in the config template. The `type` must also be changed to `paired-end`.
- The example above assumes that for each of tumor/normal DNA and tumor RNA samples, the sequencing data are contained in a single file (or paired-end file pair). However, it is often the case that sequencing data are spread across several fragments, or sequencing lanes. For this reason, the `tumor`, `normal`, and `rna` sections in the configuration file each contains a list of fragments. This allows for multiple list elements, as long as each entry has a distinct `fragment_id` value. For example, if the tumor RNA data come from multiple sequencing runs, the user could add another list element to the `rna` block with `fragment_id: L002`.

Table 2
MHC class I predictor options supported by the OpenVax pipeline

Predictor option string	Description
netmhcpan-iedb	NetMHCpan [12], version 4.0 as of this writing
netmhccons-iedb	NetMHCcons [13], version 1.1 as of this writing
smm-iedb	Stabilized matrix method [14]
smm-pbmec-iedb	SMM with a Peptide:MHC Binding Energy Covariance matrix (SMMPMBEC) [15]

Table 3
Somatic variant caller options supported by the OpenVax pipeline

Variant caller option string	Description
mutect	The original MuTect variant caller, version 1.1.7
mutect2	Mutect2, included as part of GATK version 3.7
strelka	The Strelka somatic variant caller version 1.0.14

- In the case where the pipeline is used only for somatic variant calling, the `rna` and `mhc_alleles` input sections may be omitted.

3.2.2 Pipeline Parameter Configuration

Many parameters can be shared between pipeline runs for different samples. For example, the user may want to use the same reference genome and somatic variant callers across multiple samples, in which case that section of the configuration file contents will be the same.

The reference genome paths should point to the files described in the data requirements above. Additionally, as mentioned, the `cosmic` and `capture_kit_coverage_file` entries are optional here.

Please note that for custom reference genomes that are not part of the Ensembl standard (which includes GRCh37/hg19, GRCh38, GRCm38/mm10, etc.), the pipeline cannot be used to compute ranked vaccine peptides. However, all workflow steps up to and including somatic variant calling are still supported.

The pipeline supports the use of MHC class I predictors made available through the IEDB web interface (Table 2; also *see* **Note 3**). The predictor of choice must be specified as the `mhc_predictor` in the configuration file.

Several somatic variant callers are supported in the OpenVax workflow (Table 3). The pipeline will run each variant caller specified in the `variant_callers` section of the configuration file, then use the union of all passing somatic variants as an input to the neoantigen prediction step.

Future pipeline versions will also support Strelka2 [11].

3.3 Running the Pipeline

After installation and setup of the pipeline, as well as creation of a sample-specific configuration file, the command to run the pipeline can be modeled on the following:

```
docker run -it \
-v /your/path/to/fastq/inputs:/inputs \
-v /your/path/to/pipeline/outputs:/outputs \
-v /your/path/to/reference/genome:/reference-genome \
openvax/neoantigen-vaccine-pipeline:latest \
--configfile=/inputs/sample-config-file.yaml
```

The output will be a set of ranked variants and proposed vaccine peptide results in several file formats, including basic text (ASCII) and PDF, which are further described later in this protocol.

In the following sections, we will elaborate on additional parameters which can be added to the `docker run` command. For troubleshooting the pipeline, including inspecting the Docker image and accessing relevant log files, *see* Notes **4** and **6**.

3.3.1 Reference Genome Processing

In order to use a custom reference genome, the pipeline will need to generate a number of processed index files before working with sample data. This will happen automatically the first time it is run on a sample if the requisite files do not exist, but it may be easier to run this step as standalone first. This is recommended especially if the pipeline is going to be used to process multiple samples. Standalone reference genome processing can be done with the following command:

```
docker run -it \
-v /your/path/to/fastq/inputs:/inputs \
-v /your/path/to/pipeline/outputs:/outputs \
-v /your/path/to/reference/genome:/reference-genome \
openvax/neoantigen-vaccine-pipeline:latest \
--configfile=/inputs/sample-config-file.yaml \
--process-reference-only
```

As a reminder, the `/your/path/to/reference/genome` directory must be world-writable. Additionally, the YAML config file (in this example, `sample-config-file.yaml`) must be in the `/your/path/to/fastq/inputs` directory, which is mounted as `/inputs` here.

3.3.2 Other Pipeline Options

Note that these arguments are all optional to a pipeline run:

- `--dry-run`: Print all commands that the pipeline would run on the input configuration file, without executing them. Note that this command may still result in initial data getting written to the output directory.
- `--somatic-variant-calling-only`: Use this to call somatic variants without doing any RNA processing or neoantigen prediction. The pipeline will run each variant caller specified in the configuration file.
- `--run-qc`: Run FASTQC and some Picard QC metrics on the input data. The FASTQC output will be written to `/your/path/to/pipeline/outputs/test-sample/fastqc-output`, and the digested result if any QC checks fail will be written to `/your/path/to/pipeline/outputs/test-sample/sequencing_qc_out.txt`. Note that QC check failure will not result in a pipeline crash; it is only for informational purposes, and the QC checks included in the OpenVax pipeline may not be relevant to every sequencing situation.
- `--cores`: Specify the number of threads available for pipeline execution. Default: 1 fewer than the total number of CPUs in the executing machine.
- `--memory`: Specify the total memory (in GB) available for pipeline execution. Default: 1 fewer than the total available memory on the executing machine.
- `--target`: Specify an intermediate file to be generated as the pipeline output, in which case only the necessary subset of the pipeline will execute. This is a repeated argument, so any number of final or intermediate targets (*see* **Note 2**) can be specified this way.

3.4 Interpreting the Output

All output files, both final and intermediate, are written to the `/your/path/to/pipeline/outputs/test-sample` directory. The final output of the pipeline is a set of reports, generated both as PDF (Fig. 3) and ASCII (Fig. 4) files, containing an ordered list of ranked variants with several proposed vaccine peptides for each. The filenames of these reports will start with the prefix `vaccine-peptide-report`.

In both the text and PDF format of the OpenVax vaccine report, variants are listed in order ranked by a score which combines expression and predicted MHC binding. For each variant, the report contains a basic description including its gene, predicted protein effect, and number of supporting RNA reads. After the variant description, up to three proposed long vaccine peptides are

1.

Variant	chr9 g.82927102G>T
IGV locus	chr9:82927102
Gene name	Phip
Top score	0.40075
RNA reads supporting variant allele	17
RNA reads supporting reference allele	39
RNA reads supporting other alleles	0

Predicted Effect

Effect type	Substitution
Transcript name	Phip-201
Transcript ID	ENSMUST00000034787
Effect description	p.T441N

i.

RHDNTVI H AVNNMTL	
Transcript name	Phip-201
Length	15
Expression score	4.1231
Mutant epitope score	0.097195
Combined score	0.40075
Max coding sequence coverage	15
Mutant amino acids	1
Mutation distance from edge	7
Manufacturability	
C-terminal 7mer GRAVY score	0.57143
Max 7mer GRAVY score	1.0429
N-terminal Glutamine, Glutamic Acid, or Cysteine	0
C-terminal Cysteine	0
C-terminal Proline	0
Total number of Cysteine residues	0
N-terminal Asparagine	0
Number of Asparagine-Proline bonds	0

Predicted mutant epitopes

Sequence	IC50	Score	Allele	WT sequence	WT IC50
VIHAVNNM	714.63 nM	0.088689	H-2-Kb	VITAVNNM	2329.38 nM
HAVNNMTL	1077.76 nM	0.0085062	H-2-Db	TAVNNMTL	1008.87 nM

Fig. 3 Sample section of the OpenVax PDF report for one variant and vaccine peptide

listed with associated information such as predicted MHC class I minimal epitopes.

Additionally, if the user wishes to conduct a more in-depth analysis of resulting somatic variants, the OpenVax output also includes a CSV file (whose filename starts with `all-passing-`

```
1) chr9 g.82927102G>T (Phip)
     IGV locus: chr9:82927102
     Gene name: Phip
     Top score: 0.40075
     RNA reads supporting variant allele: 17
     RNA reads supporting reference allele: 39
     RNA reads supporting other alleles: 0

     Effect type: Substitution
     Transcript name: Phip-201
     Transcript ID: ENSMUST00000034787
     Effect description: p.T441N

     Vaccine Peptides:
             i. RHDNTVI_H_AVNNMTL (score = 0.40075)
               - Transcript name: Phip-201
               - Length: 15
               - Expression score: 4.1231
               - Mutant epitope score: 0.097195
               - Combined score: 0.40075
               - Max coding sequence coverage: 15
               - Mutant amino acids: 1
               - Mutation distance from edge: 7

               Manufacturability:
               - C-terminal 7mer GRAVY score: 0.57143
               - Max 7mer GRAVY score: 1.0429
               - N-terminal Glutamine, Glutamic Acid, or Cysteine: 0
               - C-terminal Cysteine: 0
               - C-terminal Proline: 0
               - Total number of Cysteine residues: 0
               - N-terminal Asparagine: 0
               - Number of Asparagine-Proline bonds: 0

               Predicted mutant epitopes:
               Allele        IC50       Score     Sequence     WT IC50   WT sequence
               ------   ----------  ---------   --------  ----------   -----------
               H-2-Kb    714.63 nM   0.088689   VIHAVNNM  2329.38 nM      VITAVNNM
               H-2-Db   1077.76 nM  0.0085062   HAVNNMTL  1008.87 nM      TAVNNMTL
```

Fig. 4 Sample section of the OpenVax text report for one variant and vaccine peptide

`variants`) containing every somatic variant that passed all filters from its source variant caller. For each variant, this file will contain some metadata about variant filters that are conceptually applied in order:

- whether the variant is coding or not;
- for coding variants, whether it has adequate RNA support;
- for expressed coding variants, whether the resulting mutated protein contains predicted MHC ligands.

The output directory will also contain VCFs for every variant caller that is specified in the sample config—e.g., `mutect.vcf`, `mutect2.vcf`. See description of other intermediate files in **Note 2**.

4 Notes

1. *Optimal sample sequencing.* This pipeline is meant to work with high coverage short-read sequencing. It has been developed and optimized for the following characteristics of sample data:
   (a) Tumor and normal whole exome sequencing: in our trials, we target 300× tumor coverage, 150× normal.
   (b) Tumor RNA sequencing: we target 100 M reads for fresh samples (using poly-A capture) and 300 M reads for FFPE samples (using Ribo-Zero).

   While these are suggested sequencing parameters, the pipeline is generally able to handle deviations from these settings and will still generate usable vaccine peptide results.
2. *Intermediate files.* As a result of the full pipeline run, many intermediate files are generated in the output directory. It is also possible to run only those parts of the pipeline that are used to generate any of the intermediate files of interest, specifying one or more as a target to the Docker run invocation. For example,

```
docker run -it \
-v /your/path/to/fastq/inputs:/inputs \
-v /your/path/to/pipeline/outputs:/outputs \
-v /your/path/to/reference/genome:/reference-genome \
openvax/neoantigen-vaccine-pipeline:latest \
--configfile=/inputs/sample-config-file.yaml \
 --target=/outputs/test-sample/tumor_aligned_coordinate_sor-
ted_dups_indelreal_bqsr.bam \
 --target=/outputs/test-sample/normal_merged_aligned_coordi-
nate_sorted.bam
```

   This (somewhat arbitrary) example will run alignment on normal DNA and alignment + full GATK process on the tumor DNA.

   These are some of the intermediate file names one might want to use as targets, in a sample's output directory:
   (a) `{normal,tumor,rna}_merged_aligned_coordinate_sorted.bam`: after BWA alignment, merging of lanes if necessary.
   (b) `{normal,tumor,rna}_aligned_coordinate_sorted_dups.bam`: after GATK MarkDuplicates.
   (c) `{normal,tumor}_aligned_coordinate_sorted_dups_indelreal.bam`: after GATK IndelRealigner.

(d) `{normal,tumor}_aligned_coordinate_sorted_dups_indelreal_bqsr.bam`: after GATK BQSR. These BAMs are used as inputs to variant callers.

(e) `rna_aligned_coordinate_sorted_dups_cigar_N_filtered.bam`: after GATK MarkDuplicates, filtered to all tumor RNA reads with Ns in the CIGAR string (will not run IndelRealigner on these).

(f) `rna_aligned_coordinate_sorted_dups_cigar_0-9MIDSHPX_filtered.bam`: after GATK MarkDuplicates, all tumor RNA reads without Ns.

(g) `rna_cigar_0-9MIDSHPX_filtered_sorted_indelreal.bam`: tumor RNA after GATK IndelRealigner.

(h) rna_final_sorted.bam: The final RNA BAM after all processing; used as input to Vaxrank.

(i) `{mutect,mutect2,strelka}.vcf`: all-contig VCF from corresponding variant caller. Use, for example, `mutect_10.vcf`, to only call MuTect variants in chromosome 10.

3. *MHC predictors.* The pipeline we use internally differs slightly from the version described in this protocol due to licensing restrictions on the NetMHC suite of tools, which prevent their inclusion in the provided Docker image. To circumvent this issue, the open source OpenVax pipeline performs MHC binding prediction using the IEDB web interface to these tools. This may be slower but should give the same results. For users that have a license to the NetMHC tools (free for noncommercial use) and wish to run these tools locally in the pipeline, please contact the authors.

4. *Inspecting Docker image.* This command can be used to start an interactive bash session inside the container, which may be helpful for debugging, inspecting tool versions, or executing anything manually:

```
docker run -it \
-v /your/path/to/fastq/inputs:/inputs \
-v /your/path/to/pipeline/outputs:/outputs \
-v /your/path/to/reference/genome:/reference-genome \
--entrypoint /bin/bash \
openvax/neoantigen-vaccine-pipeline:latest
```

5. *Running the OpenVax pipeline outside of Docker.* While we recommend running the OpenVax pipeline using the Docker image as described in this protocol, the user can also choose to run the pipeline with a Python command on a system that contains all required software and data dependencies. Please contact the authors if you are interested in this use case or are

unable to use Docker on your system but wish to use the pipeline workflow logic another way.

6. *Pipeline errors.* If the pipeline errors, the offending task will be highlighted in red in the command-line output and will mention a log file. The first step in understanding what went wrong is to check the contents of that log file, which will be located in the `/your/path/to/pipeline/outputs/test-sample/logs` directory.

References

1. González S, Volkova N, Beer P et al (2018) Immuno-oncology from the perspective of somatic evolution. Semin Cancer Biol 52:75–85
2. Schumacher TN, Scheper W, Kvistborg P (2019) Cancer Neoantigens. Annu Rev Immunol 37:173–200
3. Li H, Durbin R (2009) Fast and accurate short read alignment with Burrows-Wheeler transform. Bioinformatics 25:1754–1760. https://doi.org/10.1093/bioinformatics/btp324
4. McKenna A, Hanna M, Banks E et al (2010) The genome analysis toolkit: a MapReduce framework for analyzing next-generation DNA sequencing data. Genome Res 20:1297–1303
5. Dobin A, Davis CA, Schlesinger F et al (2013) STAR: ultrafast universal RNA-seq aligner. Bioinformatics 29:15–21
6. Cibulskis K, Lawrence MS, Carter SL et al (2013) Sensitive detection of somatic point mutations in impure and heterogeneous cancer samples. Nat Biotechnol 31:213–219
7. Saunders CT, Wong WSW, Swamy S et al (2012) Strelka: accurate somatic small-variant calling from sequenced tumor–normal sample pairs. Bioinformatics 28:1811–1817
8. Köster J and Rahmann S (2012) Building and documenting workflows with Python-based Snakemake, In: German Conference on Bioinformatics 2012, Schloss Dagstuhl-Leibniz-Zentrum fuer Informatik
9. Sherry ST, Ward MH, Kholodov M et al (2001) dbSNP: the NCBI database of genetic variation. Nucleic Acids Res 29:308–311
10. Forbes SA, Bhamra G, Bamford S et al (2008) The Catalogue of Somatic Mutations in Cancer (COSMIC). Curr Protoc Hum Genet 10:11. https://doi.org/10.1002/0471142905.hg1011s57
11. Kim S, Scheffler K, Halpern AL et al (2018) Strelka2: fast and accurate calling of germline and somatic variants. Nat Methods 15:591–594
12. Hoof I, Peters B, Sidney J et al (2009) NetMHCpan, a method for MHC class I binding prediction beyond humans. Immunogenetics 61:1–13
13. Karosiene E, Lundegaard C, Lund O et al (2012) NetMHCcons: a consensus method for the major histocompatibility complex class I predictions. Immunogenetics 64:177–186
14. Peters B, Sette A (2005) Generating quantitative models describing the sequence specificity of biological processes with the stabilized matrix method. BMC Bioinformatics 6:132
15. Kim Y, Sidney J, Pinilla C et al (2009) Derivation of an amino acid similarity matrix for peptide: MHC binding and its application as a Bayesian prior. BMC Bioinformatics 10:394

Chapter 11

Improving MHC-I Ligand Identification by Incorporating Targeted Searches of Mass Spectrometry Data

Prathyusha Konda, J. Patrick Murphy, and Shashi Gujar

Abstract

Effective immunotherapies rely on specific activation of immune cells. Class I major histocompatibility complex (MHC-I) bound peptide ligands play a major role in dictating the specificity and activation of CD8 + T cells and hence are important in developing T cell-based immunotherapies. Mass spectrometry-based approaches are most commonly used for identifying these MHC-bound peptides, wherein the MS/MS spectra are compared against a reference proteome database. Unfortunately, the effectiveness of matching the immunopeptide MS/MS spectra to a reference proteome database is hindered by inflated search spaces attributed to a lack of enzyme restriction in searches. These large search spaces limit the efficiency with which MHC-I peptides are identified. Here, we describe the implementation of a targeted database search approach and accompanying tool, SpectMHC, that is based on a priori-predicted MHC-I peptides. We have previously shown that this targeted search strategy improved peptide identifications for both mouse and human MHC ligands by greater than two-fold and is superior to traditional "no enzyme" search of reference proteomes (Murphy et al. J Res Proteome 16:1806–1816, 2017).

Key words Mass spectrometry, Bioinformatics, MHC ligandome

1 Introduction

Antigenic peptides presented by major histocompatibility complex (MHC) molecules, also referred as human leukocyte antigens (HLA), play a major role in adaptive immunity and dictate the specificity of T-cell responses. These peptides are presented by either class I (to CD8+ cytotoxic T cells) or class II (to CD4+ helper T cells) MHCs [1, 2]. MHC-I and MHC-II molecules are associated with different types of peptide ligands. Class I MHCs are present on the surface of all nucleated cells, while class II MHC complexes are mostly present on the surface of antigen presenting cells (APCs). Class I MHCs bind to ligands usually 8–11 amino acids (AA) long, which are produced endogenously from proteins

Prathyusha Konda and J. Patrick Murphy contributed equally to this work.

Sebastian Boegel (ed.), *Bioinformatics for Cancer Immunotherapy: Methods and Protocols*, Methods in Molecular Biology, vol. 2120, https://doi.org/10.1007/978-1-0716-0327-7_11, © Springer Science+Business Media, LLC, part of Springer Nature 2020

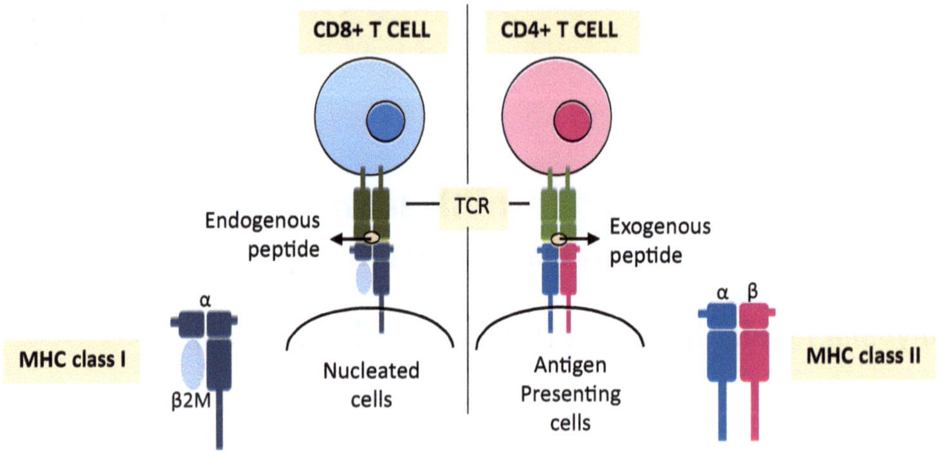

Fig. 1 Overview of MHC class I and MHC class II antigen presentation to T cells

of invading pathogens, host proteins that are considered "self," tumor-associated antigens (TAAs), or mutation-bearing tumor-specific antigens (TSAs). In contrast, Class II MHC molecules bind ligands that are several AA (9–25 AA) longer than class I peptides and are typically derived from exogenous, transmembrane, and cytosolic proteins. Class I MHC ligands are recognized by CD8 + T cells (Fig. 1a), whereas Class II MHC ligands are recognized by CD4+ T cells (Fig. 1b). Together, these peptide-bound MHC class I and class II complexes provide a contextual "window" into the constantly changing intracellular antigenic milieu and serve as a signal for surveillance by CD8+ and CD4+ T cells.

In recent times, advances in mass spectrometry have enabled the precise identification of MHC peptides, thereby contributing to a more detailed understanding of immune responses [3]. Currently, the typical experimental workflow for MHC-I peptide identification involves immunoprecipitation (IP) of MHC-bound peptides using anti-MHC I antibodies, followed by peptide elution, purification, and analysis by LC-MS/MS [4]. Traditionally, MS/MS spectra for eluted MHC-I ligands are scored against a reference proteome database with no enzymatic digestion specificity, using search algorithms such as Mascot. This lack of enzyme specificity creates vast search spaces to be explored during the searches that decrease statistical power in assigning false positive peptides and ultimately reduce the number of peptide identifications in an experiment. It has been proposed for proteomics studies that search space sizes could be reduced by limiting them to only those peptides likely to be present in the sample [5, 6]. Using a similar strategy for MHC ligandomics experiments would improve FDR estimations and increase the number of peptides identified from LC-MS/MS data.

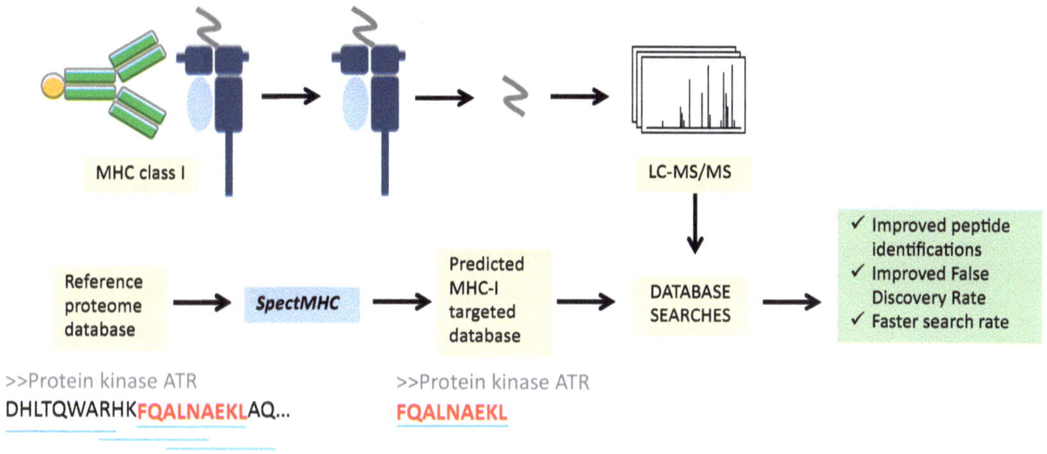

Fig. 2 Workflow of SpectMHC-based MHC-I ligand discovery

Here, we describe a solution to the limitation in MS-based MHC-I ligand identification, whereby a targeted database approach is performed, aided by our database generating tool SpectMHC (Fig. 2). The approach relies on the MHC prediction algorithm, NetMHC [7–9], to create an MHC-I peptide-specific database, thus limiting the database to the kind of peptides which are more likely to be present in the samples. We have previously shown that this targeted search strategy leads to greatly improved numbers (more than two-fold) of MHC class I peptide identifications for both mouse and human MHC-I ligandomes and, thus, is superior to traditional "no enzyme" searches of reference proteome databases [10]. As such, this approach promises to uncover otherwise missed MHC-I ligands, including those containing novel T-cell epitopes of therapeutic potential.

2 Materials

2.1 Hardware—System Requirements

In order to run SpectMHC, a Linux or Darwin (Mac OS) system, preferably with at least 6GB of RAM is required.

2.2 LC-MS/MS Data from an MHC-I Immunoprecipitation Experiment (See Note 1)

Raw LC-MS/MS data from immunoprecipitated MHC-I/HLA peptides from mouse or human tissue are required [10]. The targeted database search approach has been validated using the MHC-I ligands eluted with mouse H-2 D^b- and H-2 K^b-specific antibodies (B22.249 and Y3 hybridomas, respectively), as well as the human pan-HLA antibody (clone W6/32) (*see* **Note 2**). However, this approach can also be applied to other MHC-I alleles.

2.3 Input Reference Protein Database

To create a targeted database, input a fasta file of a reference protein database, which contains protein sequences of interest:

1. Organism-specific protein reference databases can be downloaded in fasta format from the Uniprot database (www.uniprot.org) or other reputable protein sequence repository (*see* **Note 3**).
2. Alternatively, a cell line or tissue-specific custom protein reference database can be formed from RNASeq or exome sequencing data, by translating nucleotide sequences into protein sequences and arranging in fasta file format.
3. The most common databases to access RNAseq/exome sequencing information for cell lines include European Nucleotide Archive (ENA) (https://www.ebi.ac.uk/ena), Expression Atlas (https://www.ebi.ac.uk/gxa/experiments), Array Express (https://www.ebi.ac.uk/arrayexpress/browse.html), NCBI Gene Expression Omnibus (GEO) (https://www.ncbi.nlm.nih.gov/geo/), Cancer Cell Line Encyclopedia (CCLE) (https://portals.broadinstitute.org/ccle/data), Catalogue of Somatic Mutations in Cancer (COSMIC) (https://cancer.sanger.ac.uk/cell_lines).

2.4 Dependency Software for SpectMHC

SpectMHC is dependent on the MHC/HLA ligand prediction algorithm, NetMHC. Currently, the tool supports three versions of NetMHC, NetMHCpan 4.0, NetMHC 4.0, and NetMHC 3.4. Install one of these on your local system:

1. NetMHC prediction software can be freely downloaded by academic users: NetMHCpan 4.0 (http://www.cbs.dtu.dk/services/NetMHCpan/), NetMHC 4.0 (http://www.cbs.dtu.dk/services/NetMHC-4.0/), NetMHC 3.4 (http://www.cbs.dtu.dk/services/NetMHC-3.4/).
2. Installation instructions are available in the readme file associated with the respective version of NetMHC (*see* **Note 4**).
3. Python v2.7 or above (https://www.python.org/download/releases/2.7/).

2.5 Locally Installed SpectMHC Database Compiler Tool

1. SpectMHC is available as a free download for academic users to compile fasta databases from NetMHC-predicted peptides and is available at https://github.com/Prathyusha-konda/SpectMHC-v2.
2. To download the tool, open terminal at your desired location in your system and type:

   ```
   >git clone git@github.com:Prathyusha-konda/SpectMHC-v2.git
   ```

   if using ssh authentication, or

   ```
   >git clone https://github.com/Prathyusha-konda/SpectMHC-v2.git.
   ```

   Optionally, the tool can also be directly downloaded from the github website.
3. Unzip the folder.

2.6 Proteomics Database Search Software

User-preferred database search software such as Proteome Discoverer 1.4 (or greater version) can be used. The software must be capable of implementing MS search algorithms and false discovery rate (FDR) correction. The targeted search approach has been validated using both Sequest and Mascot searches, using Percolator to control the FDR. However, the approach is also applicable to other search software such as Andromeda, MS-GF+, and X! Tandem.

3 Methods

3.1 Identifying MHC or HLA Haplotype

It is important to determine the MHC or HLA haplotypes specific to the mouse or human tissue sample to be analyzed. If uncertain, it is preferable to perform a four-digit typing on HLA A, B, and C loci to conform with the NetMHC prediction algorithms.

1. MHC-I alleles for the common C57BL/6 mouse strain are H-2 K^b and H-2 D^b.
2. If using a different mouse strain, determine the MHC haplotype using this table: http://tools.thermofisher.com/content/sfs/brochures/Mouse_Haplotype_Table.pdf.
3. If using human cancer cell lines, HLA haplotypes for most common cancer cell lines can be found at the TRON Cell Line Portal [11] (http://celllines.tron-mainz.de/).
4. For individual tissue samples, MHC haplotyping can be performed by using PCR-based HLA typing kits or from sample-specific sequencing data such as RNA-seq or exome seq data. With sequencing data, MHC haplotype can be determined using a bioinformatics tool such as Seq2HLA [12], HLAreporter [13], or HLAminer [14].

3.2 Building a Targeted MHC/HLA Database with SpectMHC

1. Move the reference fasta file (e.g., full human or mouse reference proteome) into the SpectMHC-v2 folder.
2. On your Linux/Darwin system, using a terminal window, navigate to the SpectMHC folder containing your parent or reference fasta file.
3. To understand available options, execute the following help command:

```
>bash ./spectmhc.sh -h
```

   Once executed, you will be prompted with a set of instructions about how to implement SpectMHC.

4. Implement the basic usage of SpectMHC using the command (*see* Table 1 for required and optional arguments):

Table 1
Required and optional arguments for SpectMHC

Position	Description
Required arguments	
NetMHC folder	path to NetMHC folder
MHC version	input the version of netMHC 3.4/4.0/pan
input fasta	input reference protein data in fasta format
binding cutoff	cutoff to be used (rank for netMHC 4.0/pan, binding affinity for 3.4)
allele	MHC allele to predict
length_of_peptide	Length of peptides to predict, if multiple, please separate by ",". For example, 8 or 8,9,10
number_of_split_files	only works if -s is selected. Enter a numeric value, such that every split file has between 1000 and 2000 sequences. Insert 0 if not using this
Argument	*Description*
Optional arguments	
-h	help
-r	save raw netmhc output (do not opt for this if there is not enough storage on your system)
-s	split input fasta file (if using this, number_of_split_files should be mentioned)

```
>bash ./spectmhc.sh [-r] [-s] <NetMHC folder> <MHC version> <input fasta> <binding cutoff> <allele> <length of peptides> <number of split files>
```

5. Set the number of split files, so that 1000–2000 sequences are in each file, to reduce processing time.
6. Specify a 2% rank as a cutoff for selecting the predicted peptides. Peptides with <0.5% rank are considered strong binders and 0.5–2% rank are weaker binders.
7. As an output, you will obtain a customized MHC-peptide targeted database in a fasta file format (*see* **Note 5**).

3.3 Performing a Database Search with a SpectMHC-Generated Fasta File

1. On Proteome Discover or any preferred search software, upload the raw LC-MS/MS data and newly generated targeted fasta database. Proteome Discoverer automatically appends all headers to a single header for duplicate entries. If the search software of choice does not remove the duplicates, a tool such as "dbtoolkit" can be used.
2. To assess the purity of the MHC/HLA immunoprecipitation, it is recommended to perform an initial search using no enzyme specificity against the full reference database. Ensure that

greater than 80% of these identified peptides have a <2% NetMHC rank value for the specified alleles. This can be done by using the online (or offline) version of NetMHC.

3. After confirming the purity of the immunoprecipitation experiment, perform a targeted fasta database search using "No Cleavage" enzyme specificity search options (*see* **Notes 6** and 7). In general, the approach works best with MS1 tolerances ≤5 ppm and MS2 tolerance ≤0.5 Da for collision-induced dissociation (CID) fragmentation (*see* **Note 8**). FDR correction should be done with Percolator or another comparable tool. The Percolator node in the Proteome Discoverer automatically implements the target-decoy approach [15], and so a separate reversed (decoy) database does not need to be made (*see* **Note 9**).
4. We recommend filtering the final output datasets based on q-values (Percolator FDR estimates) to acquire the peptides with "medium" (5% FDR) or "high" (1% FDR) confidence. Implementing the SpectMHC-based targeted database search approach should significantly improve the identification of peptides as compared with the "no enzyme" reference database search (*see* **Note 10**).
5. For quantitative analysis (label-free, SILAC, or tagging methods), ratios can be compared by grouping all spectra matching to a peptide to a single peptide group used for quantitation (automatically performed by Proteome Discoverer).

3.4 Applications for SpectMHC

3.4.1 TAA Analysis

Several studies have shown that cancer sera contain antibodies against the "self" or autologous cellular antigens also known as TAAs [4, 16]. To study TAAs in cancer cell lines or tissue samples using MS-based ligandome analysis, searches against cell type-specific proteome-derived targeted database can be used.

Alternatively, to obtain the protein sequences relevant to your model, nucleotide sequences from RNA-seq can be used to translate into protein sequences [17, 18]. This RNA-seq obtained proteome data can be used to generate a targeted database using SpectMHC. This approach can be used for any other disease condition.

3.4.2 TSA Analysis

Tumor mutational burden has been one emerging biomarker not only for efficient cancer-specific immune responses but also to better predict the patient response rates for cancer treatments [19–21]. To study TSAs, next-generation sequencing technologies such as RNA-seq or exome-seq are combined with MS-based ligandome analysis. For this, nucleotide sequences with mutations can be translated into protein sequences and used to generate a targeted database using SpectMHC.

3.4.3 Viral Epitope Discovery

CD8+ T cells play an important role in antiviral immunity by recognizing virus-derived peptides presented on the surface of infected cells [22–24]. Identifying these virus-derived MHC ligands provides a means for designing MHC-I peptide-based vaccines and better understanding antiviral immune responses. To detect virus-derived ligands, protein sequences of the virus of interest can be combined with the reference proteome of your cell line or tissue sample and used as an input into SpectMHC. This outputs a targeted database, which includes not only the cellular ligands but also viral ligands. This approach can be used to study any microbially-derived MHC ligands.

3.4.4 Noncanonically Translated Ligand Analysis

Advances in genomics and proteomics technologies have revolutionized our understanding of the proteome of cells. An emerging area in this field is the study of noncanonically translated proteins. New evidence shows that proteome not only contains proteins from coding genes but also includes proteins from several noncoding genes such as from long noncoding RNAs, short open reading frames, pseudogenes, and so on [25–27].

RNA sequencing or whole genome sequencing methods are employed to study MHC ligands derived from noncanonically translated proteins. The protein sequences originating from the noncanonical reading frames with non-AUG start codons can be combined with the reference proteome data and used to generate a targeted SpectMHC database.

4 Notes

1. LC-MS/MS should be collected using shotgun LC-MS/MS with high resolution (e.g., Orbitrap) MS scans (~350–700 m/z should encompass most MHC/HLA peptides), followed by data-dependent acquisition of MS/MS scans using ion trap CID, Orbitrap higher energy collisional dissociation (HCD), or Orbitrap electron transfer/higher energy collision dissociation (EThcD). The targeted search technique has been validated with CID MS2 scans.
2. Raw LC-MS/MS files from our analysis (C57BL/6 mouse MHC-I immunoprecipitate) can be accessed from https://chorusproject.org/pages/index.html, ID#1098. These can be used to confirm that the setup is working correctly.
3. Other protein sequences including those belonging to noncanonical protein forms such as isoforms, splice variants, pseudogenes, or alternative reading frames can be added to the reference database with a minimal effect on search space for

"no cleavage" targeted search compared with a "no enzyme" reference search.

4. Both NetMHC and NetMHCpan MHC/HLA binding predictions are made using artificial neural networks trained on naturally eluted and affinity binding data. NetMHC 4.0 has been trained on 81 human alleles, whereas NetMHCpan 4.0 covers 172 alleles.
5. Known common contaminant peptides (degraded proteins) observed in samples of interest can be added to the database to ensure that false spectra from these are not matched to MHC peptides.
6. In our analysis, as compared with Sequest searches with no cleavage, we find that Mascot searches with no cleavage followed by Percolator FDR correction greatly improve the number of peptides identified.
7. While using search software such as Sequest without an option "no cleavage" searches, this can be implemented by using the text "No cleavage 1 1 J" for the enzyme specificity.
8. For CID, HCD, or EThcD data collected using high resolution Orbitrap MS/MS scans, tolerances can be decreased to less than 0.5 Da and may improve specificity.
9. We have shown that using a decoy database by reversing the reference database for SpectMHC approach has no significant difference in peptide FDR estimation between the two target-decoy approaches.
10. LC-MS/MS searches using targeted databases not only increased the number of peptide identifications but also improved the search speed by ~six fold as compared with the traditional "no enzyme" reference database searches.

Acknowledgments

We gratefully acknowledge Dr. Steven Gygi (Department of Cell Biology, Harvard Medical School), Dr. Stefan Stevanovic, Dr. Dan Kowalewski, and Dr. Heiko Schuster (Department of Immunology, Institute for Cell Biology, University of Tubingen) for helpful discussions in devising the targeted database search approach. We also acknowledge financial support from the Canadian Institutes of Health Research (CIHR), Canadian Cancer Society Research Institute (CCSRI), the Beatrice Hunter Cancer Research Institute (BHCRI), and the Dalhousie Medical Research Foundation (DMRF).

References

1. Rock KL, Reits E, Neefjes J (2016) Present yourself! By MHC class I and MHC class II molecules. Trends Immunol 37(11):724–737
2. Blum JS, Wearsch PA, Cresswell P (2013) Pathways of antigen processing. Annu Rev Immunol 31:443–473
3. Caron E, Kowalewski DJ, Koh CC et al (2015) Analysis of major histocompatibility complex (MHC) immunopeptidomes using mass spectrometry. Mol Cell Proteomics 14 (12):3105–3117
4. Kowalewski DJ, Schuster H, Backert L et al (2015) HLA ligandome analysis identifies the underlying specificities of spontaneous antileukemia immune responses in chronic lymphocytic leukemia. Proc Natl Acad Sci 112(2): E166–E175
5. Noble WS (2015) Mass spectrometrists should search only for peptides they care about. Nat Methods 12(7):605–608
6. Frewen WE, Merrihew GE, Wu CC et al (2006) Analysis of peptide MS/MS spectra from large-scale proteomics experiments using spectrum libraries. Anal Chem 78:5678–5684
7. Andreatta M, Nielsen M (2015) Gapped sequence alignment using artificial neural networks: application to the MHC class I system. Bioinformatics 32(4):511–517
8. Jurtz V, Paul S, Andreatta M et al (2017) NetMHCpan-4.0: improved peptide–MHC class I interaction predictions integrating eluted ligand and peptide binding affinity data. J Immunol 199(9):3360–3368
9. Nielsen M, Lundegaard C, Worning P et al (2003) Reliable prediction of T-cell epitopes using neural networks with novel sequence representations. Protein Sci 12:1007–1017
10. Murphy JP, Konda P, Kowalewski DJ et al (2017) MHC-I ligand discovery using targeted database searches of mass spectrometry data: implications for T-cell immunotherapies. J Proteome Res 16:1806–1816
11. Scholtalbers J, Boegel S, Bukur T et al (2015) TCLP: an online cancer cell line catalogue integrating HLA type, predicted neo-epitopes, virus and gene expression. Genome Med 7 (1):118
12. Boegel S, Scholtalbers J, Löwer M et al (2015) In silico HLA typing using standard RNA-Seq sequence reads. Methods Mol Biol 1310:247–258
13. Huang Y, Yang J, Ying D et al (2015) HLAreporter: a tool for HLA typing from next generation sequencing data. Genome Med 7(1):25
14. Warren RL, Choe G, Freeman DJ et al (2012) Derivation of HLA types from shotgun sequence datasets. Genome Med 4(12):95
15. Elias JE, Gygi SP (2007) Target-decoy search strategy for increased confidence in large-scale protein identifications by mass spectrometry. Nat Methods 4(3):207–214
16. Bilich T, Nelde A, Bichmann L et al (2019) The HLA ligandome landscape of chronic myeloid leukemia delineates novel T-cell epitopes for immunotherapy. Blood 133(6):550–565
17. Smart AC, Margolis CA, Pimentel H et al (2018) Intron retention is a source of neoepitopes in cancer. Nat Biotechnol 36 (11):1056–1058
18. Laumont CM, Vincent K, Hesnard L et al (2018) Noncoding regions are the main source of targetable tumor-specific antigens. Sci Transl Med 10(470):eaau5516
19. Yadav M, Jhunjhunwala S, Phung QT et al (2014) Predicting immunogenic tumour mutations by combining mass spectrometry and exome sequencing. Nature 515 (7528):572–576
20. Sheynkman GM, Shortreed MR, Cesnik AJ et al (2016) Proteogenomics: integrating next-generation sequencing and mass spectrometry to characterize human proteomic variation. Annu Rev Anal Chem 9:521–545
21. Khodadoust MS, Olsson N, Wagar LE et al (2017) Antigen presentation profiling reveals recognition of lymphoma immunoglobulin neoantigens. Nature 543(7647):723
22. Wen J, Tang WW, Sheets N et al (2017) Identification of Zika virus epitopes reveals immunodominant and protective roles for dengue virus cross-reactive CD8+ T cells. Nat Microbiol 2(6):17036
23. Hansen SG, Wu HL, Burwitz BJ et al (2016) Broadly targeted CD8+ T cell responses restricted by major histocompatibility complex E. Science 351(6274):714–720
24. Thiele F, Tao S, Zhang Y et al (2015) Modified vaccinia virus Ankara-infected dendritic cells present CD4+ T-cell epitopes by endogenous major histocompatibility complex class II presentation pathways. J Virol 89(5):2698–2709
25. Liepe J, Marino F, Sidney J et al (2016) A large fraction of HLA class I ligands are proteasome-

generated spliced peptides. Science 354 (6310):354–358

26. Nagarajan NA, De Verteuil DA, Sriranganadane D et al (2016) ERAAP shapes the peptidome associated with classical and nonclassical MHC class I molecules. J Immunol 197 (4):1035–1043

27. Laumont CM, Daouda T, Laverdure JP et al (2016) Global proteogenomic analysis of human MHC class I-associated peptides derived from non-canonical reading frames. Nat Commun 5(7):10238

Chapter 12

The SysteMHC Atlas: a Computational Pipeline, a Website, and a Data Repository for Immunopeptidomic Analyses

Wenguang Shao, Etienne Caron, Patrick Pedrioli, and Ruedi Aebersold

Abstract

Mass spectrometry has emerged as the method of choice for the exploration of the immunopeptidome. Insights from the immunopeptidome promise novel cancer therapeutic approaches and a better understanding of the basic mechanisms of our immune system. To meet the computational demands from the steady gain in popularity and reach of mass spectrometry-based immunopeptidomics analysis, we created the SysteMHC Atlas project, a first-of-its-kind computational pipeline and resource repository dedicated to standardizing data analysis and public dissemination of immunopeptidomic datasets.

Key words Mass spectrometry, Immunopeptidome, Bioinformatics, Database, Computational proteomics

1 Introduction

Major histocompatibility complex (MHC) molecules (in humans: human leukocyte antigen [HLA] complex) are essential for the immune system to recognize abnormal cells and determinants of infectious agents. They modulate T cell immunity and are involved in generating effective antitumor and antiviral immune responses in mammals and are therefore of broad clinical relevance. Being tremendously complex, MHC molecules can present thousands of different short peptides that are collectively referred to as the immunopeptidome. The MHC class I immunopeptidome comprises for the most part peptides with length of 8–12 amino acids which are generated mainly by the degradation of intracellular proteins and are recognized by CD8+ T cells [1]. In humans, the MHC class I comprises HLA-A, HLA-B, and HLA-C molecules. The MHC class II immunopeptidome is predominantly composed of peptides of 10–25 amino acids in length, derived mainly by protease-mediated degradation of endocytosed proteins of extracellular origin and recognized by CD4+ T cells [2]. Amino acid sequences of the immunopeptidome are highly variable but not

Sebastian Boegel (ed.), *Bioinformatics for Cancer Immunotherapy: Methods and Protocols*, Methods in Molecular Biology, vol. 2120, https://doi.org/10.1007/978-1-0716-0327-7_12,

random. Through specific anchor residues that define an MHC binding motif, each peptide is usually MHC allele specific, and HLA alleles in the human population demonstrate a very high diversity [3]. For example, each individual human can express up to six different HLA class I allele types, and more than 17,100 HLA Class I alleles are expressed and found in the Immuno Polymorphism Database (June 2019). In other words, the composition of the immunopeptidome is extremely complex and variable over time and between individuals. Describing and understanding such complexity is a central task for immunology and has significant clinical implications toward the design of efficient immunotherapies and personalized medicine.

To study molecular samples of great complexity, mass spectrometry (MS) is a powerful analytical technique that allows rapid, sensitive, and robust identifications of thousands of peptides in a single measurement. The application of MS to the analysis of the immunopeptidome has been pioneered by Donald Hunt [4] and Hans-Georg Rammensee [5] and has evolved as the method of choice for immunopeptidomic studies. Over the past decade, MS-based immunopeptidomics analysis strategies steadily advanced in performance and gained in popularity [6, 7]. In particular, recent rapid developments in data-independent acquisition (DIA), exemplified as SWATH-MS [8], have further expanded the performance of analytical platforms, due to significantly improved quantitative accuracy and reproducibility [9]. As a consequence, large numbers of MS-based immunopeptidomic datasets [10–13] have been generated and continue to be generated by a rapidly growing research community of scientists using high-throughput MS techniques [14, 15]. To meet the ensuing increasing need to process, achieve, and exchange immunopeptidomic datasets, in 2017, we launched the SysteMHC Atlas project [16], a first-of-its-kind computational pipeline and resource repository dedicated to standardizing data analysis and public dissemination of immunopeptidomic datasets.

In brief, the SysteMHC Atlas [16] consists of four major components. The first is a database containing the raw MS measurement results collected from many laboratories worldwide; the second is an open source computational pipeline that standardizes and processes MS data toward the identification of the immunopeptidome and controls quality of the results; the third is a collection of sample-specific and allele-specific spectral libraries that facilitate future MS-based immunopeptidomic analyses, particularly by DIA measurements [17]; and the fourth is a web interface that enables large immunopeptidomic datasets to be explored in a transparent and user-specific manner. The current build of the SysteMHC Atlas (version 180,409) includes samples from various tissue types generated by different sample preparation protocols and MS instruments (Table 1). In total, the current version of the atlas contains 710 sample-specific spectral libraries and 39 allele-

Table 1
List of tissue types, sample preparation protocols, and mass spectrometer instruments collected in the SysteMHC Atlas

	Human	Mouse
Tissue types	Blood Brain Breast Bronchoalveolar lavage fluid Colon Skin Synovium	Adrenal Bladder Bone marrow Brain Colon Heart kidney Lewis lung carcinoma Liver Lung Lymphoma Malignant glioma Melanoma Ovary Pancreas Skin Small intestine Spinal cord Spleen Stomach Testis Thymus Uterus
Isolation types	Immunoaffinity Mild acid elution	Immunoaffinity
Antibody types	W6/32 L243 SPV-L3 (anti-DQ) 2.12.E11 (anti-DQ2) HB-145	B22 Y3
MS Instruments	LTQ-FT Orbitrap Elite Orbitrap Fusion Orbitrap Velos Orbitrap XL Q Exactive Q Exactive HF Q Exactive Plus TripleTOF 5600	Orbitrap Fusion Lumos Orbitrap XL TripleTOF 5600

specific spectral libraries (37 from human and 2 from mouse [11], respectively) (Fig. 1).

In this article, we describe the computational pipeline for immunopeptidomic analysis and illustrate the usage of the website. Finally, it should be emphasized that MS-based

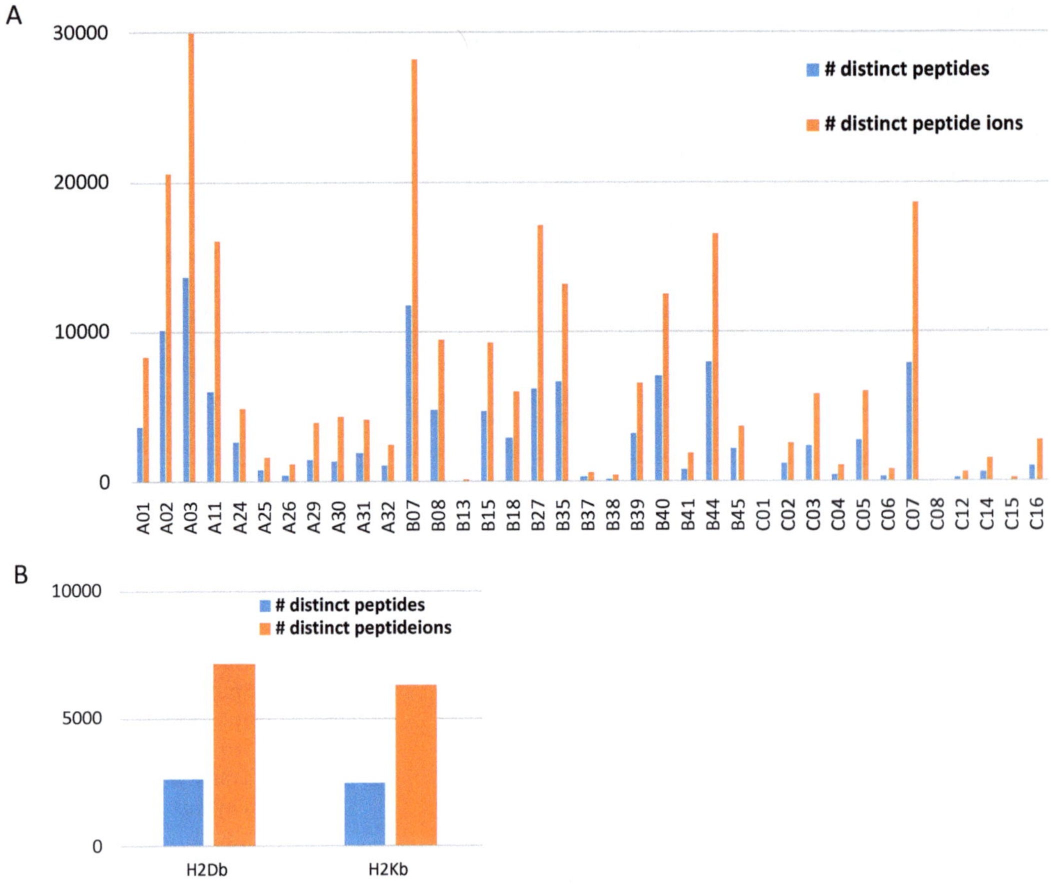

Fig. 1 The numbers of peptide ions and peptides identified in the atlas, grouped by the alleles annotated. (**a**) Human; (**b**) Mouse

immunopeptidomics is a relatively new and rapidly evolving field. As a result, many computational tools and strategies mentioned in this article are under active development, and the SysteMHC Atlas is updated periodically to include such changes and to allow for the addition of newly acquired or published immunopeptidomic datasets.

2 Materials

2.1 Software

The whole computational workflow is open source and can be freely obtained via the SysteMHC Github repository (https://github.com/SysteMHC). For detailed help information for each individual tool used in the workflow, please refer to the associated README file for installation instructions and user manual.

2.2 Hardware

The computational workflow was developed and ran on a Linux Ubuntu 14.04.6 LTS machine (GNU/Linux 3.16.0-30-generic x86_64). To run the computational workflow, a computer with a POSIX Environment, such as Linux or Mac OS, is required. To facilitate wide acceptance and broad applications in the scientific community, the whole workflow was designed and optimized to operate with a limited number of computational resources (e.g., four CPU cores; 8GB RAM) to complete the entire analysis (*see* **Note 1**).

2.3 Datasets

Grouped by SYSMHC-ID that is a unique and permanent number for each experiment, the raw mass spectrometer output files can be downloaded either directly from the SysteMHC Atlas website or externally via the hyper link to a third party repository (i.e., PRIDE column). Note that the SysteMHC Atlas also provides the converted MS files in open formats, such as mzXML files, to facilitate the downstream analysis.

3 Methods

3.1 Overall

The computational pipeline developed for the analysis of immunopeptidomic datasets is based on the Trans-Proteomic Pipeline (version 4.8.0), an open source project [18] that includes multiple key components to analyze proteomic data in an instrument-independent and highly reproducible fashion. In addition, computational tools are included for peptide annotations to specific HLA alleles, if available. This step, together with the statistical validation in the Trans-Proteomic Pipeline, provides a directly interpretable confidence of the obtained results and controls their quality (*see* **Note 2**) (e.g., spectral libraries). In the following subheading, we will describe in details these key components with the expected input and output files.

3.2 Data Conversion

The raw MS files are converted into a standardized, open, XML-based representation (i.e., mzXML files). This helps to improve compatibility of the workflow in order to support raw MS data generated by different MS instruments from different venders [19]. We use MSconvert [20] to convert raw MS data into mzXML files by default.

3.3 Database Search

The generated mzXML files are then individually searched by two widely used search engines, Comet [21], and X!Tandem [22] for peptide identifications. Using multiple search engines leads to improved sensitivity of peptide identifications. The corresponding curated protein database from UniProt is chosen for a specific organism. To estimate the false discovery rate (FDR) of the identified peptide set, a target-decoy strategy [23] is applied where a

Table 2
The key parameters used for database search

	High mass accuracy (Orbitrap/TOF)	Low mass accuracy
Precursor ion tolerance	15 ppm	
Fragment ion tolerance	0.05 Da	0.5 Da
Digestion specificity	Unconstrained	
Variable modification	Methionine oxidation	

decoy (reversed) protein sequence database is appended to the original curated protein database. The default search settings for both search engines are applied with the key parameters listed in Table 2. The search results are stored in another open format called pepXML.

3.4 Statistical Validation

After the database search, PeptideProphet [24] models the correct and incorrect peptide-spectrum match (PSM) populations and assigns a probability of being correct to each PSM. This step facilitates FDR estimation, which provides a practical confidence for error controlling. Importantly, it enables meaningful comparisons of immunopeptidome results between multiple experiments and across different laboratories. Finally, iProphet [25] is used to combine all PeptideProphet results from different search engines.

3.5 Allele Annotation

NetMHCcons (version 1.1) is applied to predict binding affinities (IC_{50}) of identified MHC Class I peptides with the default settings [26]. It is a consensus method that integrates three state-of-the-art methods applying different strategies to reach the most accurate predictions. In NetMHCcons, for each experiment or for a subset of samples, a limited number of candidate alleles are specifically selected based on the experimental design reported in literature. An annotation score [10] for each peptide is then calculated by dividing its second best IC_{50} value by its best IC_{50} value (i.e., the best predicted allele). The higher the annotation score of a peptide, the higher the chance of being correctly annotated to the best predicted allele. Only peptides with annotation scores >3 [10] are considered confidently annotated and kept for the downstream visualization and allele-specific spectral library construction.

3.6 Construction of Spectral Libraries and DIA Analysis

At a defined FDR (1% by default) estimated by iProphet [25], raw MS spectra are imported to construct raw spectral libraries with default settings in SpectraST [17, 27]. Then raw spectral libraries are further converted into consensus spectral libraries, where multiple replicate spectra generated from identical precursor ions are averaged into a consensus spectrum that best represents the associated peptide ion with the highest signal-to-noise ratio [27, 28].

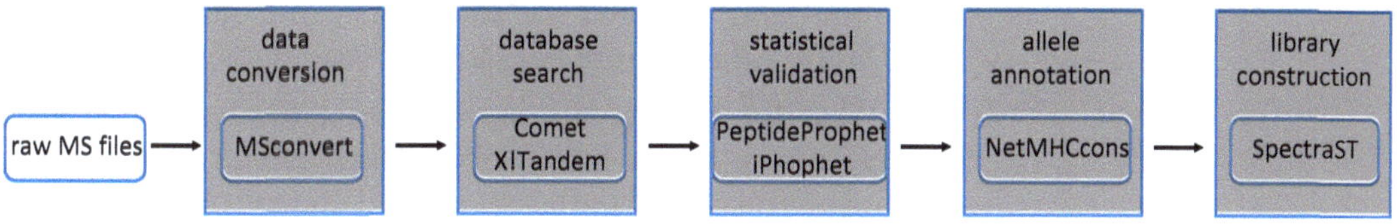

Fig. 2 An overview for the open source computational workflow

In fact, two different kinds of spectral libraries are prepared by the atlas—(1) sample-specific libraries and (2) allele-specific libraries. For each experiment, a sample-specific spectral library is generated that is context based. Moreover, after applying the filter using allele annotation scores (see above), allele-specific spectral libraries are combined from multiple samples across different experiments where the samples share the same allele. The original format of spectral library (sptxt) is further converted into the TraML format that is directly compatible with DIA/SWATH applications (*see* **Note 3**).

3.7 Web Interface

The SysteMHC Atlas offers a highly interactive and user-friendly query interface as a website (https://systemhcatlas.org), which allows researchers to browse, search, explore, and download immunopeptidomic information.

In general, the website provides two fundamental ways to explore information. In the first mode, the entire atlas functions as a repository that summarizes and visualizes all the information that has been collected from multiple experiments submitted by labs across the world. This modality can be accessed via the "EXPLORE" link to start searching any key words of interest on numerous levels, such as peptide sequences, targeted source protein names, allele classes, and types (*see* **Note 4**). In the second mode, the atlas provides all the results at each computational processing step for each individual experiment (*see* Fig. 2 and Subheading 3). In this way, the researcher can check and review each processing step in a transparent way. Overall, the SysteMHC Atlas enables large immunopeptidomics dataset to be explored in a user-specific fashion.

4 Notes

1. Note that for a production environment, a more realistic setup will comprise a computer cluster to perform the computational analysis, as well as a dedicated web server to host the results.
2. Note that all levels of data that are outputs of sequential computational steps, such as raw MS files, converted mzXML files, Prophet output files, and tandem mass spectral libraries,

are stored and prepared for researchers to download via the SysteMHC Atlas website (*see* also Subheading 3).

3. Note that for DIA applications, the retention time of each peptide ion is scaled into a normalized dimension with the use of the spiked-in landmark indexed Retention Time (iRT) peptides, if available. Thus, TraML files are directly applicable to DIA application and are available at SWATHAtlas.

4. Note that all searches are performed in a real time and interactive way, which is particularly useful to perform sequence searching with only partial peptide sequence.

Acknowledgments

The authors wish to thank the whole Aebersold lab for their support and critical discussion.

References

1. Shastri N, Schwab S, Serwold T (2002) Producing nature's gene-chips: the generation of peptides for display by MHC class I molecules. Annu Rev Immunol 20(1):463–493
2. Kambayashi T, Laufer TM (2014) Atypical MHC class II-expressing antigen-presenting cells: can anything replace a dendritic cell? Nat Rev Immunol 14(11):719
3. Falk K, Rötzschke O, Stevanović S et al (1991) Allele-specific motifs revealed by sequencing of self-peptides eluted from MHC molecules. Nature 351(6324):290
4. Hunt DF, Henderson RA, Shabanowitz J et al (1992) Characterization of peptides bound to the class I MHC molecule HLA-A2.1 by mass spectrometry. Science 255(5049): 1261–1263
5. Rammensee H-G, Friede T, Stevanović S (1995) MHC ligands and peptide motifs: first listing. Immunogenetics 41(4):178–228
6. Caron E, Kowalewski DJ, Koh CC et al (2015) Analysis of major histocompatibility complex (MHC) immunopeptidomes using mass spectrometry. Mol Cell Proteomics 14 (12):3105–3117
7. Liepe J, Marino F, Sidney J et al (2016) A large fraction of HLA class I ligands are proteasome-generated spliced peptides. Science 354 (6310):354–358
8. Gillet LC, Navarro P, Tate S et al (2012) Targeted data extraction of the MS/MS spectra generated by data-independent acquisition: a new concept for consistent and accurate proteome analysis. Mol Cell Proteomics 11(6): O111. 016717
9. Ludwig C, Gillet L, Rosenberger G et al (2018) Data-independent acquisition-based SWATH-MS for quantitative proteomics: a tutorial. Mol Syst Biol 14(8):e8126
10. Caron E, Espona L, Kowalewski DJ et al (2015) An open-source computational and data resource to analyze digital maps of immunopeptidomes. elife 4:e07661
11. Schuster H, Shao W, Weiss T et al (2018) A tissue-based draft map of the murine MHC class I immunopeptidome. Scientific data 5:180157
12. Bassani-Sternberg M, Bräunlein E, Klar R et al (2016) Direct identification of clinically relevant neoepitopes presented on native human melanoma tissue by mass spectrometry. Nat Commun 7:13404
13. Faridi P, Li C, Ramarathinam SH et al (2018) A subset of HLA-I peptides are not genomically templated: evidence for cis-and trans-spliced peptide ligands. Sci Immunol 3(28): eaar3947
14. Bassani-Sternberg M, Coukos G (2016) Mass spectrometry-based antigen discovery for cancer immunotherapy. Curr Opin Immunol 41:9–17
15. Faridi P, Purcell AW, Croft NP (2018) In immunopeptidomics we need a sniper instead of a shotgun. Proteomics 18(12):1700464

16. Shao W, Pedrioli PG, Wolski W et al (2017) The SysteMHC atlas project. Nucleic Acids Res 46(D1):D1237–D1247
17. Shao W, Lam H (2017) Tandem mass spectral libraries of peptides and their roles in proteomics research. Mass Spectrom Rev 36 (5):634–648
18. Pedrioli PG (2010) Trans-proteomic pipeline: a pipeline for proteomic analysis. In: Proteome bioinformatics. Springer, New York, NY, pp 213–238
19. Pedrioli PG, Eng JK, Hubley R et al (2004) A common open representation of mass spectrometry data and its application to proteomics research. Nat Biotechnol 22(11):1459
20. Chambers MC, Maclean B, Burke R et al (2012) A cross-platform toolkit for mass spectrometry and proteomics. Nat Biotechnol 30 (10):918
21. Eng JK, Jahan TA, Hoopmann MR (2013) Comet: an open-source MS/MS sequence database search tool. Proteomics 13(1):22–24
22. Fenyö D, Beavis RC (2003) A method for assessing the statistical significance of mass spectrometry-based protein identifications using general scoring schemes. Anal Chem 75 (4):768–774
23. Elias JE, Gygi SP (2007) Target-decoy search strategy for increased confidence in large-scale protein identifications by mass spectrometry. Nat Methods 4(3):207
24. Keller A, Nesvizhskii AI, Kolker E et al (2002) Empirical statistical model to estimate the accuracy of peptide identifications made by MS/MS and database search. Anal Chem 74 (20):5383–5392
25. Shteynberg D, Deutsch EW, Lam H et al (2011) iProphet: multi-level integrative analysis of shotgun proteomic data improves peptide and protein identification rates and error estimates. Mol Cell Proteomics 10(12):M111. 007690
26. Karosiene E, Lundegaard C, Lund O et al (2012) NetMHCcons: a consensus method for the major histocompatibility complex class I predictions. Immunogenetics 64(3):177–186
27. Lam H, Deutsch EW, Eddes JS et al (2008) Building consensus spectral libraries for peptide identification in proteomics. Nat Methods 5(10):873
28. Shao W, Lam H (2013) Denoising peptide tandem mass spectra for spectral libraries: a Bayesian approach. J Proteome Res 12 (7):3223–3232

Chapter 13

Identification of Epitope-Specific T Cells in T-Cell Receptor Repertoires

Sofie Gielis, Pieter Moris, Wout Bittremieux, Nicolas De Neuter, Benson Ogunjimi, Kris Laukens, and Pieter Meysman

Abstract

Recognition of cancer epitopes by T cells is fundamental for the activation of targeted antitumor responses. As such, the identification and study of epitope-specific T cells has been instrumental in our understanding of cancer immunology and the development of personalized immunotherapies. To facilitate the study of T-cell epitope specificity, we developed a prediction tool, TCRex, that can identify epitope-specific T-cell receptors (TCRs) directly from TCR repertoire data and perform epitope-specificity enrichment analyses. This chapter details the use of the TCRex web tool.

Key words TCR repertoire analysis, Epitope-specific T cells, Immunoinformatics, Machine learning, TCR repertoire sequencing

1 Introduction

T cells are a keystone in the adaptive immune system and indispensable in the fight against cancer. In cooperation with other immune cells, they generate targeted antitumor responses. Fundamental in the activation of T cells is their interaction with MHC-presented epitopes through their transmembrane T-cell receptor (TCR). As a result of the large variety in antigens and MHC molecules, a diverse set of epitopes can be presented by the adaptive immune system. To meet this diversity in epitope presentation, the epitope-binding regions of T-cell receptors are hypervariable. This hypervariability in TCRs is established through the process of VDJ recombination which encompasses the rearrangement of the T cells' V, D, and J genes and random insertions and deletions at the junctions [1].

To grasp a better view on the behavior of T cells in disease and healthy state, full TCR repertoire analyses are performed. Different

Pieter Moris and Wout Bittremieux contributed equally to this work.

Sebastian Boegel (ed.), *Bioinformatics for Cancer Immunotherapy: Methods and Protocols*, Methods in Molecular Biology, vol. 2120, https://doi.org/10.1007/978-1-0716-0327-7_13,

approaches are available to map the TCR repertoire including bulk TCR sequencing and single-cell TCR sequencing. Key in the analysis of TCR repertoires is the identification of epitope-specific TCRs as these play a crucial role in the induction of antitumor responses and the development of personalized cancer immunotherapies [2]. Current identification methodologies rely on the use of experimental procedures such as multimer assays [3], T cell stimulation with peptide pools [4], and the use of TCR-transduced T cells [5]. In addition to the identification of T cell epitope specificity, some approaches also allow measurement of the binding affinity between epitopes and TCRs (e.g., surface plasmon resonance) or elucidation of the structure of the pMHC-TCR complex (e.g., crystallography). In general, all these experimental procedures can be laborious and might give a biased viewpoint of the in vivo situation as they are performed in vitro.

As an alternative to these experiments, we developed a prediction tool, TCRex, for the rapid identification of epitope-specific TCRs in TCR repertoires. Our method hinges on the difference in physicochemical properties of TCRs that do and do not recognize a specific epitope. For each epitope, a machine learning classifier is trained to label TCR sequences as epitope-binding and nonbinding using their physicochemical properties. The majority of current TCR sequencing data is limited to the β chain sequence, thus at the moment, TCRex only takes TCR β sequences as input. Given a TCR β repertoire, TCRex identifies the TCRs that are able to bind with specific epitopes. Furthermore, enriched epitope specificity of the TCR repertoire can be detected by comparing the abundance of epitope-specific TCRs in the TCR repertoire to the abundance in a background TCR dataset. In this chapter, we will describe the usage of TCRex to identify epitope-specific TCRs and to perform epitope specificity enrichment analyses.

2 Materials

2.1 TCRex Tool

TCRex is available as an easy-to-use web tool at tcrex.biodatamining.be. It is free of charge for academic research.

TCRex offers both validated prediction models and the possibility to train and validate new models. All of these models are trained and validated in the same manner (Fig. 1) similar to [6]. Training prediction models requires a collection of epitope-specific TCRs and a set of nonbinding TCRs. While the first is largely present in public databases [7, 8], there is no information available on nonbinding TCRs. To circumvent this problem, we assembled a control dataset by randomly selecting TCR sequences from healthy TCR repertoires [9]. We call these TCRs background TCRs. Although there is a chance that some of these TCRs are

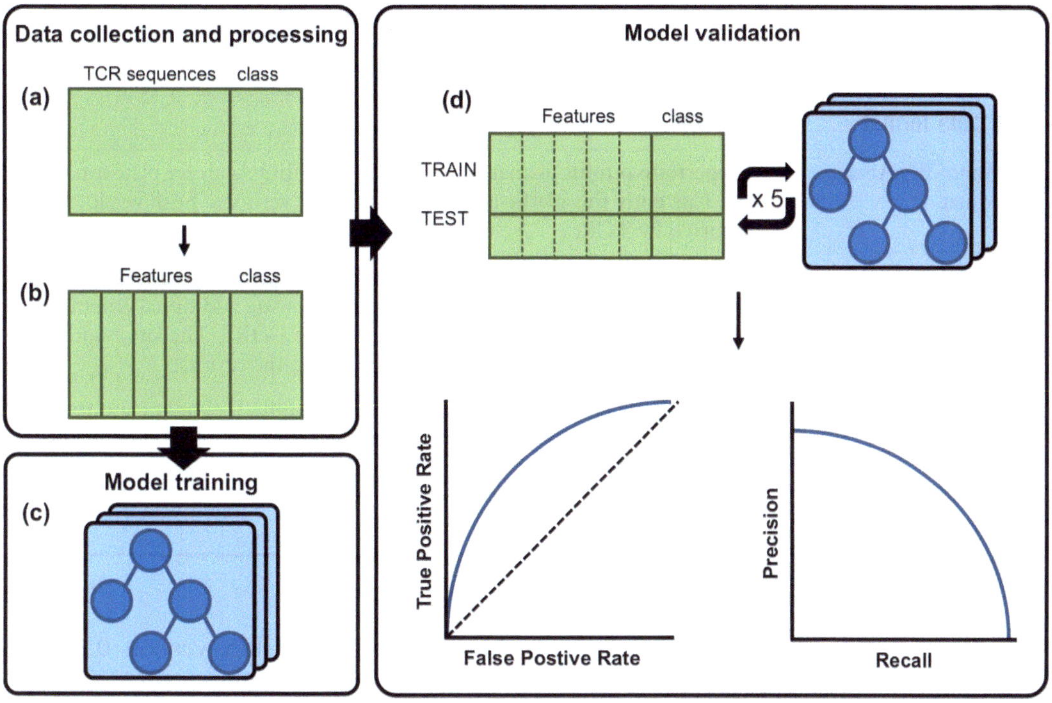

Fig. 1 Epitope-specific model training and validation in TCRex. (**a**) The training dataset for each epitope-specific model contains a minimum of 30 different epitope-specific TCRs. The number of control TCRs is set to a tenfold of the number of epitope-specific TCRs. (**b**) A feature matrix is assembled from the V/J genes and the CDR3 β sequences from all TCRs. (**c**) For each epitope-specific model, a random forest classifier of 100 trees is trained. (**d**) A fivefold cross-validation approach is used to assess the performance of the model. Models with an AUC of at least 0.7 and an average precision of at least 0.35 are kept and added to the TCRex tool. Custom trained models which do not pass these criteria are still available. Hence, it is important to check the performance metrics

binding the epitope of interest, we assume that the number of epitope-specific TCRs in this dataset is very small.

2.2 TCR Data Files

TCRex enables epitope-TCR binding predictions for human TCR β sequences. The TCR data input can be either a short list of TCR sequences or a larger TCR repertoire file. To facilitate the use of the web tool in combination with common TCR repertoire analysis pipelines, TCRex accepts different standard TCR data formats (Table 1). All input files are required to provide at least the CDR3 β amino acid sequence with the associated V/J genes for each TCR as this information is the basis for the predictions. If the V/J genes of a TCR sequence are not known, this information should be replaced by V/J family information. These are defined as the first letter of the V/J gene. For example, TRBV4 is the family of gene TRBV4-3.

Table 1
Overview of the TCR data formats supported by TCRex

TCR data format	Description
immunoSEQ Analyzer format	TCR repertoire output format from the immunoSEQ Analyzer platform [11]. Currently this platform uses two different versions. Both versions are supported by TCRex
MiXCR text file format	Text file obtained when exporting the identified MiXCR [12] clones from a .clns file using the exportClones command with following additional command line parameters: -jGene, -vGene, and -aaFeature CDR3. The latter adds the best V/J gene hits and the CDR3 β sequence to the text file
TCRex format	Tab-delimited file containing three columns listing the V/J genes and the CDR3 amino acid sequence of the TCR β chain. The column headers must be named as follows: TRBV_gene, CDR3_beta, and TRBJ_gene

3 Methods

This chapter details the operation of TCRex version 0.3.0 (*see* **Note 1**). TCRex offers epitope-TCR binding predictions for a wide range of infectious and cancerous epitopes. All epitopes, for which trained prediction models are available, are listed at the home page (tcrex.biodatamining.be). Currently the web tool contains prediction models for 44 viral and 5 cancer epitopes (Table 2). A complete overview of the steps needed to predict TCR binding to these epitopes is given in Subheading 3.1 (Fig. 2). Furthermore, TCRex enables training of new prediction models when the necessary training data are provided. This allows epitope-TCR binding predictions for "new" epitopes (i.e., epitopes for which no TCRex prediction model is at hand). A full overview of the required steps to predict binding between TCRs and epitopes using custom models is given in Subheading 3.2 (Fig. 2).

3.1 TCR–Epitope Binding Predictions Using Validated TCRex Prediction Models

1. The TCRex home page and prediction page can be accessed through tcrex.biodatamining.be.
2. On the home page, browse to the file containing the TCR sequences for which predictions are needed. One file can be uploaded and processed at a time. TCRex handles a file size limit of 50,000 TCR sequences. If your TCR data file contains more TCR sequences, you will need to split this file into smaller files.
3. Select your epitopes of interest. All available epitopes are listed at the TCRex home page. One or more epitopes can be selected by checking their associated checkboxes. Additional checkboxes for each specific virus and cancer type facilitate pathology-based selection. Before selecting your epitopes of

Table 2
List of all available epitope-specific models in TCRex (May 2019)

Viral/ Cancer	Source	Epitopes
Viral	CMV	IPSINVHHY, NLVPMVATV, QIKVRVKMV, QYDPVAALF, TPRVTGGGAM, VTEHDTLLY, YSEHPTFTSQY
	DENV1	GTSGSPIVNR
	DENV2	GTSGSPIIDK
	DENV3/4	GTSGSPIINR
	EBV	EPLPQGQLTAY, GLCTLVAML, HPVGEADYFEY, IVTDFSVIK, RAKFKQLL, YVLDHLIVV
	HCV	ARMILMTHF, ATDALMTGY, CINGVCWTV, HSKKKCDEL, KLVALGINAV
	HIV	EIYKRWII, FLKEKGGL, FPRPWLHGL, FRDYVDRFYKTLRAEQASQE, HPKVSSEVHI, IIKDYGKQM, ISPRTLNAW, KAFSPEVIPMF, KRWIILGLNK, KRWIIMGLNK, LPPIVAKEI, QVPLRPMTYK, RFYKTLRAEQASQ, RLRPGGKKK, RYPLTFGWCF, TPGPGVRYPL, TPQDLNTML
	HSV2	RPRGEVRFL
	HTLV1	SFHSLHLLF
	Influenza	GILGFVFTL, LPRRSGAAGA, PKYVKQNTLKLAT
	YFV	LLWNGPMAV
Cancer	Melanoma	AMFWSVPTV, EAAGIGILTV, ELAGIGILTV, FLYNLLTRV
	Multiple Myeloma	LLLGIGILV

interest, you can check the performances of the trained prediction models for each epitope. Standard performance metrics and curves for each of the epitope-specific models can be accessed through the statistics page (tcrex.biodatamining.be/statistics/). More information on the interpretation of these metrics is given in Table 3 and Fig. 3.

4. Although most TCR repertoire analysis pipelines make use of the IMGT format [10] to represent V/J genes, differences in the notation of these IMGT genes are common. For example, the gene TRBV1 also occurs as TRBV01-01, despite this not being valid according to the IMGT format. To ensure that all uploaded V/J genes are following the same format, TCRex can apply an IMGT parser. In short, this parser standardizes the V/J gene notation to the IMGT format as represented in the IMGT database. For example, any instance of TRBV01-01 would then be replaced by TRBV1. By default, the IMGT parser is activated. It is recommended to keep this default setting as it aligns with the format that the TCRex prediction models expect.
5. Select the enrichment threshold. After completing all TCR–epitope predictions, TCRex uses the results to perform an

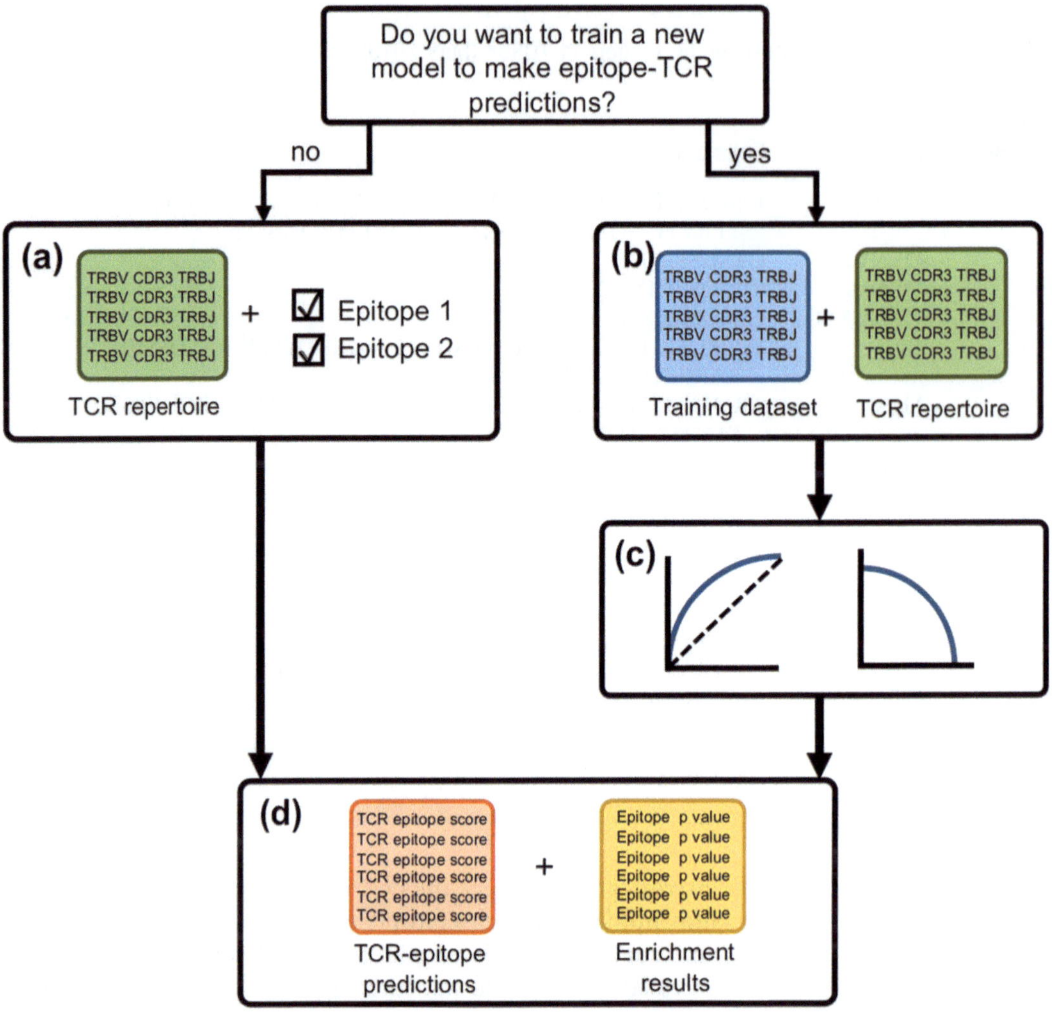

Fig. 2 TCRex workflow. TCRex enables epitope-TCR binding predictions for 49 different epitopes and provides the possibility to train and validate new models. (**a**) To use the TCRex models, upload the TCR repertoire file and select the epitopes of interest. (**b**) To train a new model, upload a training dataset with at least 30 different TCRs together with the TCR repertoire file. (**c**) Standard performance metrics are provided for all trained models. When training a new classifier, it is important to check these metrics as the results of bad performing models might not be reliable. An AUC of at least 0.7 and an average precision of at least 0.35 is recommended. (**d**) TCRex outputs both epitope–TCR predictions and epitope specificity enrichment results for the uploaded TCR repertoire. An FPR threshold can be used to control the number of false positives in the epitope–TCR predictions. Both these filtered prediction results and the complete set of predictions can be downloaded as a tab-delimited text file

epitope specificity enrichment analysis which assesses whether the uploaded TCR data file contains significantly more TCRs recognizing the studied epitope in comparison with a background TCR repertoire. This analysis requires the selection of an enrichment threshold which represents the abundance of TCRs specific for the studied epitope in a healthy repertoire.

Table 3
Explanation of the performance metrics and curves provided for each trained prediction model

Performance metric	Description	Interpretation
Number of training sequences	Number of epitope-specific sequences used to train the prediction model	TCRex requires at least 30 different epitope-specific TCR sequences (i.e., CDR3 β sequences with V/J genes) for each classifier
Balanced accuracy	Average of the fraction of epitope-specific TCRs and the fraction of control TCRs that are correctly classified	The higher the balanced accuracy, the more TCRs are likely to be classified correctly
Receiver operating characteristic (ROC) curve	Curve visualizing the identification rate of epitope-specific TCRs (i.e., true positive rate or TPR) and the fraction of wrongly classified control TCRs (i.e., false positive rate or FPR) for all confidence thresholds (Fig. 3)	A good classifier achieves a high epitope-specific TCR identification rate while having a low number of wrongly classified control TCRs. Therefore, the ideal ROC has a TPR of 1 for all FPR values. The better the performance of the model, the more the ROC curve approaches this ideal curve
AUC	A value representing the area under the ROC curve	The higher the AUC value, the better the performance of the model. This value must be greater than 0.5 (as this corresponds with random classification). TCRex considers an AUC value of 0.7 as the absolute minimum for a valid model
Precision-recall curve	Curve visualizing the fraction of identified epitope-specific TCRs that are genuinely binding the epitope of interest (i.e., precision) for different identification rates of epitope-specific TCRs (i.e., recall or TPR) (Fig. 3)	A high precision value indicates that most of the identified epitope-specific TCRs are genuinely binding the studied epitope. A high recall value indicates that we can identify most of the epitope-specific TCRs in the TCR repertoire. Therefore, the precision-recall curve of a perfect classifier has a precision of 1 for all recall values. The better the performance of the model, the more the precision-recall curve of the model approaches this ideal curve
Average precision	The estimated area under the precision–recall curve	The higher the average precision, the better the performance of the model. TCRex considers an average precision value of 0.35 as the absolute minimum for a valid model
Feature importances	A bar plot representing the 20 most important features	The higher the bar in the bar plot, the more important the feature is in the classification of TCRs into epitope-binding and nonbinding

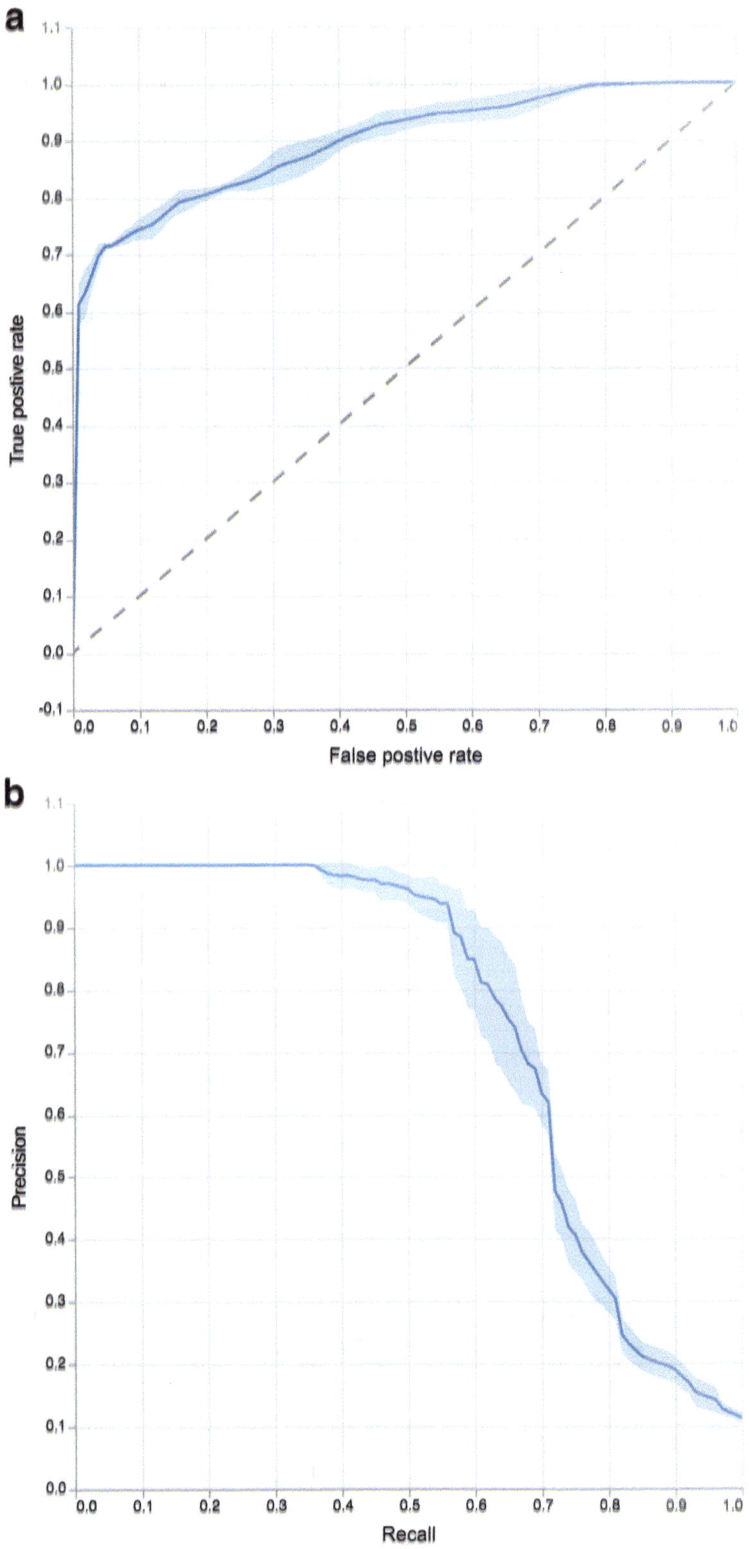

Fig. 3 Example of standard performance curves. (**a**) ROC curve for the TCRex model of the melanoma-associated epitope EAAGIGILTV. (**b**) Precision-recall curve for the TCRex model of the melanoma-associated epitope EAAGIGILTV

The enrichment threshold can take values from 0.01% to 5%. The chosen threshold is used for all selected epitopes independently. For example, when uploading a TCR repertoire, selecting three different viral epitopes and an enrichment threshold of 0.01%, TCRex evaluates for every epitope whether the abundance of TCRs specific for this epitope in the uploaded TCR repertoire is significantly higher in comparison with a background dataset where 0.01% of the TCRs bind this epitope.

6. After uploading the TCR data file, selecting the desired epitopes, and customizing the advanced settings, submit your prediction task by pressing the "Submit" button at the bottom of the home page. This will redirect you to a new web page listing all submission information including the selected epitopes, the uploaded TCR data file, the time of submission, and a unique identifier assigned to the submission called the "Task ID." Additionally, the status of the prediction task (i.e., "Completed" or "Running") is given together with a detailed progress report that lists all subtasks while they are running. Once the task is completed, the web page returns the prediction results. Using the unique Task ID and its associated unique URL, it is possible to return to the page at any time during the prediction analysis or later when the task is finished. For example, when your task is given the Task ID znaruo9qyg, the status and the prediction results can be requested using following unique URL: tcrex.biodatamining.be/task/znaruo9qyg/. All results can be viewed and downloaded up to a week after submission.

3.2 TCR–Epitope Binding Predictions Using Custom Prediction Models

In case you have access to a new set of epitope-specific TCRs, new models can be trained with TCRex and applied to any TCR dataset. Training and validation of your custom models occurs in the same manner as the epitope-specific models provided by TCRex (Fig. 1):

1. To access the module to create and apply new custom models, go to tcrex.biodatamining.be and select the "New epitopes" page.
2. Training a new epitope-specific model requires a TCR data file containing at least 30 unique TCRs (i.e., the CDR3 β sequence and corresponding V/J genes) known to bind with the epitope of interest. Files up to maximum 500 TCR sequences are allowed. To obtain a reliable prediction model, it is essential that all TCR sequences within this training file are specific to a single epitope. If you want to perform predictions for two or more epitopes, it is recommended to train and apply a prediction model for each of these epitopes separately.

3. Next, you must select the TCR data file you want to do predictions for (i.e., target dataset). One prediction file can be uploaded and processed at a time. Also here, TCRex handles a file size limit of 50,000 TCR sequences. If your TCR data file contains more TCR sequences, you will need to split this file into smaller files and repeat all steps for these smaller files.
4. The same IMGT parsing option as explained in Subheading 3.1, **step 4** is available when doing predictions with custom models. This option is recommended since training of the epitope-specific model makes use of a control dataset which is IMGT compliant. Furthermore, it is crucial that the V/J genes in the uploaded training file and the uploaded target file are formatted the same. If not, V/J genes identical to the training sequences are seen as being dissimilar by the model due to different formatting, leading to unreliable prediction results.
5. Epitope specificity enrichment analysis is also possible when using a custom prediction model. The principle of this analysis is identical to the enrichment analysis as explained in Subheading 3.1, **step 5**. Here, the chosen enrichment threshold is used to assess whether the uploaded TCR target dataset contains more epitope-specific TCRs than expected in a healthy TCR repertoire.
6. After uploading the TCR training dataset and the TCR target dataset and customizing the advanced settings, submit your prediction task by pressing the "Submit" button at the bottom of the "New epitopes" page. This will redirect you to a new web page listing all submission information, including the uploaded TCR data files, the time of submission and a unique identifier assigned to the submission called the "Task ID," and the status of the prediction task (i.e., "Completed" or "Running"). Additionally, a detailed report of the intermediate steps is provided to monitor the progress of the prediction task. Once the task is completed, the web page returns the prediction results. Using the unique TaskID and its associated unique URL, it is possible to return to the page at any time during the prediction analysis or later when the task is finished. Also here, results can be viewed and downloaded up to a week after submission.

3.3 Interpreting the Results

3.3.1 Epitope-TCR Binding Predictions

1. Epitope-TCR binding predictions are performed for every valid TCR sequence (*see* **Note 2**) in the uploaded TCR repertoire file and for every selected epitope. For each epitope-TCR prediction pair, a confidence score is given. This is a relative measure for the reliability of the prediction where higher scores indicate that the model is more certain about the prediction.

To limit the number of false positive results (i.e., TCR sequences that are falsely predicted to bind the epitope), a threshold needs to be set on this confidence score, so that only epitope-TCR pairs with a confidence score equal or higher to this threshold can be classified as binding and all epitope-TCR pairs corresponding with a confidence score below this threshold can be classified as nonbinding. This threshold is determined using the FPR value that is given with each TCR–epitope pair. This FPR value is calculated as the fraction of background TCR sequences, randomly selected from a large set of healthy TCR repertoires, having a confidence score equal or higher than the confidence score of the TCR–epitope pair. This calculation is done for each epitope model independently in a manner so that the final prediction results can be compared across models. The FPR gives an estimation of the number of false positives that is associated with the corresponding confidence score. For example, an FPR value of 0.0001 (or 0.01%) means that 1 in 10,000 uploaded TCR sequences are expected to be falsely predicted to bind the epitope of interest. The FPR threshold can be chosen at the results page. By default, a threshold of 0.1% is used.

2. The table on the results page shows the TCR–epitope pairs that are predicted to bind each other as defined by the chosen FPR threshold. For each epitope-TCR pair, the confidence score and corresponding FPR value (%) is given. As the latter is rounded to two decimal places, the number of epitope-TCR pairs listed in the table might slightly deviate from the actual results. Furthermore, this table is limited to show 5000 epitope–TCR pairs. Therefore, it is recommended to download the filtered results using the "Download results" button. This will export all filtered epitope–TCR pairs together with their confidence scores and FPR values to a tab-delimited file. It is also possible to download all prediction results without FPR filtering by clicking on "*Click here to download all results (no FPR filtering)*" underneath the "Download results" button.

3.3.2 Epitope Specificity Enrichment Analysis

The epitope specificity enrichment analysis outputs a p value for each epitope for which at least two different specific TCRs are identified. A small p value (i.e., smaller than a user-defined significance level α) corresponds with an enrichment in the number of TCRs specific for the studied epitope. This enrichment is based on a comparison with the expected number of hits in a background TCR repertoire (*see* **Note 3**). Multiple testing correction is recommended when doing predictions for multiple epitopes as this is not natively supported in the TCRex web tool.

3.3.3 Classifier Statistics

In case you trained a new epitope-specific model, the results page reports standard performance metrics and curves in addition to the prediction results. These are identical to the performance metrics for the existing models as listed on the statistics page. It is crucial to check these statistics as a poor model performance may lead to unreliable results. More explanation on how to interpret these metrics and curves is given in Table 3. In addition to the standard performance metrics, an overview of the 20 most important features of the model are provided in a bar plot.

4 Notes

1. All information regarding the working of TCRex and the models provided by TCRex relates with TCRex version 0.3.0. Future releases might contain additional features and prediction models. All updates will be communicated through the "Release notes" page. In addition, TCRex features an instruction page with an overview of all analysis steps for the latest release.
2. TCRex performs a few filtering steps on your uploaded training and target datasets: TCR sequences with CDR3 β sequences containing non-amino acid characters, unknown V/J families, or orphon genes are removed. All CDR3 β sequences are required to start with the conserved cysteine and end with the conserved phenylalanine. TCR sequences lacking these conserved residues are also removed. Hence, it is possible that some of your uploaded TCR sequences are removed before training your custom classifier or before making epitope specificity predictions.
3. The enrichment analysis relies on the use of a single background dataset, which might not be optimal for all study designs. Therefore, the enrichment results need to be interpreted with care. It is recommended to use the enrichment analysis for exploratory purposes, to get an idea of the epitope specificity of TCR repertoires. In case you have a more suitable background dataset at hand, you can use the TCRex workflow presented in this chapter to compare the identification rate of your TCR repertoire datasets to this background dataset. For this, you will need to perform epitope–TCR predictions on the background dataset using the same FPR threshold as for your other TCR repertoire datasets.

Acknowledgments

This work was funded by the University of Antwerp with a University Research Fund (BOF Concerted Research Action) and an Industrial Research Fund (IOF), by the Research Foundation Flanders (FWO) [Personal PhD grants to NDN (1S29816N), SG (1S48819N), and PMo (1141217 N); postdoctoral grant to WB (12W0418N); senior clinical investigator grant to BO (1861219 N); and research project grant (G067118 N)], and by the Belgian American Educational Foundation (BAEF) [postdoctoral grant to WB].

References

1. Roth DB (2014) V(D)J recombination: mechanism, errors, and fidelity. Microbiol Spectr 2:313–324
2. Ping Y, Liu C, Zhang Y (2018) T-cell receptor-engineered T cells for cancer treatment: current status and future directions. Protein Cell 9:254–266
3. Bentzen AK, Hadrup SR (2017) Evolution of MHC-based technologies used for detection of antigen-responsive T cells. Cancer Immunol Immunother 66:657–666
4. Chevalier MF, Bobisse S, Costa-Nunes C et al (2015) High-throughput monitoring of human tumor-specific T-cell responses with large peptide pools. Oncoimmunology 4: e1029702
5. Lorenz FKM, Ellinger C, Kieback E et al (2017) Unbiased identification of T-cell receptors targeting immunodominant peptide—MHC complexes for T-cell receptor immunotherapy. Hum Gene Ther 28:1158–1168
6. De Neuter N, Bittremieux W, Beirnaert C et al (2018) On the feasibility of mining CD8+ T cell receptor patterns underlying immunogenic peptide recognition. Immunogenetics 70:159–168
7. Shugay M, Bagaev DV, Zvyagin IV et al (2018) VDJdb: a curated database of T-cell receptor sequences with known antigen specificity. Nucleic Acids Res 46:D419–D427
8. Tickotsky N, Sagiv T, Prilusky J et al (2017) McPAS-TCR: a manually curated catalogue of pathology-associated T cell receptor sequences. Bioinformatics 33:2924–2929
9. Dean J, Emerson RO, Vignali M et al (2015) Annotation of pseudogenic gene segments by massively parallel sequencing of rearranged lymphocyte receptor loci. Genome Med 7:123
10. Lefranc MP, Giudicelli V, Duroux P et al (2015) IMGT, the international ImMunoGeneTics information system 25 years on. Nucleic Acids Res 43:D413–D422
11. Adaptive biotechnologies immunoSEQ. https://www.adaptivebiotech.com/products-services/immunoseq/
12. Bolotin DA, Poslavsky S, Mitrophanov I et al (2015) MiXCR: software for comprehensive adaptive immunity profiling. Nat Methods 12:380–381

Chapter 14

Modeling and Viewing T Cell Receptors Using TCRmodel and TCR3d

Ragul Gowthaman and Brian G. Pierce

Abstract

The past decade has seen a rapid increase in T cell receptor (TCR) sequences from single cell cloning and repertoire-scale high throughput sequencing studies. Many of these TCRs are of interest as potential therapeutics or for their implications in autoimmune disease or effective targeting of pathogens. As it is impractical to characterize the structure or targeting of the vast majority of these TCRs experimentally, advanced computational methods have been developed to predict their 3D structures and gain mechanistic insights into their antigen binding and specificity. Here, we describe the use of a TCR modeling web server, TCRmodel, which generates models of TCRs from sequence, and TCR3d, which is a weekly-updated database of all known TCR structures. Additionally, we describe the use of RosettaTCR, which is a protocol implemented in the Rosetta framework that serves as the command-line backend to TCRmodel. We provide an example where these tools are used to analyze and model a therapeutically relevant TCR based on its amino acid sequence.

Key words TCR, Rosetta, MHC, Antigen, Bioinformatics, Immunology

1 Introduction

T cell receptors (TCRs) are heterodimeric immunoglobulin proteins that play a key role in the immune system by mediating specific antigen recognition by T cells. They are composed of two different protein chains, most commonly α and β chains, while some are composed of γ and δ chains. TCRs engage foreign peptide or small molecule antigens bound to MHC (major histocompatibility complex) or MHC-like molecules, and productive binding events trigger T cell signaling, which cascades into various cellular responses from T cell development to apoptosis [1]. Due to their capability to specifically target tumor antigens, TCRs are increasingly being utilized as cell-based and soluble immunotherapeutics for cancer [2, 3], with a number of candidates under investigation by the biopharmaceutical industry and progressing to the clinic [4]. TCRs recognize different classes of ligands presented by

Sebastian Boegel (ed.), *Bioinformatics for Cancer Immunotherapy: Methods and Protocols*, Methods in Molecular Biology, vol. 2120, https://doi.org/10.1007/978-1-0716-0327-7_14,

MHC or MHC-like molecules, including peptides presented by MHC class I or class II molecules, lipid antigens presented by CD1, and metabolites bound to MHC class I-related protein (MR1). Advances in high throughput sequencing techniques allow TCR sequences to be obtained in massive numbers, providing the opportunity to use advanced computational modeling tools to gain insights into their structure, recognition, and specificity. These can capitalize on the limited but steadily growing number of experimentally determined TCR structures available in the Protein Data Bank (PDB) [5], in conjunction with sequence-based feature recognition methods [6].

We recently developed a computational method to model TCR structures in atomic resolution based on their amino acid sequences, implemented as the TCRmodel web server [7]. This algorithm consists of three main stages: template identification, grafting, and refinement and is implemented in the Rosetta framework [8].

We also developed a database, TCR3d, which provides an interface to view features of experimentally determined TCR structures and their complexes [9]. The TCR3d site is divided into three main categories: TCR complexes, TCR structures, and TCR sequences. Figure 1 shows the high-level organization of the

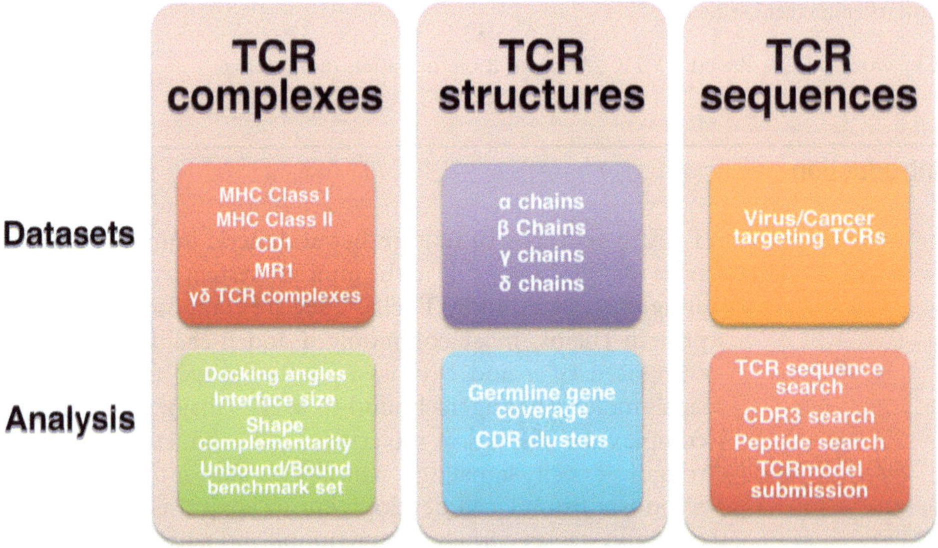

Fig. 1 The TCR3d database structure and contents. The database includes curated sets of structures and sequences ("Datasets," upper boxes), as well as derived data and search features ("Analysis," lower boxes), and is organized into three major categories: (1) TCR complexes, which include TCRs in complex with various MHC and MHC-like molecules and antigens, calculated properties of complex structures, and a docking benchmark set. (2) TCR structures, which includes details on all TCR chains in the PDB, overall structural coverage of TCR germline genes, and CDR loop clusters. (3) TCR sequences, which include virus-targeting and cancer-targeting TCR sequences from the literature, along with options to perform sequence searches against TCR structures and sequences

Table 1
TCR tools and availability

Name	Description	Website
TCRmodel web server	Web server to model TCR structures from sequence	https://tcrmodel.ibbr.umd.edu/
RosettaTCR	Protocol in the Rosetta modeling suite to model TCR structures from sequence	https://www.rosettacommons.org
TCR3d	T cell receptor structural repertoire database	https://tcr3d.ibbr.umd.edu/
TCR docking angle program	Program to calculate TCR docking angle and incident angles	https://github.com/piercelab/tcr_docking_angle

database, including its core tables. Notable features of the database include the ability to search sequences and motifs of complementarity determining region (CDR) loops against known structures, graphical views of germline gene coverage in TCR structures, representations of all known docking geometries, and sequences of cancer-targeting and viral-targeting TCRs compiled from the literature.

In this chapter, we describe information and examples on the use of the TCRmodel web server, as well as the RosettaTCR protocol for modeling TCRs, which is the command-line implementation of TCRmodel. We also highlight the interface and use of the TCR3d database, including browsing features and query-based searches. The availability of these methods is given in Table 1.

2 Materials

2.1 Hardware/Software

RosettaTCR was developed as a protocol for modeling TCRs within the Rosetta modeling suite, which can be obtained for free by academic users and installed on Windows, Linux, or Mac OS. The RosettaTCR protocol requires Rosetta to be compiled with full C++11 support. We recommend the use of GCC version 4.9 or higher, or Clang-version 3.6 or higher, to compile the Rosetta source code. The application name is "tcrmodel," which after compilation can be used as a command-line program.

The TCR modeling application is available as a web server with an easy to use interface (https://tcrmodel.ibbr.umd.edu/), and any commonly used web browsers can be used to access it. This web server provides an interface to RosettaTCR, along with an up-to-date database of template framework and loop structures for modeling, and the capability to input sets of TCRs for batch processing.

TCR3d was developed with SQLite relational database management system. The web interface for the database is developed using the Python-based microframework, Flask. The database can be accessed at https://tcr3d.ibbr.umd.edu/ and is compatible with all commonly used web browsers. The entire database can also be downloaded and accessed locally using DB Browser for SQLite (https://github.com/sqlitebrowser/sqlitebrowser). All the structures present in the databse and the complexes files can also be downloaded by the user (*see* **Note 1**).

2.2 Input Template Data

The TCR template database, which is required for running RosettaTCR, is not available by default for weekly Rosetta releases. The TCR template database directory is located at "Rosetta/database/additional_protocol_data/tcr" in the code base. Users can clone or download the template database as part of the "Rosetta/database/additional_protocol_data" directory.

For example,

git clone git@github.com:RosettaCommons/additional_protocol_data.git,

or,

git clone https://github.com/RosettaCommons/additional_protocol_data.git

The template database can also be downloaded from TCRmodel web server (*see* **Note 2**).

3 Methods

3.1 TCR3d Database

3.1.1 TCR Chains

The individual α, β, δ, and γ chains of publicly available TCR structures are provided as separate tables in TCR3d. In each table, the PDB code, chain ID, species, and the germline gene information for the TCR chains are available, as well as CDR1, CDR2, and CDR3 loop amino acid sequences.

3.1.2 Germline Genes

The current human and mouse TCR variable gene coverage by TCR structures is shown in the "Germline Genes" section. The list of PDB structures available for each gene can be accessed by clicking the gene name on the x-axes of these plots.

3.1.3 TCR Complexes

Complexes for all αβ TCRs are classified according to antigen-presenting molecules, namely MHC Class I, MHC Class II, CD1, and MR1, while γδ TCR complexes are shown separately. These tables include TCR name, MHC type, species, TCR germline gene, and the epitope information.

3.1.4 Interface Analysis

TCR complex geometries such as crossing angles and incident angles are calculated using a standalone program (Table 1) for each complex structure, while shape complementarity and interface size are calculated separately. These values can be obtained for each

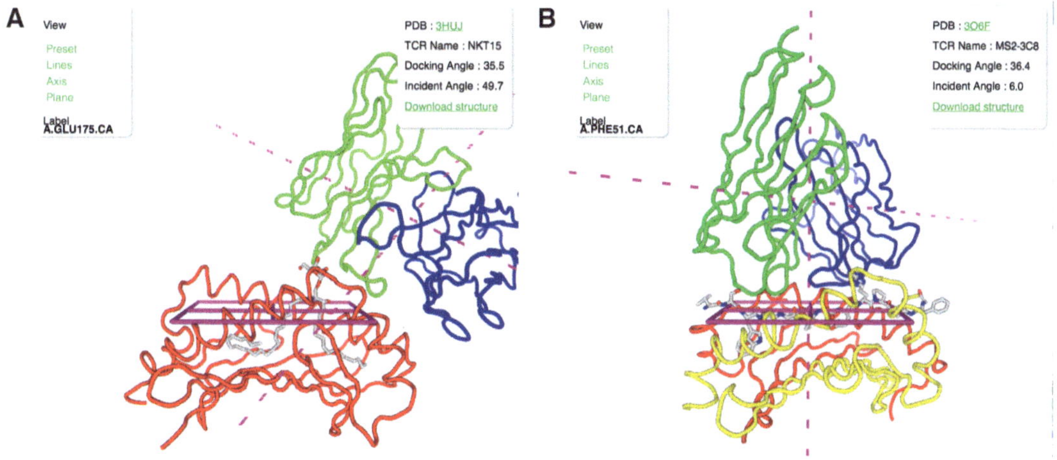

Fig. 2 TCR3d views of a complex structures for a (**a**) TCR-glycolipid-CD1d complex (PDB: 3HUJ) and (**b**) a TCR-peptide-Class II MHC complex (PDB: 3O6F). TCR chains are blue and green, CD1 d is red, Class II MHC is red and yellow, and antigenic glycolipid and peptide are shown as gray sticks. TCR orientation axes and MHC or MHC-like antigen-binding plane are magenta

TCR complex from the corresponding tables and are also plotted as figures for easy reference.

3.1.5 Structural Viewing Options

By clicking on the PDB code in the complexes table, users access the complex structure viewer showing the TCR, antigen, and MHC protein. Example structures for TCR-glycolipid-CD1d (PDB code 3HUJ) and TCR-peptide-MHC Class II (PDB code 3O6F) complexes are shown in Fig. 2.

3.1.6 CDR Clusters

TCR CDR loop structures were clustered based on backbone conformational distances and affinity propagation, analogous to a previously described method for antibody CDR loop clustering [10]. These clusters can be accessed from TCR3d under the "Structures" heading. The CDR clusters section of the database provides the cluster table that list all the clusters identified with this method, showing CDR loop type, sequence length, number of structures in each cluster, and the representative sequence logo. Each cluster can be viewed separately in more detail to show the PDB IDs, sequences, and the structural alignment of the CDR loops present in the cluster. Figure 3 shows the CDR clusters page and the cluster information and structural alignment for the cluster A_cdr3_10_C7 (CDR3 α cluster number 7 with sequence length 10).

3.1.7 TCR Sequences

Curated sets of TCR sequences that target cancer and viral peptide antigens have been collected from the recent literature and are available in the database. Given their relevance in cancer immunotherapy, sequences of TCRs that target cancer-associated neoepitopes are included in the database, and neoepitope information is

A

Cluster Name	CDR Type	Sequence length	TCR Type	Cluster center	No. of structures	Sequence logo
B_cdr1_10_C1	cdr1	10	B	5d7j_B	20	
A_cdr2_10_C2	cdr2	10	A	2eys_A	16	
A_cdr3_10_C1	cdr3	10	A	4l4v_D	15	
A_cdr1_11_C2	cdr1	11	A	5e9d_D	15	
B_cdr1_10_C2	cdr1	10	B	5eu6_E	14	

B

CDR cluster: A_cdr3_10_C7

6

Matches found

PDB ID	Chain	CDR Sequence	CDR Type	Sequence length	TCR Type
3e3q	D	AVSDPPPLLT	cdr3	10	A
4l8s	A	ASMDSNYQLI	cdr3	10	A
4l9l	A	APLDSNYQLI	cdr3	10	A
4nhu	A	AVSLHRPALT	cdr3	10	A
5isz	D	AFDTNAGKST	cdr3	10	A
6cph	D	AFKAAGNKLT	cdr3	10	A

Fig. 3 Viewing CDR clusters in TCR3d. (**a**) Table of all CDR clusters and (**b**) cluster details and structural alignment for the cluster A_cdr3_10_C7

given if available. These tables contain links to model individual TCR sequences using the TCRmodel server.

3.1.8 TCR Docking Benchmark Set

We have assembled sets of TCR complexes along with separately determined structures of unbound TCRs and peptide-MHCs, which represent an updated set of cases from a previously described predictive TCR docking benchmark [11]. There are currently 30 cases available in this set, and information on binding conformational changes is provided for each case.

3.1.9 CDR3 Sequence Search

The Sequence Search page allows users to search against TCR variable domain sequences, CDR3 sequences, and peptide antigens present in complex structures. Query CDR3 sequences are searched against all the CDR3 sequences present in the TCR structures and against the cancer−/virus-targeting TCR sequences. The input query is also searched for matches in the VDJdb database, which contains a large number of TCR sequences with known antigen specificity [12]. Sequence matches are scored and sorted by sequence similarity, using the PAM30 substitution matrix [13].

3.1.10 Subsequence or Motif Search

By default, only CDR3 sequences of the same length as the query are identified during searches. The user can optionally perform a subsequence search to identify specific sequence motifs present in CDR3 sequences. Figure 4 shows the sequence search page with results for a full-length CDR3 sequence search (Input query from the TCRmodel example described below: CDR3β sequence

Fig. 4 Sequence search page in TCR3d. Shown are (**a**) query interface, (**b**) results from CDR3 full-length sequence search (query sequence: "CASSQNYEQYF"), and (**c**) results from CDR3 motif/subsequence search (query sequence: "GNQF")

CASSQNYEQYF) and for a CDR3 subsequence search (Input query sequence: "GNQF"), representing a shared "public" CDR3α motif found in TCRs that target a peptide epitope from human cytomegalovirus (HCMV) [14].

3.1.11 Peptide Sequence Search

The user can search for peptide sequence against all the peptides present in TCR complex structures. The exact or closest matches to the peptide present in structurally characterized TCR complexes are listed along with sequence alignments. As with the CDR3 sequence search, matching peptide sequences are scored and sorted using the PAM30 substitution matrix, and subsequence searches are provided as an option.

3.1.12 TCR Variable Domain Sequence Search

Users can search using a variable domain TCR sequence against all the TCR structures available in the database.

3.2 Modeling Using the TCRmodel Web Server

TCRmodel takes α and β chain amino acid sequences as input. The input sequences must contain the variable domains of the TCR, including the CDR1, CDR2, and CDR3 loops. There are different ways that users can submit these input sequences. In addition to modeling individual TCRs, users can model multiple TCRs at once using a batch submission interface (*see* **Note 3**).

In the following example, we use a recently described TCR sequence identified in a melanoma patient ("TCR 53"), which is specific to a neoantigen CDK4 peptide [15], to demonstrate the TCRmodel method. This TCR sequence is present in the Sequences section of the TCR3d database.

3.2.1 Input Variable Domain Sequence

In TCRmodel, users can submit the variable domain sequences of α and β chains directly. The program is able to parse the variable domain sequence even if the user submits larger full-length variable and constant domain sequences.

Example α and β chain sequences:

>alpha

VTQLDSHVSVSEGTPVLLRCNYSSSYSPSLFWYVQHPNKGL
QLLLKYTSAATLVKGINGFEAEFKKSETSFHLTKPSAHM
SDAAEYFCVVSDLYNTDKLIFGTGTRLQVFP

>beta

VSQSPRYKVAKRGQDVALRCDPISGHVSLFWYQQALGQGP
EFLTYFQNEAQLDKSGLPSDRFFAERPEGSVSTLKIQRT
QQEDSAVYLCASSQNYEQYFGPGTRLTVT

3.2.2 Generate from Germline Genes

As many studies report TCRs as germline genes and CDR3 sequences, users can optionally input TCRs in this format. The user can choose the V/J genes along with input CDR3 sequence for the server to generate the full-length variable domain sequences. The following information is used to input the sequence for the same example TCR as above:

TRAV gene: TRAV8-2

TRAJ gene: TRAJ34

CDR3α sequence: CVVSDLYNTDKLIF

TRBV gene: TRBV7-8

TRBJ gene: TRBJ2-7

CDR3β sequence: CASSQNYEQYF

3.2.3 TCRmodel Server Results

Once the modeling is completed, the server redirects to the results page. On the results page, all the template information are displayed, the modeled structure can be viewed, and a download link is provided for the user to download the modeled file. Each modeling job is given a unique job ID. The job ID can also be used to retive the results later. The results are stored on the server for a month. TCRmodel server results for the example TCR sequence are shown in Fig. 5.

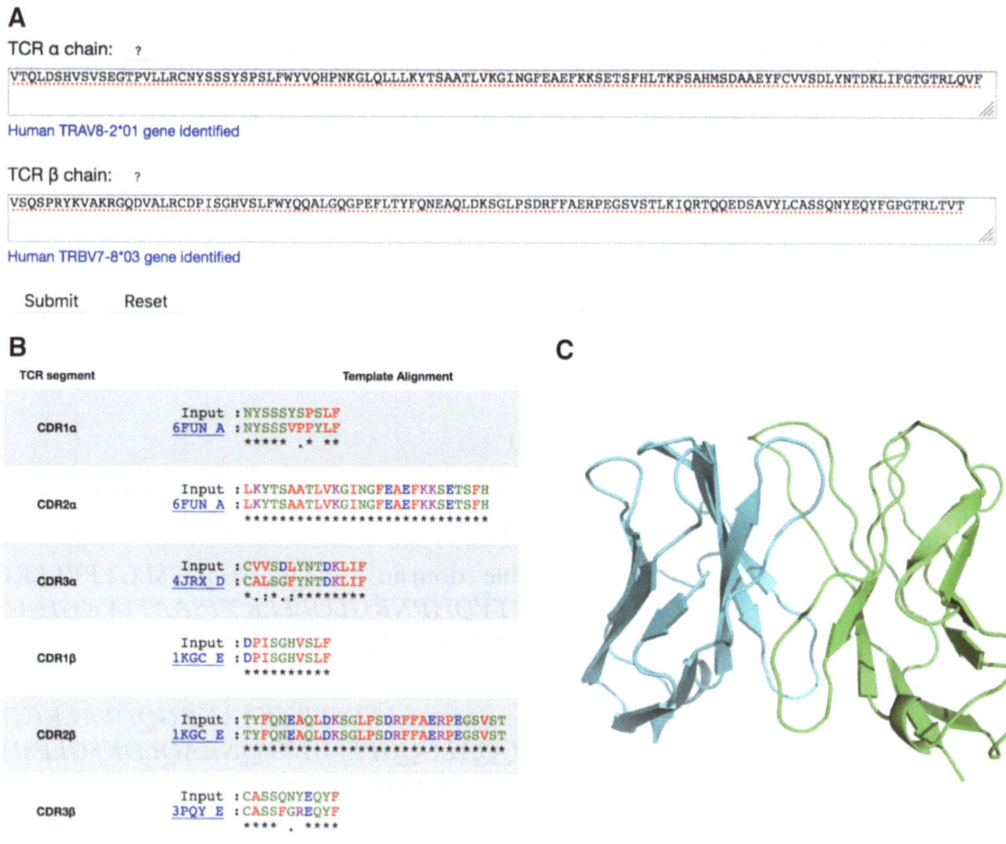

Fig. 5 TCRmodel interface and results. (**a**) Input sequences entered in the TCRmodel web server, (**b**) identified templates and alignment, and (**c**) final modeled structure

3.3 Modeling Using the RosettaTCR Protocol

3.3.1 Input Sequences

The RosettaTCR protocol is implemented as a command-line application named "tcrmodel" in Rosetta, and compared with the TCRmodel server, it has various additional options for the user to choose and control different steps in the modeling protocol. For this application, the input sequences for α and β chains need to be submitted separately. Users should include the flags "-alpha" and "-beta" to provide the sequences for α and β chains, respectively (*see* **Note 4**).

Command: $Rosetta/main/source/bin/tcrmodel.linux-gccrelease -alpha <alpha_chain_sequence> -beta <beta_chain_sequence>

3.3.2 Parsing TCR Segments

RosettaTCR has different methods to parse the input sequences into variable domain, framework, CDR1 loop, CDR2 loop, and CDR3 loop segments. The program uses regular expressions to parse the input sequences by default. For cases in which the parsing by regular expression fails, the user has other options to define the TCR segments. The TCRmodel application supports the ANARCI program for parsing the input sequence [16]. Users should use the flag "-anarci_path" to provide the path of the ANARCI program.

3.3.3 Assign TCR Segments Manually

The user can also manually assign the CDR segments. The user needs to provide the residue start and end positions for the CDR loops. The following flags are used for assigning CDR loop positions for the above example sequence:

-assign_cdr

-assign_cdr1a 23:33

-assign_cdr2a 47:74

-assign_cdr3a 91:100

-assign_cdr1b 24:33

-assign_cdr2b 47:75

-assign_cdr3b 92:100

For the above example TCR sequence, the following parsed segments were obtained:

TCR α chain variable domain: *VTQLDSHVSVSEGTPVLLRCNYSSSYSPSLFWYVQHPNKGLQLLLKYTSAATLVKGINGFEAEFKKSETSFHLTKPSAHMSDAAEYFCVVSDLYNTDKLIFGTGTRLQVF*

TCR β chain variable domain: *VSQSPRYKVAKRGQDVALRCDPISGHVSLFWYQQALGQGPEFLTYFQNEAQLDKSGLPSDRFFAERPEGSVSTLKIQRTQQEDSAVYLCASSQNYEQYFGPGTRLTVT*

TCR α chain framework sequence: *VTQLDSHVSVSEGTPVLLRCWYVQHPNKGLQLLLTKPSAHMSDAAEYFCFGTGTRLQVF*

TCR β chain framework sequence: *VSQSPRYKVAKRGQDVALRCWYQQALGQGPEFLLKIQRTQQEDSAVYLCFGPGTRLTVT.*

CDR1α: *NYSSSYSPSLF*

CDR1β: *DPISGHVSLF*

CDR2hv4α: *LKYTSAATLVKGINGFEAEFKKSETSFH*

CDR2hv4β: *TYFQNEAQLDKSGLPSDRFFAERPEGSVST*

CDR3α: *CVVSDLYNTDKLIF*

CDR3β: *CASSQNYEQYF*

3.3.4 Template Selection

Templates are selected for each segment from the template database, identifying matching segments of the same sequence length by PAM30 score. For modeling, TCR segments are grouped into one of the following sets:

Framework segment, CDR1, CDR2 and CDR3

Germline segment, CDR3

The program selects templates by default using best sequence match for germline or framework segments. If the program can find an exact template sequence for a given germline segment (including both CDR1 and CDR2 loops), then the germline template is assigned and will be used to generate the modeled variable domain in conjunction with a CDR3 template. If no exact germline template sequence is found, then individual templates are assigned for framework segment, CDR1, CDR2, and CDR3. The user can force RosettaTCR to use full germline templates using the flags "-use_alpha_germline_templates" and "-use_beta_germline_templates" for α and β chains, respectively. The important flags used for template selection are summarized in Table 2.

3.3.5 User-Provided Templates

Users can specify templates structures to use for any segment. The following flags are used to providing templates structures for the α chain sequence:

-use_alpha_germline_templates <true>

-alpha_germline_template_pdb <germline_structure.pdb>

-alpha_cdr3_template_pdb <cdr3a_structure.pdb>

The user can choose particular template structures by providing the PDB four letter code with chain ID, and the corresponding PDB structure should be present in the template

Table 2
Important flags used for template selection

Flag(s)	Description
-template_identity_cutoff	Identity cutoff to ignore template sequences from template database
-blastp_identity_cutoff	Cutoff to ignore similar template sequences from template database based on blast search
-alpha_cdr1_template_id -alpha_cdr2_template_id -alpha_cdr3_template_id -alpha_framework_template_id -alpha_germline_template_id -alpha_orientation_template_id	Template PDB ID and chain for α chain CDR1, CDR2, CDR3, Framework, germline segment, and domain orientation, respectively
-beta_cdr1_template_id -beta_cdr2_template_id -beta_cdr3_template_id -beta_framework_template_id -beta_germline_template_id -beta_orientation_template_id	Template PDB ID and chain for β chain CDR1, CDR2, CDR3, Framework, germline segment, and domain orientation, respectively
-alpha_cdr1_template_pdb -alpha_cdr2_template_pdb -alpha_cdr3_template_pdb -alpha_framework_template_pdb -alpha_germline_template_pdb -alpha_orientation_template_pdb	Template PDB structure file for α chain CDR1, CDR2, CDR3, Framework, germline segment, and domain orientation, respectively
-beta_cdr1_template_pdb -beta_cdr2_template_pdb -beta_cdr3_template_pdb -beta_framework_template_pdb -beta_germline_template_pdb -beta_orientation_template_pdb	Template PDB structure file for β chain CDR1, CDR2, CDR3, Framework, germline segment, and domain orientation, respectively

database. The following flags are used for providing template PDB names for the β chain sequence (example PDB and chain IDs shown):

-use_beta_germline_templates <true>

-beta_germline_template_id 2Q86_B

-beta_cdr3_template_id 3PQY_E

3.3.6 Blacklist Templates

The user can blacklist specific templates to prevent their use in modeling. The list of PDB IDs that should be blacklisted should be added to a file and provided to the program using the flag "-ignore_list." Users can also remove templates based on percent sequence identity between template and target sequence or ignore

Table 3
Important TCRmodel flags used for loop refinement

Flag	Description
-remodel_tcr_cdr3a_loop	remodel the CDR3 loop of α chain
-remodel_tcr_cdr3b_loop	remodel the CDR3 loop of β chain
-refine_tcr_cdr3a_loop	refine the CDR3 loop of α chain
-refine_tcr_cdr3b_loop	refine the CDR3 loop of β chain
-remodel_tcr_cdr3_loops	remodel the CDR3 loop of α and β chains
-refine_tcr_cdr3_loops	refine the CDR3 loop of α and β chains
-refine_all_tcr_cdr_loops	refine the CDR1, CDR2 and CDR3 loops of α and β chains

templates based on sequence identity. Corresponding flags are as follows:

-ignore_list <blacklist.txt>

-template_identity_cutoff <90>

-blastp_identity_cutoff Identity cutoff <85>

3.3.7 Loop Refinement

Loop modeling and refinement methods are integrated within the RosettaTCR protocol. The "remodel" flag is used to reconstruct the loop segment by sampling the backbone conformations, whereas "refine" flag is used to find the low energy conformations of the loop segment from the given starting conformation. The loop refinement can be applied to any particular CDR loop segment of the α or β chains, or only to CDR3 loops. The different options available for loop refinement are summarized in Table 3.

3.3.8 Interpreting the Output

1. Running RosettaTCR will result in the following PDB format output files, which are stored in the working directory. The -out::prefix flag can be used to add prefixes to the output file names:
   tcrmodel.pdb, the final modeled structure file.

   tcr_graftmodel.pdb, the initial grafted model without refinement.

   tcr_loopmodel.pdb, the model after optional loop modeling or refinement.
2. The various template structures used in the modeling can be obtained by using the flag "-dump_templates."
3. The final modeled structure is scored by the Rosetta scoring function, and the score is shown in the terminal output log. The score values are also included in the PDB file. The REF15

function [17] is the current scoring function used in Rosetta to score the structures. Along with the score, all the parsed TCR segment and the template information are displayed as output on the terminal.

3.4 Example Application and Results

Here, we describe an example input command and output log of the tcrmodel application. A demo for the application is included in the Rosetta software suite (*see* **Note 5**). To run the tcr model application on command line, the commands and the corresponding output log are given below:

Command:

$Rosetta/main/source/bin/tcrmodel.linuxgccrelease -alpha VTQLDSHVSVSEGTPVLLRCNYSSSYSPSLFWYVQHPN KGLQLLLKYTSAATLVKGINGFEAEFKKSETSFHLTKPS AHMSDAAEYFCVVSDLYNTDKLIFGTGTRLQVFP -beta VSQSPRYKVAKRGQDVALRCDPISGHVSLFWYQQALG QGPEFLTYFQNEAQLDKSGLPSDRFFAERPEGSVSTLK IQRTQQEDSAVYLCASSQNYEQYFGPGTRLTVT -- minimize_model true

The terminal output for the example sequence is shown below. Note that users' output details may differ due to possible differences in template library, etc.:

tcrmodel score: −316.467,

Tcr Alpha truncated Domain sequence: VTQ LDSHVSVSEGTPVLLRCNYSSSYSPSLFWYVQHPNKGLQ LLLKYTSAATLVKGINGFEAEFKKSETSFHLTKPSAHMS DAAEYFCVVSDLYNTDKLIFGTGTRLQVF

Tcr Alpha germline sequence: VTQLDSHVSVSEGTPVLLRC NYSSSYSPSLFWYVQHPNKGLQLLLKYTSAATLVKGING FEAEFKKSETSFHLTKPSAHMSDAAEYFCFGTGTRLQVF

Tcr Alpha Framework sequence: VTQLDSHVSVSEGTPVLLRC WYVQHPNKGLQLLLTKPSAHMSDAAEYFCFGTGTRLQ VF

Tcr Alpha Domain CDR1 sequence: NYSSSYSPSLF

Tcr Alpha Domain CDR2 sequence: LKYTSAATLVKGINGFEA EFKKSETSFH

Tcr Alpha Domain CDR3 sequence: VVSDLYNTDKLI

Alpha Framework template: 6fr7_A_A

Alpha CDR1 template: 6fun_A_A

Alpha CDR2 template: 6fun_A_A

Alpha CDR3 template: 4jrx_D_A

Tcr Beta truncated Domain: VSQSPRYKVAKRGQDVALRCD PISGHVSLFWYQQALGQGPEFLTYFQNEAQLDKSGLPS

DRFFAERPEGSVSTLKIQRTQQEDSAVYLCASSQNYEQY FGPGTRLTVT

Tcr Beta germline sequence: VSQSPRYKVAKRGQDVALRCDP ISGHVSLFWYQQALGQGPEFLTYFQNEAQLDKSGLPSDR FFAERPEGSVSTLKIQRTQQEDSAVYLCFGPGTRLTVT

Tcr Beta Framework sequence: VSQSPRYKVAKRGQDVALRCWYQ QALGQGPEFLLKIQRTQQEDSAVYLCFGPGTRLTVT

Tcr Beta Domain CDR1 sequence: DPISGHVSLF

Tcr Beta Domain CDR2 sequence: TYFQNEAQLDKSGL PSDRFFAERPEGSVST

Tcr Beta Domain CDR3 sequence: ASSQNYEQY

Beta germline template: 1kgc_E_B

Beta CDR3 template: 3pqy_E_B

Alpha Beta orientation template: 4ww2_A_A 4ww2_B_B

Output grafted model: tcrgraftmodel.pdb

Output final model: tcrmodel.pdb

4 Notes

1. The structures and complexes present in TCR3d are available for download. The structures of TCR chains, TCR complexes, and the docking benchmark sets can be downloaded separately. The entire database in the SQLite3d format is also available for download. All the download links can be accessed from https://tcr3d.ibbr.umd.edu/downloads.
2. The template database for the TCRmodel is updated regularly and is available for download from the TCRmodel website from this page: https://tcrmodel.ibbr.umd.edu/links. It is also available in the Rosetta software suite.
3. For advanced users and for those who may need to model a large number of TCR sequences, there is a batch process option available in the TCRmodel web server.
4. The input sequences for α and β TCR chains need to be provided by the user separately as strings using the flags "-alpha" and "-beta." All the other flags are optional flags. Other important flags used for running RosettaTCR are summarized in Table 4.
5. An example is also available as a demo in any recent Rosetta release download. The demo can be accessed from $Rosetta/demos/public/tcr_modeling. This directory contains the example commands for running the tcrmodel application along with input and output files. The various

Table 4
Summary of important TCRmodel flags

Flag	Default	Description
-alpha	None	Amino acid sequence of TCR α chain
-beta	None	Amino acid sequence of TCR β chain
-minimize_model	True	Minimize the output model
-relax_model	False	Relax the output model
-ignore_list	None	List of PDB ids to ignore as templates

optional flags used during modeling are explained and included in the demo directory.

References

1. Gaud G et al (2018) Regulatory mechanisms in T cell receptor signalling. Nat Rev Immunol 18 (8):485–497. https://doi.org/10.1038/s41577-018-0020-8
2. Hinrichs CS, Rosenberg SA (2014) Exploiting the curative potential of adoptive T-cell therapy for cancer. Immunol Rev 257(1):56–71. https://doi.org/10.1111/imr.12132
3. Liddy N et al (2012) Monoclonal TCR-redirected tumor cell killing. Nat Med 18(6):980–987. https://doi.org/10.1038/nm.2764
4. Zhang J, Wang L (2019) The emerging world of TCR-T cell trials against cancer: a systematic review. Technol Cancer Res Treat 18:1533033819831068. https://doi.org/10.1177/1533033819831068
5. Rose PW et al (2011) The RCSB protein data bank: redesigned web site and web services. Nucleic Acids Res 39(Database):D392–D401. https://doi.org/10.1093/nar/gkq1021
6. Glanville J et al (2017) Identifying specificity groups in the T cell receptor repertoire. Nature 547(7661):94–98. https://doi.org/10.1038/nature22976
7. Gowthaman R, Pierce BG (2018) TCRmodel: high resolution modeling of T cell receptors from sequence. Nucleic Acids Res 46(W1): W396–W401. https://doi.org/10.1093/nar/gky432
8. Leaver-Fay A et al (2011) ROSETTA3: an object-oriented software suite for the simulation and design of macromolecules. Methods Enzymol 487:545–574. https://doi.org/10.1016/B978-0-12-381270-4.00019-6
9. Gowthaman R, Pierce BG (2019) TCR3d: The T cell receptor structural repertoire database. Bioinformatics, In Press
10. North B et al (2011) A new clustering of antibody CDR loop conformations. J Mol Biol 406 (2):228–256. https://doi.org/10.1016/j.jmb.2010.10.030
11. Pierce BG, Weng Z (2013) A flexible docking approach for prediction of T cell receptor-peptide-MHC complexes. Protein Sci 22 (1):35–46. https://doi.org/10.1002/pro.2181
12. Shugay M et al (2018) VDJdb: a curated database of T-cell receptor sequences with known antigen specificity. Nucleic Acids Res 46(D1): D419–D427. https://doi.org/10.1093/nar/gkx760
13. Dayhoff M et al (1978) A model of evolutionary change in proteins. In: Atlas of protein sequence and structure, vol 5. National Biomedical Research Foundation Silver Spring, Waltham, MA, pp 345–352
14. Yang X et al (2015) Structural basis for clonal diversity of the public T cell response to a dominant human cytomegalovirus epitope. J Biol Chem 290(48):29106–29119. https://doi.org/10.1074/jbc.M115.691311
15. Stronen E et al (2016) Targeting of cancer neoantigens with donor-derived T cell receptor repertoires. Science 352(6291):1337–1341. https://doi.org/10.1126/science.aaf2288
16. Dunbar J, Deane CM (2016) ANARCI: antigen receptor numbering and receptor classification. Bioinformatics 32(2):298–300. https://doi.org/10.1093/bioinformatics/btv552
17. Alford RF et al (2017) The Rosetta all-atom energy function for macromolecular modeling and design. J Chem Theory Comput 13 (6):3031–3048. https://doi.org/10.1021/acs.jctc.7b00125

Chapter 15

In Silico Cell-Type Deconvolution Methods in Cancer Immunotherapy

Gregor Sturm, Francesca Finotello, and Markus List

Abstract

Several computational methods have been proposed to infer the cellular composition from bulk RNA-seq data of a tumor biopsy sample. Elucidating interactions in the tumor microenvironment can yield unique insights into the status of the immune system. In immuno-oncology, this information can be crucial for deciding whether the immune system of a patient can be stimulated to target the tumor. Here, we shed a light on the working principles, capabilities, and limitations of the most commonly used methods for cell-type deconvolution in immuno-oncology and offer guidelines for method selection.

Key words Cell-type deconvolution, Immuno-oncology, Spillover, Gene signatures

1 Introduction

Tumors consist of malignant cells that are embedded in a complex multicellular microenvironment characterized by complex and dynamic interactions [1]. Notably, this tumor microenvironment (TME) comprises a variety of different immune cells that have been associated with both positive and negative disease outcome. For instance, CD8+ T cells are the primary effector cells of antitumor immunity as they specifically recognize and kill malignant cells, while regulatory T cells (Treg) have immunosuppressive functions supporting tumor growth and immune evasion [2]. Therefore, the cellular composition of the tumor immune infiltrates can shed light on the escape mechanisms that tumor cells use to evade immunological control. Moreover, in clinical trials, knowledge of the cellular composition can help to stratify patients for the most suitable treatment option depending on the targeted cell type, hence increasing the overall chances of treatment success and ultimately accelerating access to improved treatment options [3].

Methods like fluorescence-activated cell sorting (FACS) or immunohistochemistry (IHC)-staining have been used as gold standards to estimate the immune cell content within a sample

Sebastian Boegel (ed.), *Bioinformatics for Cancer Immunotherapy: Methods and Protocols*, Methods in Molecular Biology, vol. 2120, https://doi.org/10.1007/978-1-0716-0327-7_15, © Springer Science+Business Media, LLC, part of Springer Nature 2020

[4]. However, these methods have technical limitations and might not be applicable in certain situations. More specifically, FACS requires a large amount of sample material, hence limiting its application to tumor biopsies. IHC provides an estimate from a single tumor slice, which may not be representative of a heterogeneous immune landscape of the tumor. Furthermore, both methods can utilize only a small number of cell-type-specific markers and depend on the availability of sensitive and specific antibodies. More recently, single-cell RNA sequencing (scRNA-seq) has started being used to characterize cell types and states, yet for the time being, it remains too expensive and laborious for routine clinical use. Moreover, cell-type proportions might be biased in scRNA-seq data due to differences in single-cell dissociation efficiencies [5]. At the same time, methods for gene expression profiling, RNA-seq and microarrays, have been developed and optimized to be applicable in clinical settings, resulting in a plethora of transcriptomic datasets derived from patient tumor samples such as TCGA [6]. More recently, technologies like *TruSeq RNA Access by Illumina and Nanostring* made expression profiling available even for formalin-fixed paraffin-embedded (FFPE) samples. However, these methods provide transcriptomics data from heterogeneous samples considered as a whole but offer no immediate insights into their cellular composition, which therefore has to be inferred using computational techniques.

2 Mathematical Background

Computational methods for immune-cell content estimation in tumor samples can be classified in two categories: marker gene-based approaches and deconvolution-based approaches. Marker gene-based approaches utilize a list of genes that are characteristic for a cell type. Based on these marker genes, an abundance score or enrichment score (ES) is calculated that is high when the cell-type markers are among the most highly expressed genes in the sample and low otherwise (Fig. 1a). Deconvolution methods, however, formulate the problem as a system of equations that describe the gene expression profile of a bulk sample as the weighted sum of the expression profiles of the admixed cell types (Fig. 1b). By solving the inverse problem, the unknown cell-type fractions can be inferred given a signature matrix and the bulk gene expression profile. In practice, this problem can be solved using ν-Support Vector Regression (ν-SVR; CIBERSORT) [7], constrained least square regression (quanTIseq and EPIC) [8, 9], or linear least square regression (TIMER) [10].

A limitation of the marker-based approaches is that they only allow to generate a semiquantitative score, which enables a comparison between samples but not between cell types [11].

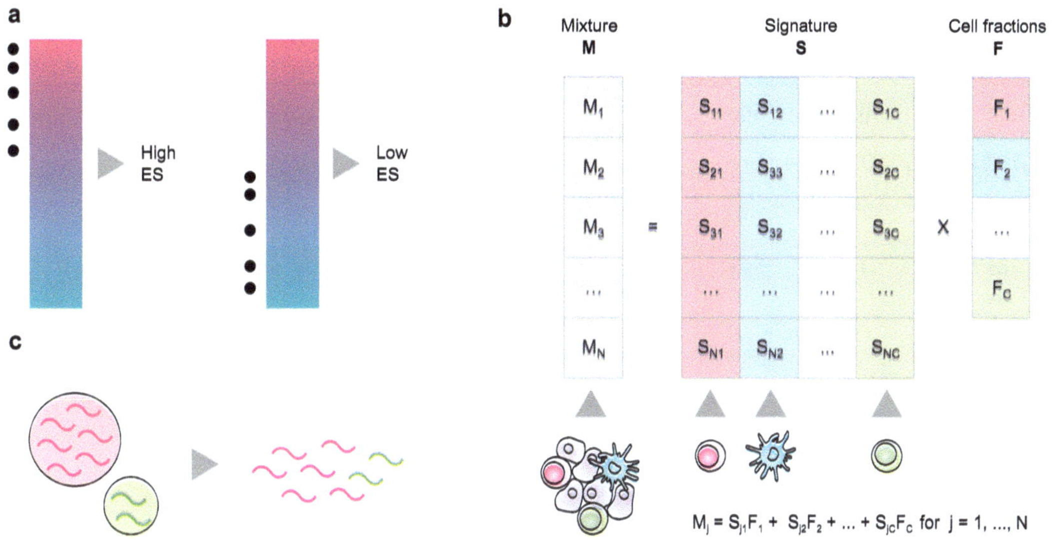

Fig. 1 Figure previously published in [12]. (**a**) Marker gene-based approaches rank genes by their expression in a sample and compute an enrichment score (ES) based on the average rank of a set of cell-type-specific marker genes (gray dots). The ES is high when the marker genes are among the top highly expressed genes (magenta) and low otherwise (cyan). (**b**) Deconvolution-based approaches fit a linear model to explain the observed expression of a gene by building a weighted sum of cell-type-specific expression profiles in a signature matrix S and relative fractions F of the cell types in the mixture. (**c**) A confounding factor is that cell types may differ in the amount of total mRNA, leading to under- or overestimation of cell-type fractions in deconvolution methods

Deconvolution allows, in principle, to generate "absolute scores" that can be interpreted as cell fractions and compared both inter- and intrasample. In practice, the properties of scores differ from method to method, as we highlight in the following section.

3 Methods for Cell-Type Deconvolution

While general-purpose deconvolution methods are available (reviewed in [12–14]), we focus on methods specifically developed for immuno-oncology that already include immune-cell-type signatures (Table 1). In this subheading, we introduce the rationale behind each method and their signatures.

Both xCell [15] and MCP-counter [16] are marker gene-based approaches. xCell comes with a compendium of 489 gene signatures for 64 immune and nonimmune cell types that have been derived from several large-scale gene expression datasets: FANTOM5 [17], ENCODE [18], Blueprint [19], Immune response in silico (IRIS) [20], Human Primary Cell Atlas (HPCA) [21], and Novershtern et al. [22]. For each cell type, xCell independently calculates an enrichment score through the following steps:

Table 1
Overview of popular methods for cell-type deconvolution in immuno-oncology

Tool	Approach	Score	Comparisons	Cell types	Reference
CIBERSORT	D	Immune cell fractions, relative to total immune cell content	Intra	22 immune cell types	[7]
CIBERSORT absolute mode	D	Score of arbitrary units that reflects the absolute proportion of each cell type	Intra, inter	22 immune cell types	[7]
EPIC	D	Cell fractions, relative to all cells in sample	Intra, inter	6 immune cell types, fibroblasts, endothelial cells	[9]
MCP-counter	M	Arbitrary units, comparable between samples	Inter	8 immune cell types, fibroblasts, endothelial cells	[16]
quanTIseq	D	Cell fractions, relative to all cells in sample	Intra, inter	10 immune cell types	[8]
TIMER	D	Arbitrary units, comparable between samples (not different cancer types)	Inter	6 immune cell types	[10]
xCell	M	Arbitrary units, comparable between samples	Inter	64 immune and nonimmune cell types	[15]

We conceptually distinguish marker-gene-based approaches (M) and deconvolution-based approaches (D). The output scores of the various methods have different properties and allow intrasample comparisons between cell types, intersample comparisons of the same cell type, or both. All methods come with a set of cell-type signatures ranging from 6 immune cell types to 64 immune and nonimmune cell types

(1) calculating a p-value for each of the 489 gene signatures using ssGSEA, (2) taking the mean enrichment score across all gene sets belonging to a cell type, (3) converting the enrichment score into an abundance score, and (4) correcting for "spillover," that is, the overprediction of a certain cell type due to the presence of a closely related cell type. The final score is indicative of the cell-type abundance but cannot be interpreted as a cell fraction. **Step 4** depends on the variance between all samples submitted to the xCell run. This implies that xCell works best if ran on many, highly heterogeneous samples. Moreover, differences in variance hamper comparability between xCell scores from different experiments and can, in extreme cases, when ran with few homogeneous samples, lead to misleading scores.

MCP-counter is based on a stringent and robust set of markers for tumor-infiltrating lymphocytes, fibroblasts, and endothelial cells derived from more than 1900 microarray samples. The MCP-counter abundance score is computed as the arithmetic mean of the gene expression of all marker genes specific for a certain cell type. The abundance score is expressed in arbitrary units and therefore only allows intersample comparisons.

CIBERSORT [7], EPIC [9], TIMER [10], and quanTIseq [8] are deconvolution-based methods, yet only EPIC and quanTIseq generate a score that can be interpreted as a cell fraction and compared both inter- and intrasample.

CIBERSORT comes with a signature of 547 genes built from microarray data containing fingerprints of 22 immune cell phenotypes including different cellular states. The method estimates the cell fractions using ν-SVR. CIBERSORT scores are expressed as a fraction of the total immune cell content of a sample and, hence, can be used to compare the abundance of different cell types within the same sample but not to compare between two samples. More recently, the authors developed an extension, called "absolute mode," that transforms the score to allow both within and between samples of the same experiment [23]. Yet, the score is still expressed in arbitrary units and does not represent a cell fraction.

TIMER allows to estimate the abundance of six immune cell types in 32 cancer types. Signature genes have been identified separately for each cancer type by selecting immune cell markers that are negatively correlated with tumor purity derived from copy number variation data. In this approach, the input samples are merged with immune-cell reference samples using COMBAT [24] to remove batch effects. Finally, a signature matrix is built from the merged immune cell profiles using the selected immune cell markers. The resulting matrix is in turn used to compute the scores via linear least squares regression. The score is expressed in arbitrary units and can be used to compare samples of the same experiment, but not between cell types in one sample. A limitation of TIMER is, similar to xCell, that the results depend on all samples submitted in a single run, that is, the results can change when submitted with a different set of samples.

EPIC ships with two signature matrices that, unlike previous tools, were derived from RNA-seq data (bulk or single-cell). The first one is derived from single-cell melanoma data describing five tumor-infiltrating immune cell types plus endothelial cells and cancer-associated fibroblasts. The second one is derived from purified PBMC samples describing six blood-circulating immune cell types. EPIC generates an absolute score and provides an estimate of the abundance of "other" cells, that is, those cell types that are not accounted for in the signature matrix. The scores, which are, thus, referred to the total cells in a sample, can be interpreted as cell fractions and be used for both inter- and intrasample comparisons.

quanTIseq is a recent deconvolution method for the analysis bulk RNA-seq data from blood or tumor samples [10]. Unlike the other methods that operate on the precomputed, normalized gene expression data, quanTIseq comes as an entire pipeline that allows to directly analyze raw RNA-seq data (i.e., FASTQ files of sequencing reads) to avoid inconsistencies due to different preprocessing approaches and to simplify the analysis. However, it is also possible to run the deconvolution module alone on gene expression data directly, at the (potential) cost of accuracy. quanTIseq is based on constrained least square regression and uses a signature matrix describing ten immune and stromal cell types. The matrix contains 171 genes and has been derived from 51 PBMC RNA-seq samples from 10 immune cell types. Similarly to EPIC, quanTIseq computes absolute scores that can be interpreted as cellular fractions and can be used for both inter- and intrasample comparisons.

4 The Impact of Gene Signatures

We performed extensive benchmarking of the methods using both single-cell data-based simulations and bulk RNA-seq samples profiled with FACS [25]. The usage of single-cell data allowed us constructing artificial bulk sequencing samples, to be used as a gold standard dataset for testing, in which we could tightly control the cell type composition to benchmark specific method properties. More specifically, we could assess the sensitivity and specificity of the tested methods by adding a slowly increasing fraction of a specific cell type. The major conclusion of our study is that RNA-seq can be utilized for estimation of tumor immune cell infiltrates robustly and with good accuracy, particularly when a cell type is well characterized. Most methods show near-perfect correlations on $CD8^+$ T cells, B cells, natural killer (NK) cells, fibroblasts, and endothelial cells. However, deconvolution of certain cell types performed was suboptimal in the benchmark. In particular, regulatory and nonregulatory $CD4^+$ T cells appear to be difficult to quantify. Moreover, we found that dendritic cells (DCs) are not sufficiently covered by currently available gene signatures. DCs comprise heterogeneous subtypes with unique functions, including plasmacytoid dendritic cells (pDCs) and myeloid dendritic cells (mDCs) [26, 27]. In our assessment, we found that none of the methods can effectively quantify total DCs, but they can quantify either mDC or pDCs, depending on the samples that have been used for generating the signatures (*see* Supplementary Table 2 in [25]).

Analysis of spillover effects, that is, erroneous prediction of a higher abundance of a cell type due to the presence of another one, revealed potential for improvement with respect to these specialized cell types. For instance, we could show that substantial spillover between DCs and B cells in several methods can be attributed to nonspecific genes that are both expressed in B cells and pDCs, but not in mDCs. By removing the nonspecific genes from the signature matrix, performance could be significantly improved.

By leveraging recently published large-scale single-cell RNA-seq datasets [5, 28–30], we envision that our understanding of immune cell types and states can be greatly improved, and by explicitly addressing spillover effects, more specific cell-type signatures can be distilled.

5 Guide

In Table 2, we summarize the results of our benchmark [25] and provide guidelines for method selection based on three criteria: interpretability of the score, overall performance, and possible limitations. For B-, CD4$^+$ T-, Treg-, CD8$^+$-, and NK-cells cells, methods with overall very good performance are available, while for CD4$^+$ non-reg. and DC, performance is limited. For choosing a

Table 2
Guidelines for method selection

Cell type	Recommended methods	Overall performance	Absolute score	No background predictions
B	EPIC	++	++	+
	MCP-counter	++	–	–
T CD4$^+$	EPIC	++	++	–
	xCell	++	–	++
T CD4$^+$ non-reg.	quanTIseq	+	++	–
	xCell	+	–	++
T reg.	quanTIseq	++	++	–
	xCell	++	–	++
T CD8$^+$	quanTIseq	++	++	+
	EPIC	++	++	–
	MCP-counter	++	–	–
NK	EPIC	++	++	+
	MCP-counter	++	–	–
Macrophages/Monocytes	EPIC	++	++	+
	MCP-counter	++	–	–
DC	None of the methods can be recommended to estimate overall DC content. MCP-counter and quanTIseq can be used to profile myeloid DCs			

For each cell type, we recommend a suitable method, highlighting advantages and possible limitations. *Overall performance*: Indicates how well predicted fractions correspond to known fractions in the benchmark. Absolute score: the method provides an absolute score that can be interpreted as a cell fraction. Methods that do not support an absolute score only support intersample comparison within one experimental dataset, that is, the score is only meaningful in relation to another sample. *Background predictions*: Indicates whether a method tends to predict a cell type although it is absent

method, it is important to understand the implications of the different scoring strategies. EPIC and quanTIseq are the only methods providing an "absolute score" that represents a cell fraction. All other methods provide scores in arbitrary units, which are only meaningful in relation to another sample of the same dataset. For this reason, and due to a robust overall performance, we recommend EPIC and quanTIseq for general-purpose deconvolution in immuno-oncology. In practice, absolute scores are not always necessary. For instance, in a clinical trial, relative scoring methods can be used to infer fold changes between treatment and control group or to monitor changes of the immune composition in longitudinal samples. In that case, using MCP-counter is a good choice, thanks to its highly specific signatures that excelled with respect to spillover. A limitation of deconvolution methods is that they are susceptible to background predictions, that is, prediction of (small) fractions for cell types that are actually absent. Therefore, when interested in presence/absence of a cell type, we suggest using xCell which is highly robust against background predictions.

6 Conclusions

As bulk RNA-seq data become increasingly ubiquitous in clinics, in silico cell-type deconvolution methods can make an important contribution to map the TME, in particular, with respect to immunotherapy. Several methods exist and show generally good performance for common cell types with few exceptions that can likely be attributed to the choice of unspecific genes in the respective signatures. Conceptual differences between marker-based and deconvolution-based methods and individual strengths and weaknesses with respect to specific cell types need to be taken into account for choosing the optimal method. We thus expect that the guidelines presented here will help users to make a well-informed choice for obtaining the best possible results in their analysis.

References

1. Fridman WH, Pagès F, Sautès-Fridman C, Galon J (2012) The immune contexture in human tumours: impact on clinical outcome. Nat Rev Cancer 12:298–306. https://doi.org/10.1038/nrc3245
2. Fridman WH, Zitvogel L, Sautès-Fridman C, Kroemer G (2017) The immune contexture in cancer prognosis and treatment. Nat Rev Clin Oncol 14:717–734. https://doi.org/10.1038/nrclinonc.2017.101
3. Friedman AA, Letai A, Fisher DE, Flaherty KT (2015) Precision medicine for cancer with next-generation functional diagnostics. Nat Rev Cancer 15:747–756. https://doi.org/10.1038/nrc4015
4. Petitprez F, Sun CM, Lacroix L (2018) Quantitative analyses of the tumor microenvironment composition and orientation in the era of precision medicine. Front Oncol 8:390. https://doi.org/10.3389/fonc.2018.00390

5. Lambrechts D, Wauters E, Boeckx B et al (2018) Phenotype molding of stromal cells in the lung tumor microenvironment. Nat Med 24:1277–1289. https://doi.org/10.1038/s41591-018-0096-5
6. Cancer Genome Atlas Research Network, Weinstein JN, Collisson EA et al (2013) The cancer genome atlas pan-cancer analysis project. Nat Genet 45:1113–1120. https://doi.org/10.1038/ng.2764
7. Newman AM, Liu CL, Green MR et al (2015) Robust enumeration of cell subsets from tissue expression profiles. Nat Methods 12:453. https://doi.org/10.1038/nmeth.3337
8. Finotello F, Mayer C, Plattner C et al (2019) Molecular and pharmacological modulators of the tumor immune contexture revealed by deconvolution of RNA-seq data. Genome Med 11:34
9. Racle J, de Jonge K, Baumgaertner P et al (2017) Simultaneous enumeration of cancer and immune cell types from bulk tumor gene expression data. elife 6:e26476. https://doi.org/10.7554/eLife.26476
10. Li B, Severson E, Pignon J-C et al (2016) Comprehensive analyses of tumor immunity: implications for cancer immunotherapy. Genome Biol 17:174. https://doi.org/10.1186/s13059-016-1028-7
11. Petitprez F, Vano YA, Becht E et al (2018) Transcriptomic analysis of the tumor microenvironment to guide prognosis and immunotherapies. Cancer Immunol Immunother 67:981–988. https://doi.org/10.1007/s00262-017-2058-z
12. Finotello F, Trajanoski Z (2018) Quantifying tumor-infiltrating immune cells from transcriptomics data. Cancer Immunol Immunother 67:1031–1040. https://doi.org/10.1007/s00262-018-2150-z
13. Avila Cobos F, Vandesompele J, Mestdagh P, De Preter K (2018) Computational deconvolution of transcriptomics data from mixed cell populations. Bioinformatics 34:1969–1979. https://doi.org/10.1093/bioinformatics/bty019
14. Newman AM, Alizadeh AA (2016) High-throughput genomic profiling of tumor-infiltrating leukocytes. Curr Opin Immunol 41:77–84. https://doi.org/10.1016/j.coi.2016.06.006
15. Aran D, Hu Z, Butte AJ (2017) xCell: digitally portraying the tissue cellular heterogeneity landscape. Genome Biol 18:220. https://doi.org/10.1186/s13059-017-1349-1
16. Becht E, Giraldo NA, Lacroix L et al (2016) Estimating the population abundance of tissue-infiltrating immune and stromal cell populations using gene expression. Genome Biol 17:218. https://doi.org/10.1186/s13059-016-1070-5
17. Forrest ARR, Kawaji H, Rehli M et al (2014) A promoter-level mammalian expression atlas. Nature 507:462–470. https://doi.org/10.1038/nature13182
18. ENCODE Project Consortium (2012) An integrated encyclopedia of DNA elements in the human genome. Nature 489:57–74. https://doi.org/10.1038/nature11247
19. Fernández JM, de la Torre V, Richardson D et al (2016) The BLUEPRINT data analysis portal. Cell Syst 3:491–495.e5. https://doi.org/10.1016/j.cels.2016.10.021
20. Abbas AR, Baldwin D, Ma Y et al (2005) Immune response in silico (IRIS): immune-specific genes identified from a compendium of microarray expression data. Genes Immun 6:319–331. https://doi.org/10.1038/sj.gene.6364173
21. Mabbott NA, Baillie JK, Brown H et al (2013) An expression atlas of human primary cells: inference of gene function from coexpression networks. BMC Genomics 14:632. https://doi.org/10.1186/1471-2164-14-632
22. Novershtern N, Subramanian A, Lawton LN et al (2011) Densely interconnected transcriptional circuits control cell states in human hematopoiesis. Cell 144:296–309. https://doi.org/10.1016/j.cell.2011.01.004
23. Newman AM, Liu CL, Green MR, Gentles AJ, Feng W, Xu Y, Hoang CD, Diehn M, Alizadeh AA CIBERSORT website. In: CIBERSORT. https://cibersort.stanford.edu/. Accessed 20 Oct 2018
24. Johnson WE, Li C, Rabinovic A (2007) Adjusting batch effects in microarray expression data using empirical Bayes methods. Biostatistics 8:118–127. https://doi.org/10.1093/biostatistics/kxj037
25. Sturm et al (2019) Comprehensive evaluation of computational cell-type quantification methods for immuno-oncology. Bioinformatics 35(14):i436–i445. https://doi.org/10.1093/bioinformatics/btz363
26. Collin M, McGovern N, Haniffa M (2013) Human dendritic cell subsets. Immunology 140:22–30. https://doi.org/10.1111/imm.12117
27. Villani A-C, Satija R, Reynolds G et al (2017) Single-cell RNA-seq reveals new types of human blood dendritic cells, monocytes, and progenitors. Science 356. https://doi.org/10.1126/science.aah4573

28. Sade-Feldman M, Yizhak K, Bjorgaard SL et al (2019) Defining T cell states associated with response to checkpoint immunotherapy in melanoma. Cell 176:404. https://doi.org/10.1016/j.cell.2018.12.034
29. Azizi E, Carr AJ, Plitas G, et al Single-cell immune map of breast carcinoma reveals diverse phenotypic states driven by the tumor microenvironment. https://doi.org/10.1101/221994
30. Guo X, Zhang Y, Zheng L et al (2018) Global characterization of T cells in non-small cell lung cancer by single-cell sequencing. Nat Med 24:978–985. https://doi.org/10.1038/s41591-018-0045-3

Chapter 16

Immunedeconv: An R Package for Unified Access to Computational Methods for Estimating Immune Cell Fractions from Bulk RNA-Sequencing Data

Gregor Sturm, Francesca Finotello, and Markus List

Abstract

Since the performance of in silico approaches for estimating immune-cell fractions from bulk RNA-seq data can vary, it is often advisable to compare results of several methods. Given numerous dependencies and differences in input and output format of the various computational methods, comparative analyses can become quite complex. This motivated us to develop *immunedeconv*, an R package providing uniform and user-friendly access to seven state-of-the-art computational methods for deconvolution of cell-type fractions from bulk RNA-seq data. Here, we show how *immunedeconv* can be installed and applied to a typical dataset. First, we give an example for obtaining cell-type fractions using quanTIseq. Second, we show how dimensionless scores produced by MCP-counter can be used for cross-sample comparisons. For each of these examples, we provide R code illustrating how *immunedeconv* results can be summarized graphically.

Key words Cell-type deconvolution, R package, Immuno-oncology, Benchmark, Comparative analyses

1 Introduction

The composition of the tumor microenvironment has a profound influence on treatment response and outcome [1, 2]. Knowledge of the immune-cell content in tumor samples is therefore invaluable for understanding the immune response against tumors and the escape mechanisms put in place by malignant cells.

Methods like fluorescence-activated cell sorting (FACS) or immunohistochemistry (IHC)-staining have been used as gold standards to estimate the immune cell content within a sample [3]. More recently, single-cell RNA-seq is used to characterize cell types and states within tumor samples. However, estimates of cell-type abundances with one of these technologies are often not available, due to technical limitations and/or the high cost of the assay. On the contrary, bulk RNA-seq has been established as a cost-efficient standard assay for molecular characterization of bulk

Sebastian Boegel (ed.), *Bioinformatics for Cancer Immunotherapy: Methods and Protocols*, Methods in Molecular Biology, vol. 2120, https://doi.org/10.1007/978-1-0716-0327-7_16, © Springer Science+Business Media, LLC, part of Springer Nature 2020

tumors. Recently, several computational methods have been proposed to infer cell-type proportions from bulk RNA-seq data generated from cell mixtures like blood or tumor samples (*see* review in Chapter 15).

Here, we present *immunedeconv*, an R package providing unified access to the deconvolution methods described in Chapter 15. We have previously highlighted how the selected methods are conceptually different and excel in different cell-types and use-cases [4]. Therefore, it can be desirable to run different methods on the same dataset to compare and/or complement results. *Immunedeconv* simplify this process as it does not require the user to download and install all methods separately, or to manually coerce input data into the required formats.

2 Materials

2.1 Available Methods

Immundeconv currently contains seven state-of-the-art methods (Table 1), which can be conceptually separated into marker-gene-based approaches (M) and deconvolution-based approaches (D). The output scores of the methods have different properties and allow intrasample comparisons between cell types, intersample comparisons of the same cell type, or both (see review in Chapter 15).

2.2 Installation

There are two supported ways for installing the *immunedeconv* package, namely using the `install.packages` command in `R` or through the `conda` package manager. The former works on all platforms, the latter only on MacOS and Linux. *immunedeconv* has many dependencies owing to the respective dependencies of the integrated deconvolution methods. Consequently, `install.packages` can be slow and tends to be unreliable due to potential version conflicts. We, therefore, strongly recommend to use `conda` for the installation.

Table 1
Tools integrated in *immunedeconv* with the corresponding name used in the R package

Tool	Approach	Method name in *immunedeconv*	Reference
CIBERSORT	D	`cibersort`	[6]
CIBERSORT absolute mode	D	`cibersort_abs`	[6]
EPIC	D	`epic`	[7]
MCP-counter	M	`mcp_counter`	[8]
quanTIseq	D	`quantiseq`	[9]
TIMER	D	`timer`	[10]
xCell	M	`xcell`	[11]

2.2.1 Installation Through Conda Package Manager

System requirements: Linux or MacOS.

Installation time: typically within few minutes

1. If you do not have `conda` available on your system, download Miniconda from https://conda.io/miniconda.html. You can install Miniconda in your home directory without root privileges.
2. We recommend creating a dedicated environment for deconvolution (Optional):

```
conda create -n deconvolution
conda activate deconvolution
```

   You can skip this step if you prefer to install *immunedeconv* in your global environment.
3. Install the *immunedeconv* package.

```
conda install -c bioconda -c conda-forge r-immunedeconv
```

   Conda will automatically install the R package and all its dependencies. You can now open an instance of R within the environment and use the package.

2.2.2 Installation as Standard R Package (install.packages)

System requirements: R $\geq$ 3.5.0 with the `devtools` and `BiocManager` packages installed (available via install.packages from CRAN).

Installation time: typically around 30 min.

1. First, manually install the following dependencies that are not available on CRAN. If you have a very recent version of `devtools`, it will automatically resolve these dependencies and you can skip to **step 2**.

```
Bioconductor
  BiocManager::install("preprocessCore")
  BiocManager::install("Biobase")
  BiocManager::install("GSVA")
  BiocManager::install("sva")
  BiocManager::install("GSEABase")
GitHub
  library(devtools)
  install_github('dviraran/xCell')
  install_github('GfellerLab/EPIC')
  install_github('ebecht/MCPcounter/Source')
```

2. Now, install *immunedeconv* and all CRAN dependencies using.

```
devtools::install_github('icbi-lab/immunedeconv')
```

2.2.3 Installing CIBERSORT

Due to licensing restrictions, we could not include CIBERSORT into the package. However, you can still use CIBERSORT through *immunedeconv*'s unified interface, if you download the source code from the authors' website:

1. Download the cibersort source code from https://cibersort.stanford.edu/. The source code package contains two files that are required for running CIBERSORT through *immunedeconv*.

```
CIBERSORT.R
LM22.txt
```

2. Extract the source code to a directory of your choice, for example,

```
/home/deconvolution/cibersort
```

3. Specify the paths within R:

```
library(immunedeconv)
set_cibersort_binary("/home/deconvolution/cibersort/CIBER-
SORT.R")
set_cibersort_mat("/home/deconvolution/cibersort/LM22.txt")
```

2.3 Issues and Support

If you have any issues installing or running *immunedeconv*, please open an issue on GitHub (https://github.com/icbi-lab/immunedeconv/issues).

3 Methods

3.1 Estimation of Immune Cell Contents

To estimate immune cell contents from transcriptomics samples, use the `deconvolute` function. In the simplest case, the function requires only two arguments: the gene expression data and the method to use for estimation. Depending on the selected method, optional arguments can be supplied. TIMER uses tumor type-specific reference profiles (*see* **Note 1**). quanTIseq and EPIC differentiates between tumor and peripheral blood samples (*see* **Note 2**). CIBERSORT and quanTIseq differentiate between RNA-seq and microarray data (*see* **Note 3**). It is also possible to exclude certain genes in the signatures (*see* **Note 4**).

3.1.1 Input

The input can be formatted as either

1. a `matrix` or `data.frame` or,
2. a Bioconductor `ExpressionSet`.

The matrix must be formatted as genes *x* samples, with HGNC gene symbols as row-names and sample identifiers as column-names:

```
>my_gene_expression_matrix
    SampleA     SampleB    ...
    CD8A         8.2         1.7
    CD4         0          4.2
    PDCD1       3.3        0.1
    ...         ...        ...
    >deconvolute(matrix, method="mcp_counter")
```

The ExpressionSet needs to have a column in `featureData` that contains HGNC gene symbols. Default parameters assume this column to be called `gene_symbol`; however, it can be adjusted using the `column` parameter:

```
>deconvolute(eset, method="mcp_counter", column="gene_symbol")
```

3.1.2 Output

The `deconvolute` function produces a `data.frame` with the first column containing the cell-type names and the other columns containing the estimates for each sample:

```
    >my_result
    cell_type        SampleA    SampleB  ...
    T cell CD8+      0.1        0.22
    B cell           0.07       0.31
    NK cell          0.53       0.0
    ...              ...        ...
```

To interpret the results, it is very important to understand the different scoring strategies of the methods. EPIC and quanTIseq are the only methods providing an "absolute score" that represents a cell fraction. All other methods provide scores in arbitrary units, which are only meaningful in relation to another sample of the same dataset. Please refer to Chapter 15 for more details about the scoring strategies of the individual methods.

3.2 Case Study

Immunedeconv ships with an example dataset containing purified immune cells extracted from three healthy donors. For demonstration purposes, let us create a subset of this dataset and estimate immune cell fractions:

```
>library(immunedeconv)
my_matrix <- example_gene_expression_matrix[, c
(1,2,12,13,19,20)]
```

The matrix has HGNC gene symbols as rownames:

```
>rownames(my_matrix)[1:5]
[1] "A1BG"      "A1BG-AS1" "A1CF"      "A2M"       "A2ML1"
```

And sample names as column names:

```
>colnames(my_matrix)
[1] "CD4-positive T Cells, donor1" "CD4-positive T Cells,
donor2"
[3] "CD8-positive T Cells, donor1" "CD8-positive T Cells,
donor2"
[5] "Natural Killer Cells, donor1" "Natural Killer Cells,
donor2"
```

To estimate immune cell fractions, we have to invoke the `deconvolute` function. It requires the specification of one of the following methods for deconvolution (Table 1):

```
>deconvolution_methods
  MCPcounter              EPIC        quanTIseq              xCell
"mcp_counter"           "epic"      "quantiseq"            "xcell"
  CIBERSORT CIBERSORT (abs.)              TIMER
"cibersort"  "cibersort_abs"            "timer"
```

For this example, we use quanTIseq [7]. As a result, we obtain a *cell_type x sample* data frame with cell-type scores for each sample:

```
>res_quantiseq <- deconvolute(my_matrix, "quantiseq")
```

Now, `res_quantiseq` contains a `data.frame` with the results. quanTIseq generates scores that can be interpreted as cell-type fractions. Therefore, a visualization where the plot area is proportional to cell-type fractions is suitable. In Figs. 1–2 we show example plots created with ggplot2 from the tidyverse R-package collection (https://www.tidyverse.org/; to install the tidyverse using conda, use `conda install -c conda-forge r-tidyverse`), but other tools or functions can be used. The following command first reformats the results from quanTIseq using the command gather to create key-value pairs for sample and fraction. This is followed by a ggplot command to create a bar chart with flipped axis (coord_flip) to produce a horizontal version with a matching color scheme (scale_fill_brewer). Individual commands are connected through a pipe operator % > %:

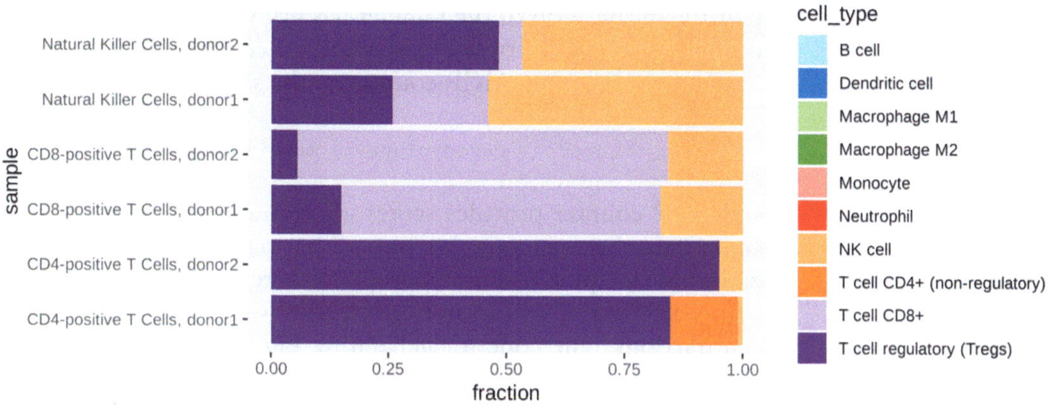

Fig. 1 Example for cell-type fractions reported by quanTIseq and plotted with ggplot2

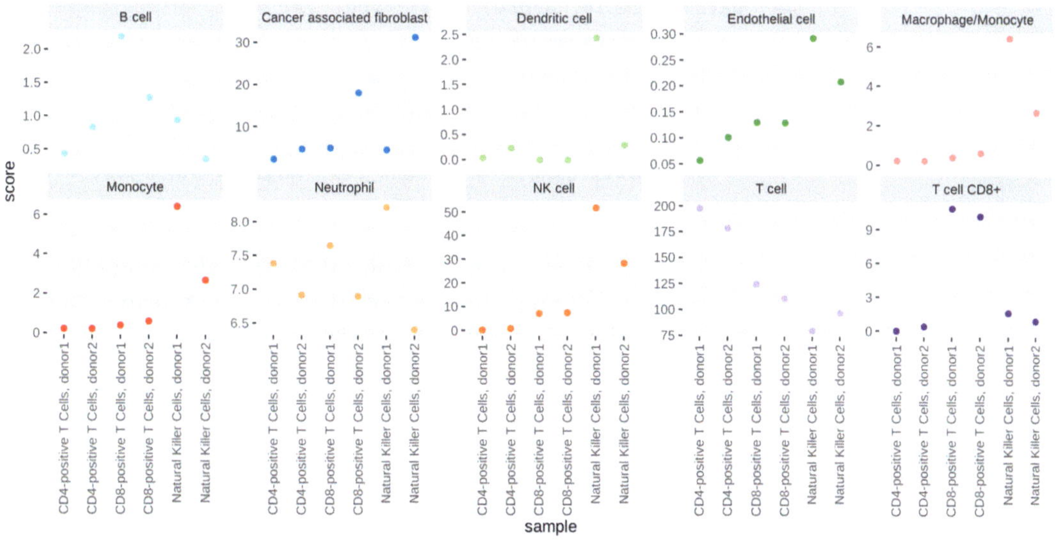

Fig. 2 Example for a ggplot created dot plot visualizing the scores reported by MCP-counter

```
>library(tidyverse)
>res_quantiseq %>%
  gather(sample, fraction, -cell_type) %>%
  ggplot(aes(x=sample, y=fraction, fill=cell_type)) +
    geom_bar(stat='identity') +
    coord_flip() +
    scale_fill_brewer(palette="Paired")
```

In Fig. 1, we observe that the correct cell types are recovered; however we observe small fractions of NK cells and CD4$^+$ predicted to be present in CD8$^+$ T cells and considerable fractions of CD8$^+$ and regulatory T cells in NK cells. This can either be due to the fact that the purification of the samples has not worked reliably or due to spillover-effects between closely-related cell types [4].

Let us now apply MCP-counter to the same dataset:

```
>res_mcp_counter <- deconvolute(my_matrix, "mcp_counter")
```

MCP-counter provides scores in arbitrary units that are only comparable between samples, but not between cell types. A visualization as dot plot is more appropriate here to highlight that the scores are relative and to not erroneously suggest the scores were cell fractions. The code for generating Fig. 2 is similar to that of Fig. 1, except that we are generating a dot plot (geom_point) and generate one facet for each cell type (facet_wrap):

```
>res_mcp_counter %>%
  gather(sample, score, -cell_type) %>%
  ggplot(aes(x=sample, y=score, color=cell_type)) +
    geom_point() +
    facet_wrap(~cell_type, scales="free_y", nrow=2) +
    scale_color_brewer(palette="Paired", guide=FALSE) +
    theme(axis.text.x = element_text(angle = 90, vjust = 0.5,
hjust=1))
```

With the scores in arbitrary units, the results shown in Fig. 2 are not useful for judging whether a cell type is present in an individual sample, but they allow comparing relative quantities between samples, for example, we can spot samples where NK cells and CD8+ T cells are more abundant.

3.3 Conclusion

A variety of in silico cell-type deconvolution methods for immuno-oncology are now available, but each one has its individual strengths and limitations. In addition, one can expect deconvolution performance to vary due to the properties and quality of the dataset under investigation. So far, a major hurdle in comparing results between different methods was the lack of a unified and easy-to-use interface. The *immunedeconv* R package closes this gap and, besides providing simplified access to deconvolution methods also to non-expert users, it is suited to benchmark the performance of cell-type deconvolution methods on new datasets or to compare a newly developed method to the state of the art.

4 Notes

1. Setting the tumor-type.

   TIMER ships with profiles for 33 TCGA cancer types [5]. To be able to run TIMER, you have to provide the proper tumor-type to the `deconvolute` function. The full list of abbreviations is available on the TCGA wiki (https://gdc.cancer.gov/resources-tcga-users/tcga-code-tables/tcga-study-abbreviations) and stored in.

```
>timer_available_cancers
```

For instance, if `gene_expression` contains two breast cancer and two liver cancer samples,

```
>deconvolute(gene_expression, method="timer", indications = c
("BRCA", "BRCA", "LIHC", "LIHC"))
```

The length of the indication vector must correspond to the number of samples in the gene expression matrix.

2. Setting the sample-type (tumor vs. nontumor).

   quanTIseq and EPIC provide specialized modes for tumor and peripheral blood samples. If running in tumor-mode, quanTIseq excludes signature genes with spurious, high expression in tumor-samples. EPIC ships with two signature matrices, one of them optimized for blood samples, one for tumor-samples. Per default, *immunedeconv* runs the methods in tumor-mode. If you wish to run them in nontumor mode, set the option `tumor` to `FALSE`:

```
>deconvolute(gene_expression, method="epic", tumor=FALSE)
```

3. Running on Microarray samples.

   CIBERSORT has been specifically developed for microarray samples, but the authors suggest that it can be readily applied to RNA-seq data, if quantile normalization is disabled. quanTIseq, while developed for RNA-seq data, provides an optimized mode for microarrays. All other methods do not distinguish between the two platform types. In microarray mode, CIBERSORT will perform quantile normalization of the input data as an additional processing step. quanTIseq includes additional genes in the signature matrix, which are excluded in RNA-seq mode. As default, *immundeconv* runs the methods in RNA-seq mode. To switch to microarray-mode, set the option `arrays` to `TRUE`:

```
>deconvolute(gene_expression, method="cibersort",arrays=TRUE)
```

4. Excluding genes.

   We have suggested that excluding genes with low cell-type specificity can reduce spillover-effects and improve overall results [4]. If you find a gene to impair the results, you can exclude it using the `rmgenes` parameter. A possible reason for a gene to hamper the analysis can be an high expression of the gene in unrelated cell types that are present in your sample but have not been considered in the signature matrices of the deconvolution methods. For instance, we identified five genes that are expressed by both B cells and plasmacytoid dendritic cells (pDCs). As the methods considered in our benchmark have not been trained on pDCs, these genes can lead to an overestimation of B cells in the presence of pDCs. To exclude these five genes, use.

```
>deconvolute(gene_expression, method="quantiseq", rmgenes = c
("TCL1A", "TCF4", "CD37", "SPIB", "IRF8"))
```

References

1. Fridman WH, Pagès F, Sautès-Fridman C, Galon J (2012) The immune contexture in human tumours: impact on clinical outcome. Nat Rev Cancer 12:298–306
2. Fridman WH, Zitvogel L, Sautès-Fridman C, Kroemer G (2017) The immune contexture in cancer prognosis and treatment. Nat Rev Clin Oncol 14:717–734
3. Petitprez F, Sun CM, Lacroix L (2018) Quantitative analyses of the tumor microenvironment composition and orientation in the era of precision medicine. Front Oncol 8:390. https://doi.org/10.3389/fonc.2018.00390
4. Sturm G, Finotello F, Petitprez F et al (2019) Comprehensive evaluation of computational cell-type quantification methods for immuno-oncology. Bioinformatics 35:i436–i445
5. Zhang J, Baran J, Cros A et al (2011) International cancer genome consortium data portal—a one-stop shop for cancer genomics data. Database 2011:bar026
6. Newman AM, Liu CL, Green MR et al (2015) Robust enumeration of cell subsets from tissue expression profiles. Nat Methods 12:453
7. Racle J, de Jonge K, Baumgaertner P et al (2017) Simultaneous enumeration of cancer and immune cell types from bulk tumor gene expression data. elife 6:e26476
8. Becht E, Giraldo NA, Lacroix L et al (2016) Estimating the population abundance of tissue-infiltrating immune and stromal cell populations using gene expression. Genome Biol 17:218
9. Finotello F, Mayer C, Plattner C et al (2019) Molecular and pharmacological modulators of the tumor immune contexture revealed by deconvolution of RNA-seq data. Genome Med 11:34
10. Li B, Severson E, Pignon J-C et al (2016) Comprehensive analyses of tumor immunity: implications for cancer immunotherapy. Genome Biol 17:174
11. Aran D, Hu Z, Butte AJ (2017) xCell: digitally portraying the tissue cellular heterogeneity landscape. Genome Biol 18:220

Chapter 17

EPIC: A Tool to Estimate the Proportions of Different Cell Types from Bulk Gene Expression Data

Julien Racle and David Gfeller

Abstract

Gene expression profiling is nowadays routinely performed on clinically relevant samples (e.g., from tumor specimens). Such measurements are often obtained from bulk samples containing a mixture of cell types. Knowledge of the proportions of these cell types is crucial as they are key determinants of the disease evolution and response to treatment. Moreover, heterogeneity in cell type proportions across samples is an important confounding factor in downstream analyses.

Many tools have been developed to estimate the proportion of the different cell types from bulk gene expression data. Here, we provide guidelines and examples on how to use these tools, with a special focus on our recent computational method EPIC (Estimating the Proportions of Immune and Cancer cells). EPIC includes RNA-seq-based gene expression reference profiles from immune cells and other nonmalignant cell types found in tumors. EPIC can additionally manage user-defined gene expression reference profiles. Some unique features of EPIC include the ability to account for an uncharacterized cell type, the introduction of a renormalization step to account for different mRNA content in each cell type, and the use of single-cell RNA-seq data to derive biologically relevant reference gene expression profiles. EPIC is available as a web application (http://epic.gfellerlab.org) and as an R-package (https://github.com/GfellerLab/EPIC).

Key words Gene expression analysis, Cell fraction predictions, Computational biology, RNA-seq deconvolution, Immunoinformatics, Tumor immune microenvironment

1 Introduction

Most gene expression analyses of clinically relevant samples are performed on tissue materials that contain a mixture of different cell types. This is best exemplified in cancer research, where the tumor microenvironment is composed of various cell types [1, 2]: cancer cells form the main part of it, but different immune cells (e.g., CD4+ T cells, CD8+ T cells, macrophages), stromal cells, and endothelial cells are also present. Evidence has been accumulated over the past years showing that the complex interplay between these various cell types is a major driver of cancer progression, response to therapy, and patient survival [3, 4]. Knowing the proportions of the cell types in a tumor is therefore of crucial

Sebastian Boegel (ed.), *Bioinformatics for Cancer Immunotherapy: Methods and Protocols*, Methods in Molecular Biology, vol. 2120, https://doi.org/10.1007/978-1-0716-0327-7_17,

Table 1
Main repositories containing gene expression data

Name	Description	URL	Ref.
Gene Expression Omnibus (GEO)	Public genomics data repository, containing many microarray and RNA-seq datasets	https://www.ncbi.nlm.nih.gov/geo/	[36]
ArrayExpress	Public genomics data repository, containing many microarray and RNA-seq datasets	https://www.ebi.ac.uk/arrayexpress	[37]
cBioPortal for Cancer Genomics	Portal containing large-scale cancer genomics data sets, which can be analyzed on the portal directly or downloaded	https://www.cbioportal.org/	[33, 34]
International Cancer Genome Consortium (ICGC)	A consortium providing comprehensive genomics data for many cancer types	https://icgc.org/	[38]
Genomic Data Commons Data portal	A data portal from the National Cancer Institute that hosts all the data that had been generated by the former Cancer Genome Atlas project (TCGA), as well as some additional datasets	https://portal.gdc.cancer.gov/	[39]

These repositories often also host additional types of data. GEO and ArrayExpress contain data from many experiments, not only cancer-related. The three other repositories contain only cancer-related datasets

importance, both for fundamental understanding of cancer and for translational research. Moreover, the presence of distinct cell types in different tissue samples is a major confounding factor for differential expression and correlative analyses of gene expression data [5, 6].

In particular, knowing the different cell fractions would enable more robust analyses of cancer genomics data, where bulk tumor samples were analyzed without isolating the various cell types. Unfortunately, these proportions are not always measured experimentally (e.g., with flow cytometry or immunohistochemistry). Bearing in mind the large amount of gene expression data deposited in publicly available repositories (Table 1), researchers have developed many computational methods to reanalyze such data and estimate the abundance of various cell types found in these samples [7–23]. In many cases, these methods have been used to estimate the abundance of various immune cell types, but the methods can also be applied to other cell types. These methods can be classified as "gene marker-based" or "deconvolution-based." Some of the main gene marker-based methods are ESTIMATE [8], MCPcounter [13], TIminer [17], xCell [18], and the approach of Danaher and colleagues [15] using a set of gene markers derived from TCGA data. These methods are usually based on gene set enrichment analysis [24, 25]; they rely therefore only on a list of genes describing the various cell types. In general, they are more robust to noise in the data, and they could detect cell types present at lower abundance, but they are only semiquantitative and can only describe the relative abundances per cell type between the samples. As such, it is not possible to obtain quantitative values for the true abundance of cell types, which limits comparisons between cell types and validation on external data such as flow cytometry. In contrary, deconvolution-based methods are more quantitative, but they rely on the knowledge of reference gene expression profiles for the various cell types to deconvolve. Some of the main deconvolution methods are EPIC [16], CIBERSORT [12], and TIMER [14], although many new methods have been developed in the last few years [19–23]. Reviews giving more details about these methods are available elsewhere [26, 27].

Many of these methods were developed based on microarray data and are still using reference gene expression profiles obtained from microarrays, despite being often applied to bulk RNA-seq data. In addition, several of these methods use reference gene expression profiles derived from blood (which may not be relevant for solid tissues such as tumors), do not account for variable mRNA content per cell, and require information (i.e., reference profiles) about each cell type present in the sample. Additional challenges include the lack of validation for some cell types included in some methods, or the use of in silico or artificial in vitro mixtures for benchmarking, which may not represent biologically relevant proportions and does not account for the variability induced by the in vivo mixing of different cell types (e.g., T cells are known to modify their gene expression profiles in the presence of antigens expressed on tumor cells).

EPIC method [16] was developed to address these shortcomings, with a special focus on cell fraction predictions in tumor samples. Specifically, it includes reference gene expression profiles (RNA-seq) from blood and from tumor-infiltrating cells for all main nonmalignant cell types found in human tumor samples (including various immune cell types, as well as stromal and endothelial cells). As a source for the reference gene expression profiles for absolute cell fraction predictions, we showed, for the first time to our knowledge, that these could be obtained from single-cell RNA-seq data. This idea was then reused in more recent publications [21, 23, 28]. Similar to the concept of "effective RNA fraction" introduced for the deconvolution of microarray data by Liebner and colleagues [29], EPIC includes a step of renormalization by the cell-type-specific mRNA content which, as we showed, is improving the cell proportion prediction accuracy in RNA-seq deconvolution [16]. This improvement in RNA-seq deconvolution was later on confirmed in subsequent studies [19, 20, 23]. A third improvement introduced in EPIC is the ability to consider other "uncharacterized" cell types. This enabled us to estimate the proportion of the cells for which reference gene expression profiles were not available (see also a recent study exploring the same mathematical framework [19]). In practice, these cells correspond mostly to the cancer cells when working with bulk tumor samples because the other main cell types found in a tumor have their reference gene expression profiles defined in EPIC. EPIC thus does not define any reference gene expression profile for the cancer cells per se, since these are known to be much more variable between patients and tumor types than the rest of the cell types present in a tumor. Importantly, an independent recent benchmark study showed that EPIC was the most accurate method to predict most cell types present in bulk tumor RNA-seq data [30].

Here, we present an example application of EPIC, showing step by step how to apply this tool on a publicly available RNA-seq dataset. This example is inspired by the study of Angelova and colleagues [31], where they observed differences in immune infiltration of colorectal cancers that harbor a microsatellite instability or not. We will describe in parallel how to do such an analysis through EPIC web application or through EPIC R-package. Figure 1 illustrates the schematic workflow of EPIC.

2 Materials

In order to be available for a broad audience, EPIC is written as a package for *R* (available for Windows, Mac OS, or Linux) and is also available as a web application. Here, we will outline both uses of EPIC. Note however that the R-package is preferred for large datasets, both in terms of speed and the availability of additional advanced options.

2.1 Hardware

There is no limitation on the hardware or operating system as EPIC can be run through a web application.

2.2 Software

2.2.1 Web Application

For the web application, no software installation is needed. It is simply needed to have an internet browser available (it has been tested to work with the most common browsers).

2.2.2 R-package

If using the R-package, it is needed to first install the free software for statistical computing and graphics: *R* (version 3.2 or newer, https://www.r-project.org).

In addition, the R-package *devtools* (version 1.13 or newer) need to be installed; this can simply be done from within *R* with the following command:

```
install.packages("devtools")
```

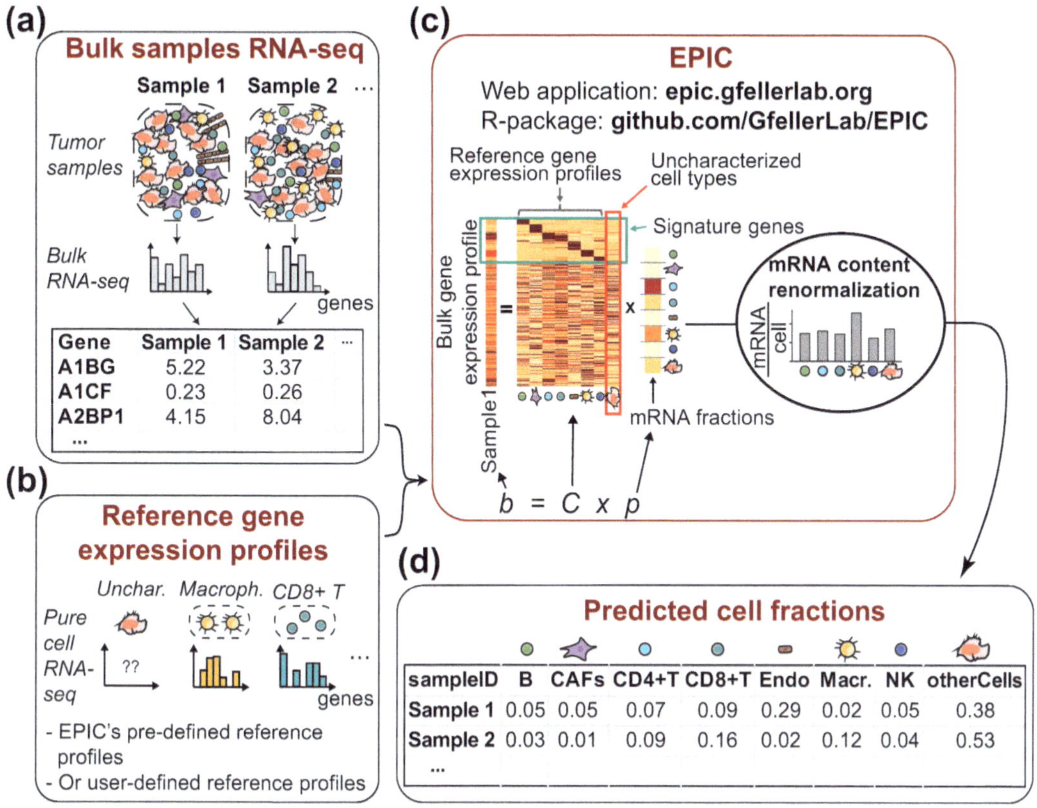

Fig. 1 Workflow from EPIC. (**a**) Input data are a matrix of the RNA-seq gene expression data measured in bulk samples. (**b**) Typically, predefined cell reference gene expression profiles are used for the deconvolution. Alternatively, users can define their own reference profiles representing any cell type. These could be obtained from RNA-seq data of pure cell types or from averaging single-cell RNA-seq data of a given cell type. (**c**) EPIC estimates the fraction of mRNA contributed to the bulk by each cell type. This is based on a constrained weighted least square optimization on the set of signature genes representing the different cell types. As a final step, EPIC renormalizes the mRNA fractions based on the mRNA content from each cell type, resulting in cell fractions. (**d**) The result returned by EPIC is a matrix containing the predicted cell fractions for each sample. It includes the *otherCells* (i.e., cancer cells when applied to tumor samples) for which no reference gene expression profile is defined. (Figure adapted from [16])

2.3 Getting EPIC

2.3.1 Web Application

To run EPIC web application, go to the url: http://epic.gfellerlab.org.

2.3.2 R-package

The source code of EPIC is available via a GitHub repository (https://github.com/GfellerLab/EPIC). To install this package, open the software *R*, and from within this software, enter the following command:

```
devtools::install_github("GfellerLab/EPIC", build_vignet-
tes=TRUE)
```

This will install the package in the default repository, building additionally the vignette of the help documentation.

2.4 Get RNA-Seq Data

In this case study, we will use colorectal cancer data obtained by The Cancer Genome Atlas (TCGA) project [32], downloaded through cBioPortal [33, 34]. EPIC can however be used on any other RNA-seq data, either generated in-house or obtained from public datasets (*see* Table 1 for a list of the main repositories containing such data).

1. Go to the url: https://www.cbioportal.org/datasets.
2. Download the data from the study named "*Colorectal Adenocarcinoma (TCGA, Nature 2012)*" (click on the small icon representing a download symbol next to the study name).
3. Uncompress the data after they have been downloaded (on Windows computers, it might be needed to first install a software to decompress such data, for example, 7-Zip: https://www.7-zip.org/).
4. For this example, we only need to keep the files called "*data_clinical_sample.txt*" and "*data_RNA_Seq_expression_median.txt.*" The first one tells, among else, which sample is from a microsatellite stable tumor (MSS), or with a low or high microsatellite instability (MSI-L and MSI-H, respectively). The other file ("*data_RNA_Seq_expression_median.txt*") gives the RNA-seq gene expression value in Reads Per Kilobase Million (RPKM).

If using other RNA-seq datasets, please note that the mathematics behind EPIC assume that the input RNA-seq data are normalized as Transcripts Per Million (TPM), Reads Per Kilobase Million (RPKM), or Fragments Per Kilobase Million (FPKM) (*see* **Notes 1** and **2**), and this is the value that must be used in the input, not a log-transformed value.

3 Methods

3.1 Running EPIC

3.1.1 Web Application

1. The web application needs as input a tab-delimited text file containing the RNA-seq data. This file must have $N_{samples} + 1$ columns: the first column gives the gene names (*see* **Note 3**), and each other column represents a sample, giving the gene expression found in it. The first row must be a header indicating *gene_name* (or anything) for the first column and the name of each sample for the remaining columns. See Fig. 1a. In the example file "*data_RNA_Seq_expression_median.txt*," instead of one column for the gene names, there are two ("*Hugo_Symbol*" and "*Entrez_Gene_Id*" columns). You should therefore open this file (e.g., with Microsoft Excel or any other spreadsheet editor), you should then delete the second column ("*Entrez_Gene_Id*") and save the updated file under the name "*data_RNA_Seq_expression_median_singleName.txt*", for example.

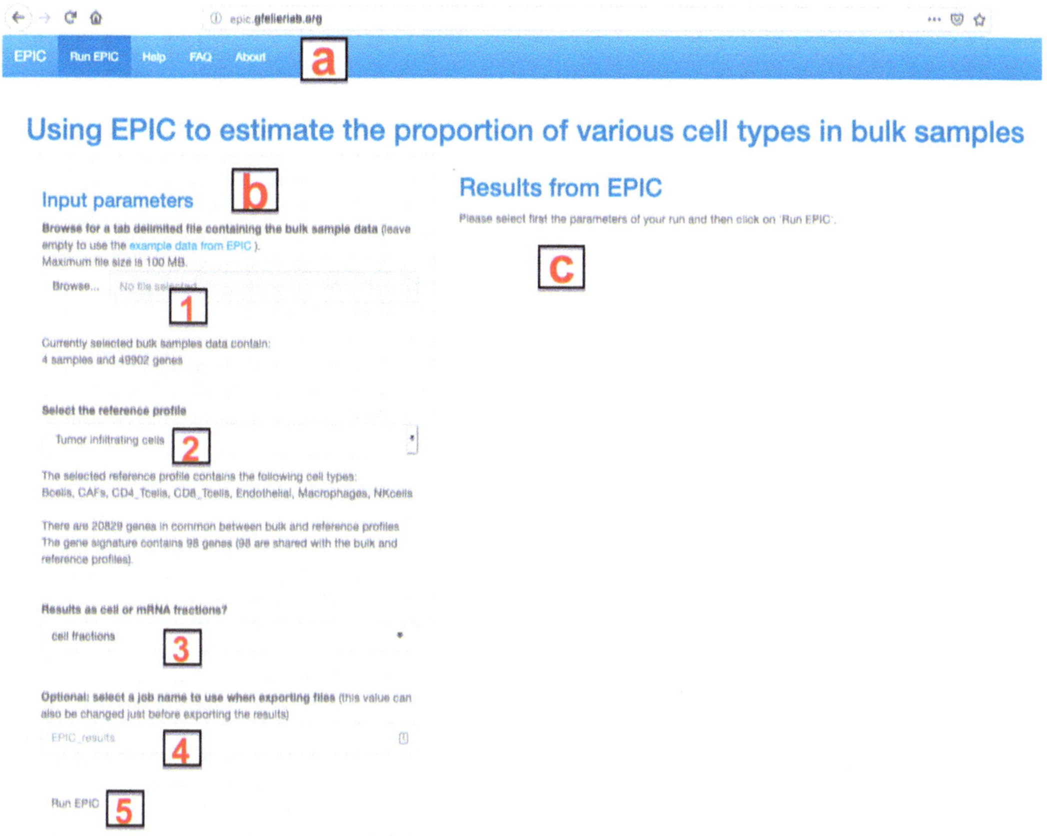

Fig. 2 Screenshot of EPIC web application. It is composed of three regions lettered (**a–c**): (**a**) A top navigation bar to switch between the application and help documents; (**b**) the input area; (**c**) the results area (filled once EPIC has been launched)

2. Go then to EPIC web application page (http://epic.gfellerlab.org) and define the input parameters of the run: (1) Upload the file "*data_RNA_Seq_expression_median_singleName.txt*" (Fig. 2—input (1)). This can take some time due to file size. (2) Give a job name to this run, for example, "*TCGA_colorectal*" (Fig. 2—input (4)). (3) The other parameters are the best recommended values and can remain as is (*see* **Notes 4** and **5**). (4) Click then on "Run EPIC" to launch the computations (Fig. 2—input (5)). This can also take some time depending on data size.
3. Once the run is finished, a table giving the percentage of each cell type in each sample appears (filling the area (c) of Fig. 2). Note that some warning messages might appear in yellow at the bottom-right corner of the page. These are usually not important and are explained in the FAQ page of the web application (*see* also **Notes 5** and **6**).
4. One can display the results either as a "Table" of the cell fractions or as summary "Figures" (Fig. 3), depending on the

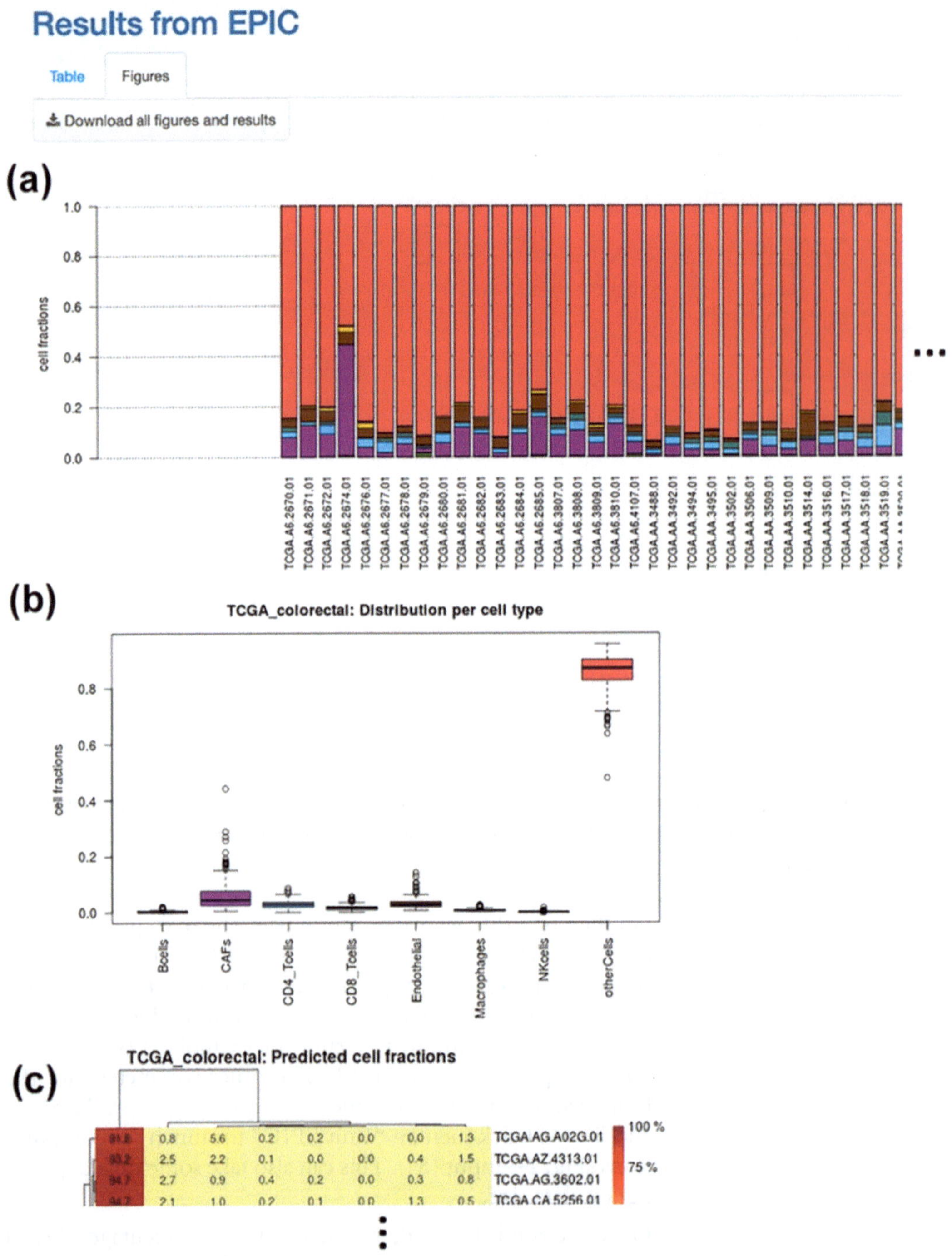

Fig. 3 Results page from EPIC web application. This corresponds to a zoom into the area (c) from Fig. 2 after running EPIC. (**a**) Bar plot of the fraction of each cell type present in each sample. (**b**) Boxplot of the distribution of each cell type between the samples—colors are the same than in above bar plot. (**c**) Heatmap of the fraction of each cell type per sample after hierarchical clustering of the cell types and samples

tab that is selected in the results section from EPIC (area (c) of Fig. 2). Instead of the cell fractions, all the results can also be displayed as mRNA fractions (based on what is selected in the input (3) from Fig. 2; *see* **Note 5**).

5. The results can next be downloaded by clicking on the "*Download all figures and results*" button.

3.1.2 R-package

1. Open the software *R*.
2. Load the RNA-seq data into a matrix "*bulk*" of dimensions "N_{genes} x $N_{samples}$" containing the gene expression values for each gene in each sample. This matrix needs to have row names corresponding to the gene names (*see* **Note 3**) and optionally column names giving the sample names. Assuming that the example data is present in the folder "*path/to/data/*," we can get this matrix with the following commands in *R*:

```
tdata <- read.delim(file="path/to/data/data_RNA_Seq_expres-
sion_median.txt", as.is=T, check.names=F)
bulk <- as.matrix(tdata[,-(1:2)])
rownames(bulk) <- tdata[,1]
```

Note that the first two columns of "*tdata*" are not included in the "*bulk*" matrix because these are two columns of gene names. Contrary to what was needed for the web application, we do not need to first open the data with some external spreadsheet editor to remove the second gene name column because this can be done directly from within *R*.

3. Run EPIC using the recommended parameters (*see* also **Note 4**):

```
res <- EPIC::EPIC(bulk)
```

Some warning messages might appear during the run of EPIC. These are usually not important and are explained in the FAQ section of EPIC documentation (accessible, for example, in *R* through the command "*vignette('EPIC')*"—*see* also **Notes 5** and **6**).

4. The main results from EPIC, giving the cell fractions, are then available under "*res$cellFractions.*" This is a matrix of dimensions "$N_{samples}$ x $N_{cell\text{-}types}$" giving the fraction of each cell type in each sample. The mRNA fractions are also available under "*res$mRNAProportions*" (*see* **Note 5**).

3.2 Interpreting the Output

The result from EPIC is a matrix giving the most probable cell fractions of each cell type present in each sample (instead of the cell fractions, the mRNA fractions can also be outputted, as explained above and in **Note 5**). These values correspond to fractions or proportions, not percentages (i.e., they are always smaller or equal to 1).

The estimation is obtained in each sample with help of a constrained weighted least square regression, searching for the best set of cell fractions that explain the observed gene expression values of a list of signature genes (Fig. 1c). The default reference profile (named "*TRef*" in the R-package) contains reference gene expression profiles for *B cells*, *cancer-associated fibroblasts* (*CAFs*), *CD4+ T cells*, *CD8+ T cells*, *endothelial cells*, *macrophages*, and *NK cells*, profiles that have been obtained from tumor-infiltrating cell samples (*see* **Note 7**). In addition to these cell types, the fraction of "*otherCells*" is returned by EPIC: this represents the remaining cells (making that the sum all cell fractions is always equal to 1 per sample); these cells mainly correspond to cancer cells when applying EPIC on tumor samples.

3.3 Example of Further Processing EPIC Outputs

The results from EPIC can then be integrated into a broader analysis. For example, in this case study, we are interested to see if there are differences in cell fractions between MSI and MSS tumors.

But first, from the summary boxplot obtained in the web application (Fig. 3b), we can already note that the largely most-abundant cell type is the *otherCells* in all samples. This was expected because these cells represent the cancer cells and only tumor samples containing at least 80% tumor nuclei were processed in TCGA.

To analyze the difference in infiltration based on microsatellite instability, we will assume that we used EPIC R-package and will use *R* to then draw figures showing these differences. Below are the various additional steps needed from within *R*:

1. Load the previously downloaded clinical data and rework it slightly to remove the cases not annotated to MSI or MSS:

```
clinData <- read.delim(file="path/to/data/data_clinical_-sample.txt", skip=4)
clinData <- clinData[match(colnames(bulk), as.character(clinData$SAMPLE_ID)),] # Put this data in the same order than bulk, keeping only samples with RNA-seq.
levels(clinData$MSI_STATUS)[levels(clinData$MSI_STATUS) %in% c("", "Not Evaluable")] <- NA # Remove unknown MSS/MSI status.
```

2. Make multiple boxplots of the fractions predicted for each cell type, grouping the samples per microsatellite instability status:

```
dev.new(width=10, height=6, noRStudioGD=T)
par(mfrow=c(2,4), oma=c(0,0,1,0))
for (cCell in colnames(res$cellFractions)){
  boxplot(res$cellFractions[,cCell] ~ clinData$MSI_STATUS,
```

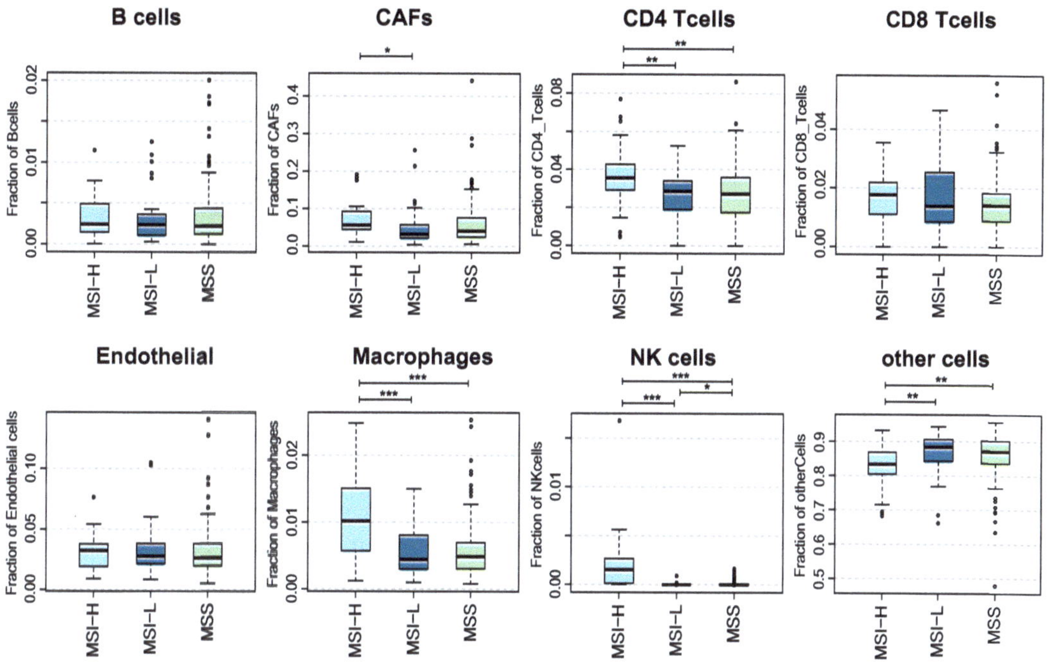

Fig. 4 Comparison of the cell fractions based on microsatellite instability. Boxplots indicate the median, upper, and lower quartiles of the predicted cell fractions for the cell type indicated in the title. Two-sided Wilcoxon signed rank test is performed and adjusted for multiple testing based on Holm's procedure [35] ($*$: p-value <0.05, $**$: p-value <0.01, $***$: p-value <0.001)

```
    ylab=paste("Fraction of", cCell), col=c("lightskyblue2",
       "deepskyblue4", "darkseagreen2"), las=3, main=cCell)
}
title(main=paste("Comparison of the fraction of cells present in
tumors",
   "based on the MSI status"), outer=T, line=-0.5)
```

These results first show that the MSS and MSI-L tumors display a similar immune contexture (Fig. 4). Then, similar to what was observed by Angelova and colleagues [31], we observe a different immune contexture for the MSI-H tumors: there are significantly higher infiltrations of CD4+ T cells, macrophages, and NK cells in these tumors (Fig. 4). CD8+ T cells follow the same trend, while B cells do not show differences. With respect to the nonimmune cells, we observe that the CAFs show a trend of higher infiltration in the MSI-H tumors and the endothelial cells do not show differences. Finally, as discussed above, the most-abundant cells are the *otherCells*, with a proportion above 0.8 in each sample, and this proportion is significantly lower in the MSI-H tumors (Fig. 4).

Even though the fraction of each immune cell type was very low in these samples (<0.05 in most samples), we could see that EPIC was able to recapitulate the subtle differences in infiltration existing between microsatellite highly instable versus stable tumors. This was just an example case study, but thanks to its ease of use and the ability to easily include other reference gene expression profiles (*see* **Note 4**), EPIC can be used in a broad variety of applications.

4 Notes

1. EPIC does internally a scaling of the gene expression, allowing RPKM (or FPKM) to be invisibly transformed to TPM without the need to specify which unit was used for the gene expression. For this reason, the input bulk gene expression matrix should contain all the genes measured in the RNA-seq experiment, not only the genes that are present in the signatures of the various cell types. If needed, the R-package has an option not to do such a normalization, which is only recommended if both the bulk and reference gene expression profiles are inputted and if a consistent normalization was performed between them.
2. If custom reference gene expression profiles are used as input to EPIC, the gene expression values can also be given in other units than TPM/RPKM/FPKM (e.g., directly as counts), as long as the expression values are in the same units for the input bulk gene expression matrix and reference gene expression profiles. Note however that the mathematics behind EPIC assume that the bulk gene expression can be written as a linear relation between the cell proportions and their reference gene expression profiles (Fig. 1c); therefore, it is not recommended to use log-transformed data even when using custom reference gene expression profiles.
3. EPIC does not do any gene name conversion. Therefore, when using the predefined reference gene expression profiles, the gene names in the bulk sample need to correspond to the official names (also known as HUGO symbols). Further, all genes measured should be included instead of a subset (*see* **Note 1**). As a manual check that the correct gene names are used, the number of genes in common between the bulk and reference profiles is indicated by EPIC (there should be in principle approximately 20,000 genes in common or more). The number of signature genes in common is also indicated (if only few signature genes are missing, this should be fine, but ideally all signature genes should be present in the bulk). When using user-defined reference gene expression profiles, the gene

names can be anything, as long as the same names are used for the bulk and reference profiles and for the signature gene names.

4. EPIC comes along with two predefined reference gene expression profiles, and it is possible to define other profiles to use with EPIC. (1) The main reference profile (called "*TRef*" in the R-package) is based on human tumor-infiltrating cells obtained from single-cell RNA-seq data. It includes the following cell types: *B cells*, *cancer-associated fibroblasts* (*CAFs*), *CD4+ T cells*, *CD8+ T cells*, *endothelial cells*, *macrophages*, and *NK cells*. (2) Another reference profile (called "*BRef*" in the R-package) is based on human blood circulating immune cells obtained from bulk RNA-seq data of sorted immune cells. It includes the following cell types: *B cells*, *CD4+ T cells*, *CD8+ T cells*, *monocytes*, *neutrophils*, and *NK cells*. When studying tumor samples, it is recommended to use the *TRef* profile instead because it contains more cell types found in tumors, and tumor-infiltrating cells may express multiple of their genes at different levels than in resting conditions. (3) In addition to these reference profiles, users can build their own reference gene expression profiles and use them either with the web application or the R-package. This could be used for example when studying other cell types or to do the deconvolution in other organisms. Please read EPIC's help documents to know the format needed for such a user-defined reference profile.

5. The mathematics behind EPIC (and most other cell type deconvolution methods) assume that we can write the bulk gene expression data as a linear combination of the reference gene expression profiles (Fig. 1c). Moreover, RNA-seq returns relative gene expression values per sample instead of absolute values, such that some form of data normalization is needed. We additionally observed that different cell types have different abundances of mRNA per cell [16]. For all these reasons, we showed that the linear combination coefficients estimated during the deconvolution correspond in fact to the proportion of mRNA that each cell type contributes to the bulk sample instead of directly representing the cell fractions [16]. EPIC accounts for that: it first estimates the mRNA proportions contributed by each cell type and then renormalizes the results based on the mRNA content from each cell type in order to return the cell fractions (Fig. 1c), which is often the most useful information. Both values are nevertheless returned by EPIC, and users can select to use the one or the other for further processing. In case the mRNA content from some cell types is unknown, a warning message appears during the run of EPIC and an average value is used for these cell types (which is currently the case for the *CAFs* and *endothelial cells*).

If the user uses his own reference gene expression profiles, defining different cell types than those found in the predefined reference profiles, he should ideally also define the mRNA per cell values for each cell type. If the user did not measure the mRNA content per cell in such a case, then the mRNA proportions can still be used for further analyses, keeping in mind that these values might not correspond exactly to what a FACS experiment would return, for example.

6. During a run of EPIC, some warning message might indicate that there was not a full convergence of the optimization for some samples. These could be outlier samples, in which the expression of some signature genes did not fully follow the expected distribution based on the reference profiles, possibly due to some outlier genes or maybe due to a too low infiltration of the reference cell types in the given sample. From our experience, it seems that the cell proportions are also well predicted in these outlier samples, but care should be taken when analyzing them.
7. As explained in **Note 4**, EPIC comes along with two sets of predefined reference gene expression profiles, but it can be run with any user-defined reference profile describing the cells of interest. Concerning the predefined profiles, we focused on the main nonmalignant cell types observed in tumors and validated our predictions on biologically relevant samples. For the blood circulating immune cells, we could validate the predictions from each defined cell type in bulk samples of blood from human donors. With respect to the tumor-infiltrating cell reference profiles, we could first validate the predictions of *B cells*, *CD4 T cells*, *CD8+ T cells*, *macrophages*, *NK cells*, and the uncharacterized cancer cells with help of human bulk tumor samples. We could additionally validate the predictions for all these cell types and for the *CAFs* and *endothelial cells*, with help of cross-validation within single-cell RNA-seq data of human tumors. The addition of more cell subtypes (i.e., CD4+ T helper and T reg) was discussed in our work [16], showing that, currently, deconvolution was not sufficiently accurate to include them in the results. Therefore, users attempting to work with reference gene expression profiles from other cell types in EPIC (or with other tools that may include them without necessarily providing direct experimental validations) should be aware that predictions may not be accurate and need to be experimentally validated. Our work also indicated that cell types present at very low proportions are difficult to predict [16] and conclusions drawn from such predictions should be handled with care and always experimentally validated.

References

1. Hanahan D, Weinberg RA (2011) Hallmarks of cancer: the next generation. Cell 144:646–674. https://doi.org/10.1016/j.cell.2011.02.013
2. Joyce JA, Fearon DT (2015) T cell exclusion, immune privilege, and the tumor microenvironment. Science 348:74–80. https://doi.org/10.1126/science.aaa6204
3. Fridman WH, Pagès F, Sautès-Fridman C, Galon J (2012) The immune contexture in human tumours: impact on clinical outcome. Nat Rev Cancer 12:298–306. https://doi.org/10.1038/nrc3245
4. Croci DO, Zacarías Fluck MF, Rico MJ et al (2007) Dynamic cross-talk between tumor and immune cells in orchestrating the immunosuppressive network at the tumor microenvironment. Cancer Immunol Immunother 56:1687–1700. https://doi.org/10/fj4m65
5. Shen-Orr SS, Gaujoux R (2013) Computational deconvolution: extracting cell type-specific information from heterogeneous samples. Curr Opin Immunol 25:571–578. https://doi.org/10.1016/j.coi.2013.09.015
6. Hagenauer MH, Schulmann A, Li JZ et al (2018) Inference of cell type content from human brain transcriptomic datasets illuminates the effects of age, manner of death, dissection, and psychiatric diagnosis. PLoS One 13:e0200003. https://doi.org/10.1371/journal.pone.0200003
7. Repsilber D, Kern S, Telaar A et al (2010) Biomarker discovery in heterogeneous tissue samples -taking the in-silico deconfounding approach. BMC Bioinformatics 11:1. https://doi.org/10.1186/1471-2105-11-27
8. Yoshihara K, Shahmoradgoli M, Martínez E et al (2013) Inferring tumour purity and stromal and immune cell admixture from expression data. Nat Commun 4:2612. https://doi.org/10.1038/ncomms3612
9. Zhong Y, Wan Y-W, Pang K et al (2013) Digital sorting of complex tissues for cell type-specific gene expression profiles. BMC Bioinformatics 14:1. https://doi.org/10.1186/1471-2105-14-89
10. Quon G, Haider S, Deshwar AG et al (2013) Computational purification of individual tumor gene expression profiles leads to significant improvements in prognostic prediction. Genome Med 5:29. https://doi.org/10.1186/gm433
11. Gong T, Szustakowski JD (2013) DeconRNASeq: a statistical framework for deconvolution of heterogeneous tissue samples based on mRNA-Seq data. Bioinformatics 29:1083–1085. https://doi.org/10.1093/bioinformatics/btt090
12. Newman AM, Liu CL, Green MR et al (2015) Robust enumeration of cell subsets from tissue expression profiles. Nat Methods 12:453–457. https://doi.org/10.1038/nmeth.3337
13. Becht E, Giraldo NA, Lacroix L et al (2016) Estimating the population abundance of tissue-infiltrating immune and stromal cell populations using gene expression. Genome Biol 17:218. https://doi.org/10.1186/s13059-016-1070-5
14. Li B, Severson E, Pignon J-C et al (2016) Comprehensive analyses of tumor immunity: implications for cancer immunotherapy. Genome Biol 17:174. https://doi.org/10.1186/s13059-016-1028-7
15. Danaher P, Warren S, Dennis L et al (2017) Gene expression markers of Tumor Infiltrating Leukocytes. J Immunother Cancer 5:18. https://doi.org/10.1186/s40425-017-0215-8
16. Racle J, Jonge K, de Baumgaertner P et al (2017) Simultaneous enumeration of cancer and immune cell types from bulk tumor gene expression data. elife 6:e26476. https://doi.org/10.7554/eLife.26476
17. Tappeiner E, Finotello F, Charoentong P et al (2017) TIminer: NGS data mining pipeline for cancer immunology and immunotherapy. Bioinformatics 33:3140–3141. https://doi.org/10.1093/bioinformatics/btx377
18. Aran D, Hu Z, Butte AJ (2017) xCell: digitally portraying the tissue cellular heterogeneity landscape. Genome Biol 18:220. https://doi.org/10.1186/s13059-017-1349-1
19. Finotello F, Mayer C, Plattner C et al (2019) Molecular and pharmacological modulators of the tumor immune contexture revealed by deconvolution of RNA-seq data. Genome Med 11:34. https://doi.org/10.1186/s13073-019-0638-6
20. Monaco G, Lee B, Xu W et al (2019) RNA-seq signatures normalized by mRNA abundance allow absolute deconvolution of human immune cell types. Cell Rep 26:1627–1640.e7. https://doi.org/10/gft8b3
21. Frishberg A, Peshes-Yaloz N, Cohn O et al (2019) Cell composition analysis of bulk genomics using single-cell data. Nat Methods 16:327–332. https://doi.org/10/gfw8kp
22. Hunt GJ, Freytag S, Bahlo M, Gagnon-Bartsch JA (2019) dtangle: accurate and robust cell type deconvolution. Bioinformatics

35:2093–2099. https://doi.org/10.1093/bioinformatics/bty926

23. Wang X, Park J, Susztak K et al (2019) Bulk tissue cell type deconvolution with multi-subject single-cell expression reference. Nat Commun 10:380. https://doi.org/10/gfxgjq
24. Subramanian A, Tamayo P, Mootha VK et al (2005) Gene set enrichment analysis: a knowledge-based approach for interpreting genome-wide expression profiles. Proc Natl Acad Sci 102:15545–15550. https://doi.org/10.1073/pnas.0506580102
25. Barbie DA, Tamayo P, Boehm JS et al (2009) Systematic RNA interference reveals that oncogenic KRAS-driven cancers require TBK1. Nature 462:108–112. https://doi.org/10.1038/nature08460
26. Finotello F, Trajanoski Z (2018) Quantifying tumor-infiltrating immune cells from transcriptomics data. Cancer Immunol Immunother 67:1031–1040. https://doi.org/10.1007/s00262-018-2150-z
27. Petitprez F, Sun C-M, Lacroix L et al (2018) Quantitative analyses of the tumor microenvironment composition and orientation in the era of precision medicine. Front Oncol 8:390. https://doi.org/10.3389/fonc.2018.00390
28. Schelker M, Feau S, Du J et al (2017) Estimation of immune cell content in tumour tissue using single-cell RNA-seq data. Nat Commun 8:2032. https://doi.org/10.1038/s41467-017-02289-3
29. Liebner DA, Huang K, Parvin JD (2014) MMAD: microarray microdissection with analysis of differences is a computational tool for deconvoluting cell type-specific contributions from tissue samples. Bioinformatics 30:682–689. https://doi.org/10.1093/bioinformatics/btt566
30. Sturm G, Finotello F, Petitprez F et al (2019) Comprehensive evaluation of cell-type quantification methods for immuno-oncology. Bioinformatics 35:i436–i445. https://doi.org/10.1093/bioinformatics/btz363
31. Angelova M, Charoentong P, Hackl H et al (2015) Characterization of the immunophenotypes and antigenomes of colorectal cancers reveals distinct tumor escape mechanisms and novel targets for immunotherapy. Genome Biol 16:64. https://doi.org/10.1186/s13059-015-0620-6
32. The Cancer Genome Atlas Network (2012) Comprehensive molecular characterization of human colon and rectal cancer. Nature 487:330–337. https://doi.org/10.1038/nature11252
33. Cerami E, Gao J, Dogrusoz U et al (2012) The cBio cancer genomics portal: an open platform for exploring multidimensional cancer genomics data. Cancer Discov 2:401–404. https://doi.org/10.1158/2159-8290
34. Gao J, Aksoy BA, Dogrusoz U et al (2013) Integrative analysis of complex cancer genomics and clinical profiles using the cBioPortal. Sci Signal 6:pl1. https://doi.org/10.1126/scisignal.2004088
35. Holm S (1979) A simple sequentially rejective multiple test procedure. Scand J Stat 6:65–70
36. Edgar R, Domrachev M, Lash AE (2002) Gene expression omnibus: NCBI gene expression and hybridization array data repository. Nucleic Acids Res 30:207–210. https://doi.org/10.1093/nar/30.1.207
37. Kolesnikov N, Hastings E, Keays M et al (2015) ArrayExpress update—simplifying data submissions. Nucleic Acids Res 43:D1113–D1116. https://doi.org/10.1093/nar/gku1057
38. Zhang J, Baran J, Cros A et al (2011) International cancer genome consortium data portal—a one-stop shop for cancer genomics data. Database J Biol Databases Curation 2011:bar026. https://doi.org/10.1093/database/bar026
39. Grossman RL, Heath AP, Ferretti V et al (2016) Toward a shared vision for cancer genomic data. N Engl J Med 375:1109–1112. https://doi.org/10/gfxgjx

Chapter 18

Computational Deconvolution of Tumor-Infiltrating Immune Components with Bulk Tumor Gene Expression Data

Bo Li, Taiwen Li, Jun S. Liu, and X. Shirley Liu

Abstract

Tumor-infiltrating immune cells play critical roles in immune-mediated tumor rejection and/or progression, and are key targets of immunotherapies. Estimation of different immune subsets becomes increasingly important with the decreased cost of high-throughput molecular profiling and the rapidly growing amount of cancer genomics data. Here, we present **T**umor **IM**mune **E**stimation **R**esource (TIMER), an in silico deconvolution method for inference of tumor-infiltrating immune components. TIMER takes bulk tissue gene expression profiles measured with RNA-seq or microarray to evaluate the abundance of six immune cell types in the tumor microenvironment: B cell, CD4+ T cell, CD8+ T cell, neutrophil, macrophage, and dendritic cell. We further introduce its associated webserver for convenient, user-friendly analysis of tumor immune infiltrates across multiple cancer types.

Key words Tumor immune interaction, Infiltrating immune cells, Cancer immunotherapy, Deconvolution, RNA-seq, Interactive website

1 Introduction

Tumor microenvironment usually consists of diverse immune cell types [1–3], including both lymphocytes and myelocytes as a consequence of tumor antigen recognition [Boon, Coul, Eynde] and immune infiltration [4–7]. Recent development in cancer immunotherapies necessitates the understanding of different immune subsets that co-localize with the tumor during cancer progression [8–12]. Clinical efforts to boost the antitumor immune responses by reinvigoration of the infiltrating cytotoxic T cells [13] or elimination of the regulatory T cells [14] and/or myeloid-derived suppressive cells [15] have demonstrated curative potentials for some late-stage cancer patients. Therefore, learning the components of the infiltrating immune cells is critical to understanding tumor immune interaction and designing effective immunotherapies targeting different immune subsets.

Sebastian Boegel (ed.), *Bioinformatics for Cancer Immunotherapy: Methods and Protocols*, Methods in Molecular Biology, vol. 2120, https://doi.org/10.1007/978-1-0716-0327-7_18,

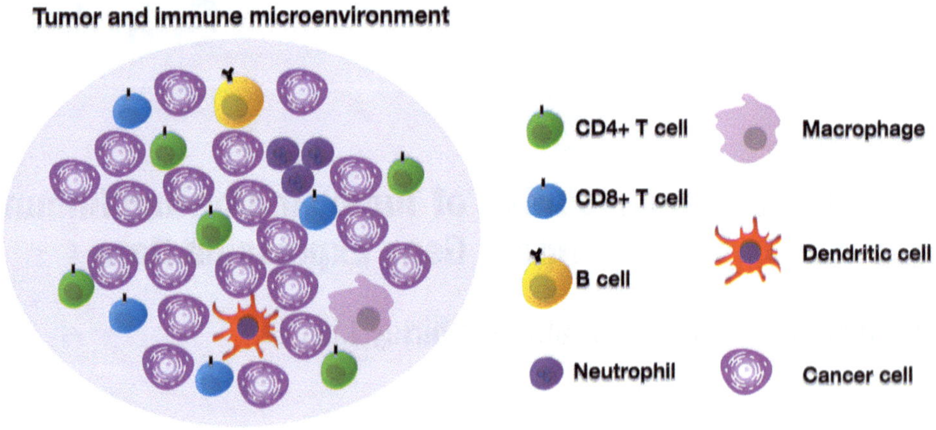

Fig. 1 Overview of the immune components in the tumor microenvironment. Diverse immune subsets have been identified in the tumor microenvironment, and the heterogeneity of immune infiltrates across individuals have been associated with the clinical outcome for diverse cancer types

Bulk tumor tissue is a mixture of diverse cell types, including cancer cells, immune cells, endothelial cells (blood vessel), fibroblasts, and so on (Fig. 1). Experimental procedures to dissociate and individually process each cell type are currently only available for small study cohorts due to cost and logistic considerations. Alternatively, with the rapid accumulation of tumor gene expression data, especially the large cancer consortium studies such as TCGA [16–19], ICGC [8, 20–22], and TARGET [23, 24], it is highly desirable to implement a computational deconvolution algorithm to infer immune cell compositions from bulk tissue gene expression data.

High-throughput molecular profiling takes whole transcriptome as input and generates a measurement of gene expression. Currently, two platforms are commonly used for gene expression profiling: massive parallel RNA sequencing, or RNA-seq, and whole-genome gene expression microarray. In general, RNA-seq produces more reliable and unbiased measurements, where sometimes microarray is less robust for lowly expressed transcripts and saturates for high gene expression [25]. Nonetheless, when taking bulk tissue sample, which is a mixture of different cell types, as input, both platforms produce gene expression measurement for all the cell types. It is expected that the observed expression level for each gene is a weighted linear combination of all different cell types in the sample:

$$\Upsilon_i = \sum_j f_j \times X_i^j + \varepsilon_i \quad (1)$$

where Υ_i is the observed gene expression level of gene i, f_j the relative abundance of cell type j in the tissue, X_i^j the unobserved expression level for gene i in cell type j, and ε_i the measurement error. This equation is a general mathematical representation of deconvolution problems with finite number of hidden components. It is straightforward that when neither f_j nor X_i^j is known, the above inference is computationally unidentifiable: for any hypothesized solution of f_j and X_i^j, $m \times f_j$ and $\frac{1}{m} \times X_i^j$ will be a valid solution as well, where m is any finite positive number.

Depending on the nature of the biological problem, sometimes it is feasible to apply a normalization constraint of f_j to make the above problem identifiable:

$$\sum_j f_j = 1 \tag{2}$$

The constraint dictates that the relative abundances for all the cell types in consideration must sum up to 1. The hidden hypothesis behind it is that even though the tissue is an unknown mixture, the collection of cell types is finite and known. For example, when studying the expression levels of normal brain tissue, it is usually valid to assume that the tissue is a mixture of several well-studied cell types in the central neural system, such as neurons, oligodendrocytes, astrocytes, glia, and microglia cells.

This strategy, however, usually does not apply to tumor tissues, as there remain uncharacterized cell types in the microenvironment. Alternatively, it is sometimes feasible to obtain X_i^j for cell types of interest, to make the equation identifiable. This is true to immune infiltrates, as previous studies have profiled many pure immune subsets through flow cytometry. The major challenge to study tumor tissue gene expression is, due to genome instability, gene expression patterns of tumor cells are usually unknown. Thus, the deconvolution problem for bulk tumor tissue can be generally written as:

$$Y_i = T_i + \sum_{j \in \text{immune cells}} f_j \times X_i^j + \sum_{s \in \text{other cell types}} f_s \times X_i^s + \varepsilon_i \tag{3}$$

In the above equation, T_i is the unknown tumor expression for gene i, with the two summations representing immune and other cell types, and for most cancer types, the latter remains poorly understood. The goal in the TIMER algorithm is to estimate f_j for the immune subsets using bulk tumor gene expression data.

2 Materials

TIMER [26] is written in the R programming language [27]. We applied TIMER to estimate abundances of different immune cell types using the TCGA data and presented the TIMER webserver for users to conveniently explore the associations between immune cells and a wide spectrum of patient clinical features. We will explain the rationale of TIMER methodology in this subheading and introduce the usage of the webserver in the next.

2.1 Tumor Purity Estimation

Tumor purity is defined as the fraction of malignant cells in the tumor tissue in a genomic sequencing experiment. Due to genome instability, it is expected that cancer cells usually carry copy number alterations (CNAs), which distinguish them from normal somatic cells. Purity is estimated using R package CHAT [28], which takes allele-specific SNP array data to infer the fraction of cells with aneuploidy genome (Fig. 2a). Detailed usage of CHAT and its input data format is available at:

```
https://sourceforge.net/p/clonalhetanalysistool/
wiki/CHAT/
```

For each sample, CHAT outputs an estimation of tumor purity (AGP) and its associated quality score (PoP). PoP is the percentage of genome with copy number changes, which is important in purity estimation. PoP smaller than 0.05 indicates that the inference is unreliable due to limited information and needs to be excluded from downstream analysis. AGP is a value ranging from 0 to 1. Purity equals to 1 means that the sample contains almost none of noncancerous cells. A complete set of purity inference for all the TCGA samples is available at:

```
http://cistrome.org/TIMER/misc/AGPall.zip
```

2.2 Selection of Informative Genes and Model Simplification

As it is usually infeasible to learn the true values of T_i for different cancer types, we select genes that are lowly expressed in the cancer cells to exclude this component in the model. Specifically, for each cancer type, we calculated the Pearson's correlation (r) between the expression levels of each gene and tumor purity. Genes with negative r is higher expressed in the tumor microenvironment than in the tumor. We selected genes with $r < -0.2$ as informative markers (Fig. 2b). When the inference is restricted to these genes, Eq. (3) can be rewritten as:

$$Y_i = \sum_{j \in \text{immune cells}} f_j \times X_i^j + \sum_{s \in \text{other cell types}} f_s \times X_i^s + \varepsilon_i \quad (4)$$

We further selected markers with expression enriched or restricted within the immune cell lineage [8] from purity-selected genes (Fig. 2c), to exclude the effects of other cell types in the microenvironment. Eq. (4) is then simplified as:

$$Y_i = \sum_{j \in \text{immune cells}} f_j \times X_i^j + \varepsilon_i \quad (5)$$

2.3 Selection of Immune Cell Types

Hematopoietic stem cells give rise to all the immune cells, including the lymphocytes and myelocytes [29]. Major immune cell types in the tumor include T/B cell, natural killer (NK) cell, monocyte, macrophage, neutrophil, dendritic cell, and so on. Each of the major cell type may also be further divided into subgroups. For

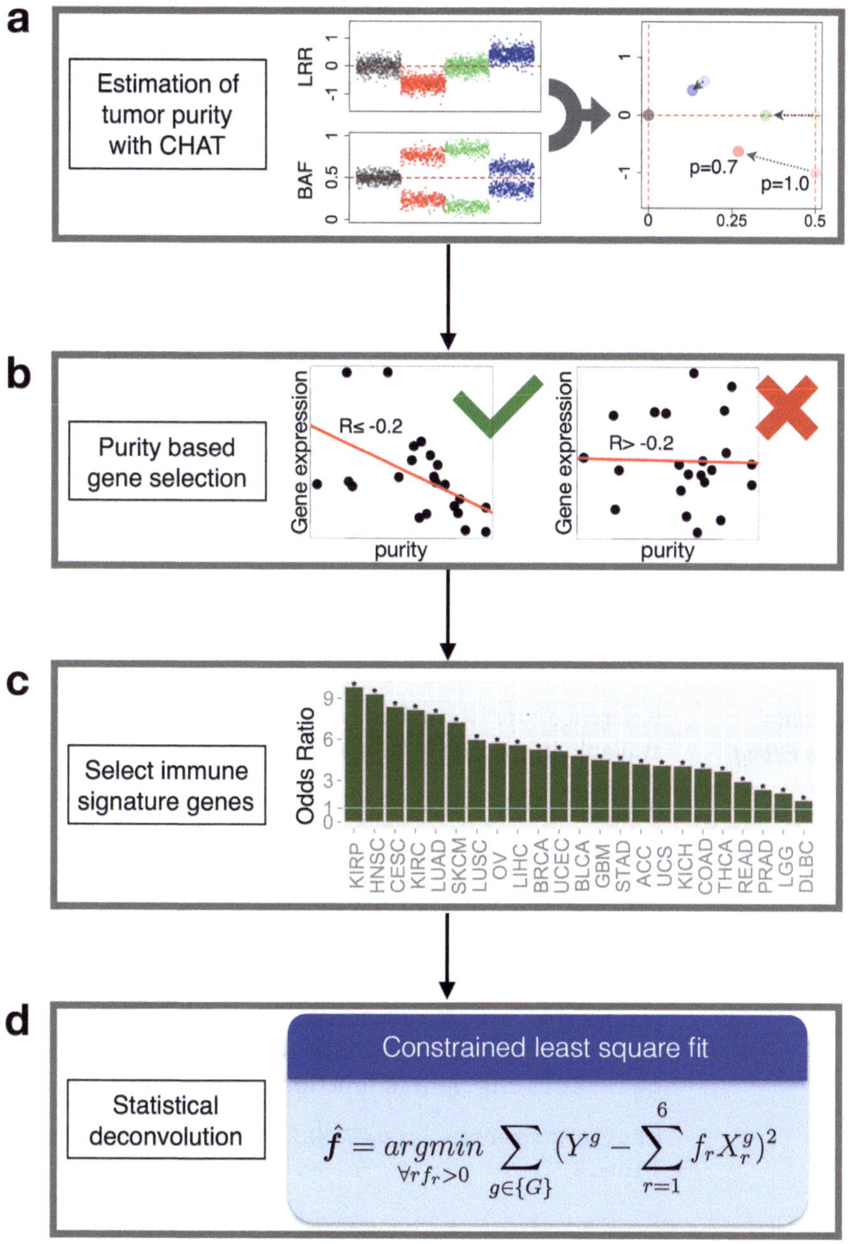

$$\hat{f} = \underset{\forall r f_r > 0}{argmin} \sum_{g \in \{G\}} (Y^g - \sum_{r=1}^{6} f_r X_r^g)^2$$

Fig. 2 Flowchart of the TIMER methodology. TIMER uses genomic estimation of tumor purity (**a**) to select genes with higher expression in the microenvironment. These genes typically show negative correlations with tumor purity (**b**). A substantial fraction of the purity-selected genes are enriched for immune signature (**c**), which are further selected to estimate the immune components with constraint least square fitting (**d**). (This flowchart is a modified version of Fig. 1 in [26])

example, the T-cell population contains CD4+ and CD8+ T cells and macrophage contains M1 or M2 polarized subtypes. Sometimes these subtypes can be split into even finer subsets. CD4+ T cells consist of a mixture of naïve, regulatory, helper cells, where CD8+ T cells in the tumor usually harbor effector, memory, and exhausted cell types. While it is attractive to deconvolve all the known immune cell subsets, due to computational limitations, it is practical to include only selected cell types in the model.

The first limitation is *identifiability.* Since it is impossible to include all immune cell types, the normalization constraint Eq. (2) usually does not apply. To make Eq. (5) identifiable, one needs to known $X_{\bullet}^{j}$, that is, the expression profile for cell type j. Such data can be obtained from previous experiments on sorted immune cells from human blood samples [30] but limited for only a few cell types. The second limitation is *statistical colinearity.* Theoretically, introducing highly correlated terms in regression models will generate unstable estimations of the coefficients. For example, if $X_{\bullet}^{k}$ and $X_{\bullet}^{l}$ represent two immune cell types with similar gene expression pattern, the estimation for f_k and f_l will be inaccurate [31]. Due to shared lineage or similar functions, different immune cell types can be very similar, for example, NK and CD8+ T cell, regulatory and helper CD4+ T cells, monocyte and macrophage, and so on. Therefore, we selected six linearly-separable cell types: B cell, CD4+ T cell, CD8+ T cell, macrophage, neutrophil, and dendritic cell for deconvolution.

2.4 Constrained Least Square Fitting

Reference expression levels of selected immune cell types, $X_{\bullet}^{j}$ were available in the public domain, and downloadable at

```
http://cistrome.org/TIMER/misc/HPCTimmune.Rdata
```

For tumor samples profiled from either RNA-seq or Affymetrix HGU133plus2 microarray, the immune components can be inferred using Eq. (5) with the reference immune cell expression data. Coefficients f_j for each cell type were estimated with least square fitting with constraint $f_j \geq 0$ (Fig. 2d). Coefficients are comparable across individuals but not between different cell types (*see* **Note 1**). This method can be implemented with the `getFractions.Abbas` function in the following R codes:

```
http://cistrome.org/TIMER/codes/CancerImmunePipe
line.R
```

3 Methods

As discussed above, unbiased estimation of selected immune cell types from bulk tissue data is practical but only accurate when tumor purity can be estimated for informative gene selection. The Cancer Genome Atlas provided both DNA and RNA profiling data for this task. Based on the inference of immune cell abundance, we developed the TIMER website [32] for users to explore different aspects of tumor immune interactions:

```
https://cistrome.shinyapps.io/timer/
```

3.1 Overview of TIMER Website

TIMER consists of four modules for correlative analysis of immune infiltrates estimated from the TCGA data, including `Gene`, `Survival`, `Mutation`, and `SCNA` modules. There are two additional modules for convenient investigation of differentially expressed genes between tumor and adjacent normal tissues (`DiffExp`), or the correlations between a pair of genes (`Correlation`). In the following are detailed instructions to the website, and there is also a step-by-step guide on YouTube:

https://youtu.be/94v8XboCrXU

3.2 `Gene` Module

This module is intended to study the correlation between gene expression and the abundance of given immune cell type(s). Three input boxes, **Gene Symbol**, **Cancer Types**, and **Immune Infiltrates** are presented on the webpage. Users need to type in the official symbol for one gene. The website will suggest alternative spells if the input was not found in the database. Users may select one or more cancer types by either typing in the cancer disease names or selecting from the drop down list in the **Cancer Types** input box. All TCGA cancer name and abbreviations are available in Table 1. By default, all six immune cell types are included in the **Immune Infiltrates**, and users can delete one or more cell type from the analysis. After the parameters are selected, clicking the "Submit" button will start the analysis, where the partial Spearman's correlation is calculated for each pair of gene/infiltrate. Tumor purity is automatically corrected for each analysis since it is a known confounding factor for both gene expression and immune infiltration levels.

For each cancer type, 1 + **X** scatter plots are returned (Fig. 3), with the first one being gene expression against tumor purity, and the following figures for **X** selected immune cell types ($0 \leq \mathbf{X} \leq 6$). On top of each plot, the Lowess smooth curve with confidence interval estimations is overlaid to visualize the trend of correlation. The figures can be directly downloaded as JPG or PDF format.

3.3 `Survival` Module

This module builds flexible Cox proportional regression models, with diverse variable options. First, users can select one or more diseases in the **Cancer Types** input field. Three categories of covariates are allowed in the model, including clinical factors (**Clinical**), infiltrating immune cell abundance (**Immune Infiltrates**), and the expression levels of given gene(s) (**Gene Symbols**). For each category, one or more covariates can be included. After all the fields have been set, the resulting Cox model will be immediately displayed on the right side of the webpage. Estimations of parameters are displayed as a table, with the following columns: coef (log hazard ratio), HR (hazard ratio), 95%CI_l (lower boundary of 95% confidence interval for HR), 95%CI_u (upper boundary of 95% confidence interval for HR), *p*.value (*p* value for individual

Table 1
Disease full name and abbreviations of 33 cancer types covered in TCGA and TIMER website

Abbreviation	Disease full name
ACC	Adrenocortical carcinoma
BLCA	Bladder urothelial carcinoma
BRCA	Breast invasive carcinoma
CESC	Cervical and endocervical cancers
CHOL	Cholangiocarcinoma
COAD	Colon adenocarcinoma
DLBC	Lymphoid neoplasm diffuse large B-cell lymphoma
ESCA	Esophageal carcinoma
GBM	Glioblastoma multiforme
HNSC	Head and neck squamous cell carcinoma
KICH	Kidney chromophobe
KIRC	Kidney renal clear cell carcinoma
KIRP	Kidney renal papillary cell carcinoma
LAML	Acute myeloid leukemia
LGG	Brain lower grade glioma
LIHC	Liver hepatocellular carcinoma
LUAD	Lung adenocarcinoma
LUSC	Lung squamous cell carcinoma
MESO	Mesothelioma
OV	Ovarian serous cystadenocarcinoma
PAAD	Pancreatic adenocarcinoma
PCPG	Pheochromocytoma and paraganglioma
PRAD	Prostate adenocarcinoma
READ	Rectum adenocarcinoma
SARC	Sarcoma
SKCM	Skin cutaneous melanoma
STAD	Stomach adenocarcinoma
TGCT	Testicular germ cell tumors
THCA	Thyroid carcinoma
THYM	Thymoma
UCEC	Uterine corpus endometrial carcinoma
UCS	Uterine carcinosarcoma
UVM	Uveal melanoma

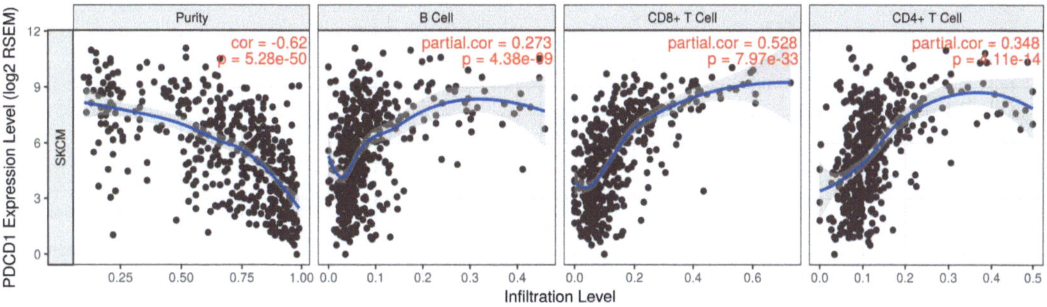

Fig. 3 Example of Gene module in the TIMER website. When choosing PDCD1 as Gene Symbol, metastatic melanoma (SKCM) as Cancer Type, and B cell, CD4+, and CD8+ T cells as immune infiltrates, the website returns 4 scatter plots. The first one is always comparison of gene expression and tumor purity, with the following panels for each of the selected immune cell type

Table 2
Example of Survival module in the TIMER website

Model: Surv(LUAD) ~ Age + Purity + B_cell + CD19						
455 patients with 161 dying						
	coef	**HR**	**95% CI_l**	**95% CI_u**	***p*.value**	**sig**
Age	0.009	1.009	0.992	1.025	0.303	
Purity	−0.163	0.849	0.416	1.736	0.655	
B_cell	−2.638	0.071	0.007	0.701	0.023	*
CD19	−0.069	0.933	0.845	1.03	0.17	
R square = 0.043 (max possible = 9.75e-01)						
Likelihood ratio test: $p = 5.12e04$						
Wald test: $p = 1.31e-03$						
Score (log rank) test: $p = 1.04e-03$						

When choosing lung adenocarcinoma (LUAD) as Cancer Type, B cell as immune infiltrates, Age as clinical factor, and CD19 as Gene Symbol, the website returns a summary table of the Cox proportional hazard built with the given variables

covariate), and sig (significance levels). Test results for the complete model are displayed on the bottom. For example, when selecting lung adenocarcinoma as cancer type, age and purity as clinical cofactors, B cell as immune infiltrates, and CD19 as gene symbol, the website returns Table 2.

With the above parameters, users can also generate Kaplan-Meier curves (Fig. 4) for each selected covariate (excluding clinical cofactors) using the "Plot KM Curve" button. Survival of patients with the upper X% and lower X% of the covariate will be compared, with X ranging from 5 to 50 (**Split Percentage of Patients** slider). Users may also choose to view a subset of the data by limiting

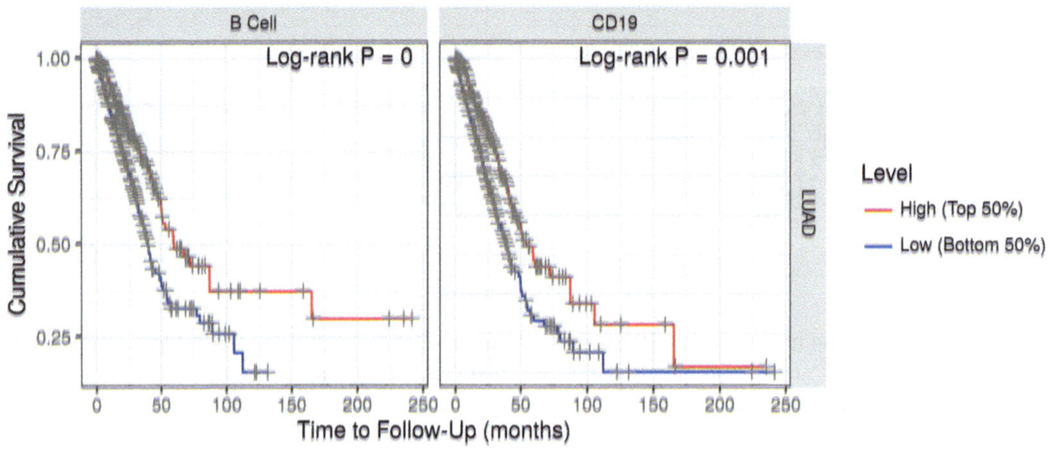

Fig. 4 Example of Survival module in the TIMER website. When choosing lung adenocarcinoma (LUAD) as Cancer Type, B cell as immune infiltrates and CD19 as Gene Symbol, the website returns two Kaplan-Meier curves, each for one covariate. By default, the upper and lower 50% of samples for each variable are displayed in the two groups and compared with Log-rank test for statistical significance

survival time within a certain period (**Survival Time Between** slider). It should be noted that truncating data with an empirical time window is for visualization purposes only and shall not be used for statistical significance estimations.

3.4 `Mutation` and SCNA Modules

Cancer somatic single nucleotide mutations (SNVs) or copy number alterations (SCNAs) may induce immune responses either via signaling pathways within the cancer cells or providing novel antigenic targets for the adaptive immune system. The `Mutation` and SCNA modules allow users to explore the heterogeneity of immune infiltrates in tumors with different mutational backgrounds. In both modules, the cancer type needs to be selected first. As most somatic SNVs are rare events [33], in the `Mutation` module, we focus on most frequent mutations to ensure statistical power in the analysis. Specifically, if a cancer type has less than 50 mutations with frequency greater than 10%, we select the top 50 mutations. Otherwise mutations with frequency $\geq$ 10% are selected. Users may choose any SNVs in the **Gene with Mutation** drop down list to visualize the results. Clicking on the "Submit" button will return a Boxplot with 6 pairs of distributions, each for one immune cell type. The infiltration levels between mutated or wild type samples are compared with two-sided Wilcoxon rank sum test, with significance level labeled on the top of each pair. For example, when cancer type is selected as ovarian cancer (OV) and mutation as CSMD3, the website returns Fig. 5, where CD4+ T cell and dendritic cell showing significantly higher levels in mutated samples.

In the SCNA module, users can input the official symbol for the gene of interest, and the copy number values estimated from

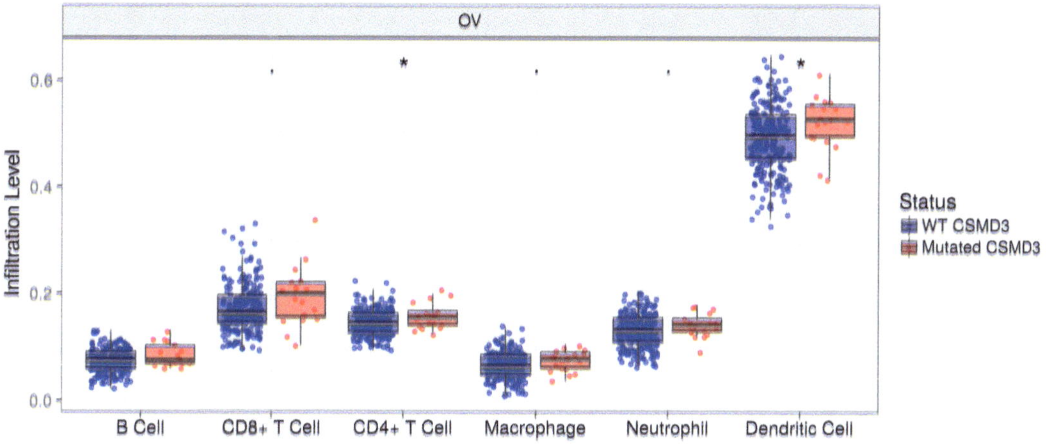

Fig. 5 Example of Mutation module in the TIMER website. When choosing ovarian cancer (OV) as Cancer Type and CSMD3 mutation, the website returns six pairs of boxplots, each for one immune cell type. Infiltration levels in mutated or wild type (WT) individuals are compared with Wilcoxon rank sum test and the statistical significance levels are labeled above the boxes

GISTIC 2.0 [34] will be used for the analysis. There are five types of copy number events defined in the GISTIC method: focal deletion ($n = -2$), arm-level deletion ($n = -1$), normal diploid ($n = 0$), arm-level deletion ($n = 1$), and focal amplification ($n = 2$). Once parameters are selected, clicking the "Submit" button will return a Boxplot with six sets of distributions, each for one cell type. Samples with copy number variations ($n \neq 0$) are compared with diploid samples with Wilcoxon rank sum test, and significance levels are labeled on top of each box.

3.5 `DiffExp` Module

Perhaps one of the most commonly implemented analyses in cancer research is to compare gene expression levels between tumor and the matched normal samples. In TIMER, we provide a convenient solution for quick visualization of such comparisons for all the genes in the database. In this module, users simply input the gene symbol, and click "Submit." The website will return a comprehensive Boxplot showing the distributions of the gene expression levels (in RSEM [35]). In total, 33 cancer types and 17 normal tissue types are investigated, and Wilcoxon rank sum test is applied to all the cancer/normal pairs to estimate statistical significance. Major subtypes of selected cancers are also displayed as individual columns on the plot for quick comparisons.

3.6 `Correlation` Module

This module explores the dependency between pairs of genes. Users may type in m genes in the **Gene Symbols: (Y-axis)** input field, and n genes in the **Gene Symbols: (X-axis) field**. One founding factor (either tumor purity or age) can be included in the analysis using the drop down list in the **Correlation Adjusted by** field. If no confounding factor is selected, clicking "Submit"

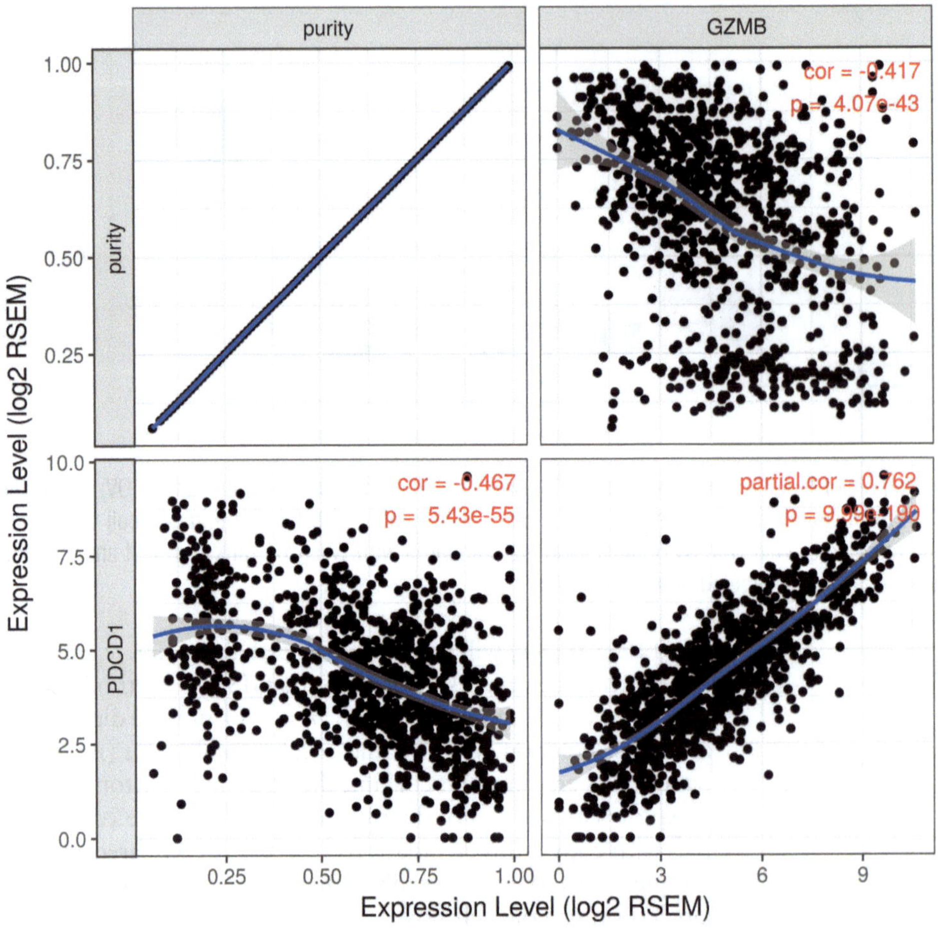

Fig. 6 Example of Correlation module in the TIMER website. With Y-axis gene selected as PDCD1 and X-axis gene as GZMB, when correcting for purity, the website returns a set of four scatter plots. The upper left panel is noninformative, with the rest the correlations between gene expression and either purity or the expression levels of the other gene. Gene–gene correlation (lower-left) is corrected by purity and estimated by partial Spearman's correlation

button will return *m*-by-*n* scatter plots, each displaying the correlation of the corresponding pair of genes.

If one of the factor is selected for adjustment, $(m + 1)$-by-$(n + 1)$ plots will be returned, with the first row and column of figures showing the dependencies of the gene expression levels and the confounding factor. Correlations between genes are calculated with partial Spearman's correlation corrected for the selected factor. For example, when choosing metastatic melanoma as cancer type, PDCD1 as Y-axis, and GZMB as X-axis, without purity adjusted, the website returns Fig. 6, and it is clear that the two genes are significantly correlated. This is in line with the observation that effector T cells secreting GZMB in the tumor are likely to become exhausted [36].

4 Notes

1. Since we chose not to apply the normalization constraint in Eq. (2), the estimated fractions usually do not sum up to 1. In addition, different immune cell type may express informative marker genes at different levels, which will impact the scales of f_j. As a result, comparison between two immune cell types of the same sample, that is, f_j and f_k, is meaningless. We have demonstrated that enforcing normalization constraint to make f_j comparable across immune cell types will falsely impose negative correlations between the estimated infiltration levels [31]. Therefore, all the analysis performed in the TIMER website are restricted to comparisons across individuals for the same immune cell type.

References

1. Binnewies M, Roberts EW, Kersten K et al (2018) Understanding the tumor immune microenvironment (TIME) for effective therapy. Nat Med 24:541–550
2. Junttila MR, de Sauvage FJ (2013) Influence of tumour micro-environment heterogeneity on therapeutic response. Nature 501:346–354
3. Quail DF, Joyce JA (2013) Microenvironmental regulation of tumor progression and metastasis. Nat Med 19:1423–1437
4. Dushyanthen S, Beavis PA, Savas P et al (2015) Relevance of tumor-infiltrating lymphocytes in breast cancer. BMC Med 13:202
5. Eruslanov E, Neuberger M, Daurkin I et al (2012) Circulating and tumor-infiltrating myeloid cell subsets in patients with bladder cancer. Int J Cancer 130:1109–1119
6. Marvel D, Gabrilovich DI (2015) Myeloid-derived suppressor cells in the tumor microenvironment: expect the unexpected. J Clin Invest 125:3356–3364
7. Pages F, Galon J, Dieu-Nosjean MC et al (2010) Immune infiltration in human tumors: a prognostic factor that should not be ignored. Oncogene 29:1093–1102
8. Abbas AR, Baldwin D, Ma Y et al (2005) Immune response in silico (IRIS): immune-specific genes identified from a compendium of microarray expression data. Genes Immun 6:319–331
9. Gubin MM, Zhang X, Schuster H et al (2014) Checkpoint blockade cancer immunotherapy targets tumour-specific mutant antigens. Nature 515:577–581
10. Hamid O, Robert C, Daud A et al (2013) Safety and tumor responses with lambrolizumab (anti-PD-1) in melanoma. N Engl J Med 369:134–144
11. Tran E, Turcotte S, Gros A et al (2014) Cancer immunotherapy based on mutation-specific CD4+ T cells in a patient with epithelial cancer. Science 344:641–645
12. Van Allen EM, Miao D, Schilling B et al (2015) Genomic correlates of response to CTLA-4 blockade in metastatic melanoma. Science 350:207–211
13. Rosenberg SA, Restifo NP, Yang JC et al (2008) Adoptive cell transfer: a clinical path to effective cancer immunotherapy. Nat Rev Cancer 8:299–308
14. Simpson TR, Li F, Montalvo-Ortiz W et al (2013) Fc-dependent depletion of tumor-infiltrating regulatory T cells co-defines the efficacy of anti-CTLA-4 therapy against melanoma. J Exp Med 210:1695–1710
15. Ko JS, Zea AH, Rini BI et al (2009) Sunitinib mediates reversal of myeloid-derived suppressor cell accumulation in renal cell carcinoma patients. Clin Cancer Res 15:2148–2157
16. The Cancer Genome Atlas Research Network (2012) Comprehensive molecular portraits of human breast tumours. Nature 490:61–70
17. The Cancer Genome Atlas Research Network (2008) Comprehensive genomic characterization defines human glioblastoma genes and core pathways. Nature 455:1061–1068
18. The Cancer Genome Atlas Research Network (2012) Comprehensive genomic

characterization of squamous cell lung cancers. Nature 489:519–525

19. The Cancer Genome Atlas Research Network (2011) Integrated genomic analyses of ovarian carcinoma. Nature 474:609–615
20. Nik-Zainal S, Davies H, Staaf J et al (2016) Landscape of somatic mutations in 560 breast cancer whole-genome sequences. Nature 534:47–54
21. Richter J, Schlesner M, Hoffmann S et al (2012) Recurrent mutation of the ID3 gene in Burkitt lymphoma identified by integrated genome, exome and transcriptome sequencing. Nat Genet 44:1316–1320
22. Tirode F, Surdez D, Ma X et al (2014) Genomic landscape of Ewing sarcoma defines an aggressive subtype with co-association of STAG2 and TP53 mutations. Cancer Discov 4:1342–1353
23. Bolouri H, Farrar JE, Triche T Jr et al (2018) The molecular landscape of pediatric acute myeloid leukemia reveals recurrent structural alterations and age-specific mutational interactions. Nat Med 24:103–112
24. Ma X, Liu Y, Liu Y et al (2018) Pan-cancer genome and transcriptome analyses of 1,699 paediatric leukaemias and solid tumours. Nature 555:371–376
25. Zhao S, Fung-Leung WP, Bittner A et al (2014) Comparison of RNA-Seq and microarray in transcriptome profiling of activated T cells. PLoS One 9:e78644
26. Li B, Severson E, Pignon JC et al (2016) Comprehensive analyses of tumor immunity: implications for cancer immunotherapy. Genome Biol 17:174
27. R Core Team: R: A language and environment for statistical computing. 2014
28. Li B, Li JZ (2014) A general framework for analyzing tumor subclonality using SNP array and DNA sequencing data. Genome Biol 15:473
29. Orkin SH, Zon LI (2008) Hematopoiesis: an evolving paradigm for stem cell biology. Cell 132:631–644
30. Mabbott NA, Baillie JK, Brown H et al (2013) An expression atlas of human primary cells: inference of gene function from coexpression networks. BMC Genomics 14:632
31. Li B, Liu JS, Liu XS (2017) Revisit linear regression-based deconvolution methods for tumor gene expression data. Genome Biol 18:127
32. Li T, Fan J, Wang B et al (2017) TIMER: a web server for comprehensive analysis of tumor-infiltrating immune cells. Cancer Res 77: e108–e110
33. Martincorena I, Campbell PJ (2015) Somatic mutation in cancer and normal cells. Science 349:1483–1489
34. Mermel CH, Schumacher SE, Hill B et al (2011) GISTIC2.0 facilitates sensitive and confident localization of the targets of focal somatic copy-number alteration in human cancers. Genome Biol 12:R41
35. Li B, Dewey CN (2011) RSEM: accurate transcript quantification from RNA-Seq data with or without a reference genome. BMC Bioinformatics 12:323
36. Crespo J, Sun H, Welling TH et al (2013) T cell anergy, exhaustion, senescence, and stemness in the tumor microenvironment. Curr Opin Immunol 25:214–221

Chapter 19

Cell-Type Enrichment Analysis of Bulk Transcriptomes Using xCell

Dvir Aran

Abstract

Tissues are a complex milieu of cell types of different lineages and subtypes, each with its own unique transcriptomic profile. Bulk transcriptome profiling is therefore the sum of the cell-type-specific gene expression weighted by cell-type proportion in the given sample. Deconvolution of gene expression profiles allows to reconstruct the cellular composition of tissues. xCell is a robust computational method that converts gene expression profiles to enrichment scores of 64 immune and stroma cell types across samples. Here, we described the method, discuss correct usage, and demonstrate an analysis of a cohort of peripheral blood mononuclear cells (PBMC).

Key words Deconvolution, Cell types, Gene expression, Gene signatures, Immune cells

1 Introduction

In the past two decades, around one million gene expression profiles of human samples have been deposited in the public domain (Fig. 1). Most of those are from bulk tissues, meaning that they are composed of a mixture of cell types. Tissues are highly heterogenous in the proportions of cell types, and these proportions may change drastically in pathologies. Thus, a difference in gene expression between conditions may not be a result of a change in the gene's expression, but a difference in the cellular composition between the conditions. The cellular heterogeneity is therefore a confounding factor, which strongly affects differential gene expression analyses, and may lead to false downstream investigations, such as inferring genetic effects, assembling co-expression networks and detecting regulatory factors [1].

However, acknowledging that bulk tissues are cellularly heterogenous provides also an opportunity. The cell-type composition of different tissues has been found to be perturbed in certain disease states and in aging and to be predictive and associated with response to certain treatments [2–4]. Thus, characterization of

Sebastian Boegel (ed.), *Bioinformatics for Cancer Immunotherapy: Methods and Protocols*, Methods in Molecular Biology, vol. 2120, https://doi.org/10.1007/978-1-0716-0327-7_19,

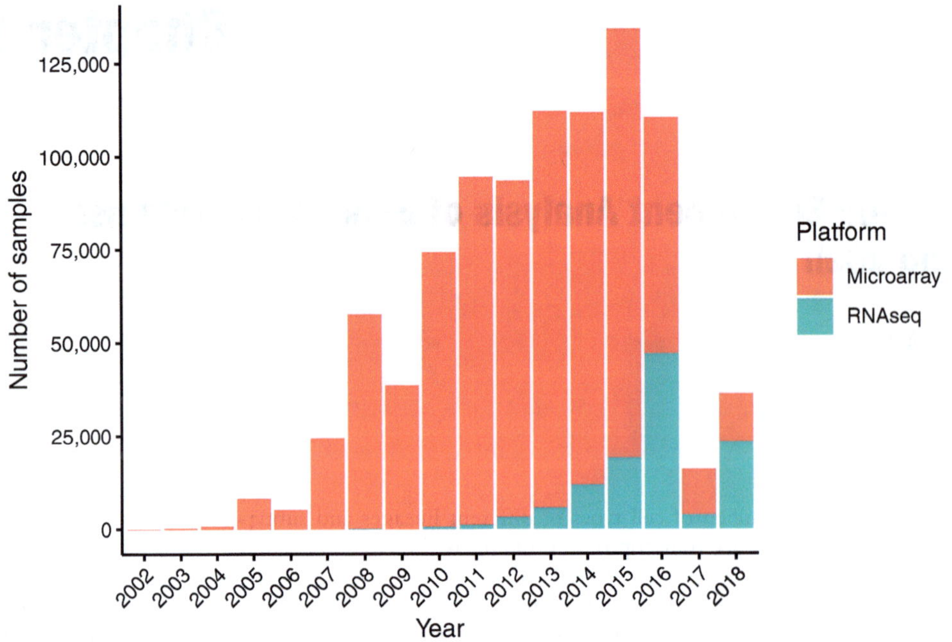

Fig. 1 Bulk gene expression samples by year. Counts of gene expression microarray (red) and RNA-seq samples of deposited in GEO and ArrayExpress by year

the variation in the composition of cell types across subjects can identify cellular targets for diseases and suggest novel therapeutic strategies. Moreover, adjusting for these variations can allow to detect real-gene expression differences and enhance the interpretation of downstream analyses.

In the past decade, many computational methods have been suggested for performing such deconvolution of the cellular heterogeneity from bulk gene expression profiles [5]. Generally, these methods are based on associating a "reference" gene set that have been learned from purified cell types with bulk transcriptomic profiles. We have recently developed xCell, a novel computational method for inferring the cellular composition using bulk transcriptomic profiles [6]. xCell is a gene signature-based method, which was learned from comprehensive and diverse pure cell-type transcriptomes from various sources. xCell was developed using a machine-learning approach for learning the best signatures for 64 immune and stroma cell types. To account for unwanted affects between closely related cell types, xCell employs a spillover compensation technique, inspired by such correction applied in flow cytometry analyses. By comparing the xCell scores to cytometry immunophenotyping of hundreds of samples, we showed its superiority over other currently available methods.

In this chapter, we describe the xCell pipeline which is composed of three steps. We also describe how to use the xCell R package by investigating a set of microarray dataset of peripheral blood mononuclear cells (PBMC). We will discuss the limitations of xCell and suggest how to correctly perform analyses with it.

2 Materials

xCell was developed in R, and an R package for running xCell is available as an open-source code in a GitHub repository (https://github.com/dviraran/xCell).

Installation of R is a prerequisite (https://www.r-project.org/). RStudio is the recommended environment for running R scripts (https://www.rstudio.com/).

To install the xCell R package, all commands below should be typed in an R environment:

1. If the **devtools** package not previously installed, first install it:

```
install.packages('devtools')
```

2. Install the current xCell version from Github:

```
devtools::install_github('dviraran/xCell')
```

The xCell package is dependent on the following packages:
Bioconductor packages—GSVA, GSEABase.

```
if (!requireNamespace("BiocManager", quietly = TRUE))
    install.packages("BiocManager")
BiocManager::install(c("GSVA","GSEABase"), version = "3.8")
```

CRAN packages—pracma, utils, stats, MASS, digest, curl, quadprog.

```
install.packages('pracma', 'utils', ' stats ', ' MASS ',
                          'digest', 'curl', 'quadprog')
```

3 Methods

3.1 Loading xCell

```
library(xCell)
```

When loading the xCell package, an object named "xCell.data" is loaded. This is a list containing the spill over and calibration parameters, the signatures, and the list of genes it uses.

3.2 Data Input

The input for xCell is a gene expression matrix from human mixed samples. It should be read prior to running the xCell functions. The matrix should contain HUGO gene symbols as row names and the columns are samples.

If the gene expression is a tab delimited file, it can be read using the following call:

```
expr = read.table(file.name, header=TRUE, row.names=1, as.
is=TRUE, sep='\t')
```

If the gene expression data are from a microarray, no normalization is required. If the gene expression data are from a sequencing platform, values must be normalized to gene length (i.e., RPKM, TPM, FPKM). xCell uses the expression levels ranking and not the actual values, thus further normalization does not have an effect. *See* **Note 1** for more details about normalization.

3.3 xCell Pipeline

The xCell pipeline is composed of three steps, which are also represented as function in the R package:

1. rawEnrichmentAnalysis,

   This function reads a gene expression matrix and returns an enrichment score for each of the 64 cell types across the input gene expression samples. For its calculations, xCell considers 10,808 common genes. The list of genes can be found in:

   ```
   xCell.data$genes
   ```

   As a minimal requirement, at least 5000 genes are required to be available in the input gene expression matrix; however, low number of shared genes may affect the accuracy of the results.

   xCell uses multiple signatures for each cell type. Altogether, there are 489 signatures that correspond to the 64 cell types. The full list of signatures is available in:

   ```
   xCell.data$signatures
   ```

   The scores are computed using single-sample gene set enrichment analysis (ssGSEA) [7]. For each cell type, the average of the multiple scores from the multiple corresponding signatures is calculated. Finally, average scores are shifted such that the minimal score for each cell type is zero.

   Hence, xCell performs best with heterogenous dataset. If a cell type has similar levels across all samples, the scores will be low and will not correspond well with the expected proportions. Thus, it is recommended to use all available data combined in one run and not to break down to smaller subsets.

Note that scores of samples that were analyzed in different runs are therefore not comparable. *See* **Note 2** for further discussion about variation.

The function usage is as follows:

```
scores = rawEnrichmentAnalysis(expr, signatures,
genes, file.name, parallel.sz, parallel.type = "SOCK")
```

'expr' is the gene expression matrix as described above; 'signatures' and 'genes' are the signatures and genes as described above; 'file.name' is optional and can be used to specify to save the raw scores to a tab delimited text file; 'parallel.sz' is the number of threads to use (default is 4); 'parralel.type' is the type of cluster architecture to use: can be 'SOCK' (default) or 'FORK'; Fork is faster, but is not supported in Windows.

2. transformScores,

   This function is used to transform scores from raw enrichment scores to a linear scale that resembles percentages, xCell uses precomputed calibration parameters for the transformations. xCell uses different set of parameters for sequencing-based gene expression values and microarray-based values (*see* **Note 3** for information about adjusting the calibration parameters). The parameters for sequencing-based values can be found in:

```
xCell.data$spill$fv
```

   and for microarray-based values:

```
xCell.data$spill.array$fv
```

   The function usage is as follows:

```
tscores = transformScores(scores, fit.vals, scale, fn)
```

   'scores' are the output of rawEnrichmentAnalysis; 'fit.vals' are the calibration parameters as described above; scale if a logical, of whether to scale the transformed scores (default is TRUE and recommended). 'fn' is optional and can be used to specify a file name to save the transformed scores to a tab delimited text file.

3. spillOver.

   This function is used to perform the spill-over compensation applied by xCell to reduce dependency between scores of closely related cell types. The spill over is performed using a precomputed dependency matrix 'K'. Different matrices are used for sequencing-based and microarray-based inputs. The matrix for sequencing-based inputs can be found in:

```
xCell.data$spill$K
```

and for microarray-based inputs:

```
xCell.data$spill.array$K
```

The removal of spill over may have an effect that is too strong and may remove real signal. Thus, an alpha parameter is used to calibrate the level of compensation to perform. Alpha equals 0 means no compensation, and 1 means full compensation. In our experiments, a value of 0.5 reduces the dependencies and increases real signals (this is also the default value):

```
spillOver(transformedScores, K, alpha = 0.5, file.name = NULL)
```

3.4 Significance Analysis

In addition, xCell provides significance assessment of the null hypothesis that the cell type is not in the mixture. It is important to understand this null hypothesis—the p-values that are calculated do not provide significance for the estimate itself, only if the estimate is significantly different from zero. Thus, this is a tool that can allow to remove cell types that are not in the mixture. We will discuss how removing cell types that are not in the mixture from the analysis may improve the inferences.

The available functions for significance are as follows:

1. xCellSignifcanceBetaDist,

   This function uses predefined beta distribution parameters from random mixtures generated using the reference data sets. This is the method used in the xCell manuscript, and it is recommended.

2. xCellSignifcanceRandomMatrix.

   uses random matrix and calculates beta distribution parameters.

3.5 xCell Usage

xCell contains a wrapper function to run all three steps using one function:

```
xCellAnalysis(expr, signatures = NULL, genes = NULL, spill
= NULL, rnaseq = TRUE, file.name = NULL, scale = TRUE, alpha =
0.5, save.raw = FALSE, parallel.sz = 4, parallel.type = "SOCK",
cell.types.use = NULL)
```

Here, the inputs are similar to the inputs in the three functions above. If 'signatures', 'genes', and 'spill' are NULL (default), then xCell uses the xCell.data objects. If 'rnaseq' is true, then use xCell.data$spill, else use xCell.data$spill.array. 'cell.types.use' can be a subset of the 64 cell types analyzed by xCell, see below considerations for best practice.

See the xCell R package documentation for more details on each function.

3.6 Example of Using xCell

Below, we describe in detail an application of xCell to analyze data a study which contains gene expression from peripheral blood mononuclear cells. This study also contains counts of certain cell types using spectrometry-based immunoprofiling, allowing to compare the results generated with xCell from the gene expression data with an independent measurement for the cellular composition. The results from these datasets were presented in the original xCell publication.

3.6.1 Data Gathering

From the ImmPort web portal [8], we downloaded the study file of SDY420 [9]. The study has a file named *fcs_analyzed_result.txt* which contains counts of cells. We extracted the number of cells from all live cells to create a matrix of the fraction of cells for each sample. In addition, we downloaded the gene expression data, also available with the ImmPort study files. This is an Illumina array, and it was normalized using Matlab functions. The object is available here: https://github.com/dviraran/xCell/tree/master/vignettes.

To read the object type,

```
sdy = readRDS('sdy420.rds')
```

Altogether, this study contains 104 samples with gene expression and immunoprofiling counts.

3.6.2 Generating xCell Scores

As described, there are three steps to create xCell scores. First, we use the *rawEnrichmentAnalysis* function to generate the raw scores:

```
library(xCell)
raw.scores = rawEnrichmentAnalysis(as.matrix(sdy$expr),
                                   xCell.data$signatures,
                                       xCell.data$genes)
```

The next step is transforming the raw scores and applying the spill-over compensation. To get best results, it is best to run the spill-over compensation only on relevant cell types (e.g., if we know there are no macrophages in the mixtures, it is best to remove them from the analysis). Thus, we subset the scores matrix to only cell types that are also measured in the CyTOF dataset:

```
cell.types.use = intersect(colnames(xCell.data$spill$K),
                           rownames(sdy$fcs))
```

```
transformed.scores = transformScores(raw.scores[cell.types.
use,],
                                     xCell.data$spill.array$fv)
```

The last step is compensating the scores for spill-over effects:

```
scores = spillOver(transformed.scores,xCell.data$spill.array
$K)
```

Note that we use here xCell.data$spill.array data since the expression data were generated with microarrays.

The pipeline detailed above can be similarly performed using the *xCellAnalysis* wrapper function:

```
scores = xCellAnalysis(sdy$expr, rnaseq=F,
                       cell.types.use = cell.types.use)
```

3.6.3 Correlating xCell Scores and CyTOF Immunoprofilings

Using these scores, we can now find the correlation between the xCell scores and the cell types fractions from the CyTOF immunoprofilings:

```
library(psych)
library(ggplot2)
fcs = sdy$fcs[rownames(scores),colnames(scores)]
res = corr.test(t(scores),t(fcs),adjust='none')
qplot(x=rownames(res$r),y=diag(res$r),
      fill=diag(res$p)<0.05,geom='col',
      main='SDY420 association with immunoprofiling',
     ylab='Pearson R', xlab='') + labs(fill = "p-value<0.05") +
      theme_classic() +
      theme(axis.text.x = element_text(angle = 45, hjust = 1))
```

This code generates a bar plot for the correlation of the xCell scores with the expected fractions from the immunoprofiling (Fig. 2a). We see significant correlations (p-value <0.05) in 13 of 18 cell types, and high correlations ($R > 0.5$) in 7 cell types. It is important to note that the xCell produces enrichment scores and not proportions of cell types, thus it is not expected that the scores will be similar to the CyTOF proportions, just that there will be linear correlation between the measurements (*see* **Note 4** for more information).

In the above analysis, we ran xCell with a subset of cells and not for all 64 cell types. In some cases, this may improve the accuracy since the spill-over compensation procedure may be overcompensating. Thus, we can run the same analysis for all cell types and compare the associations with the immunoprofiling:

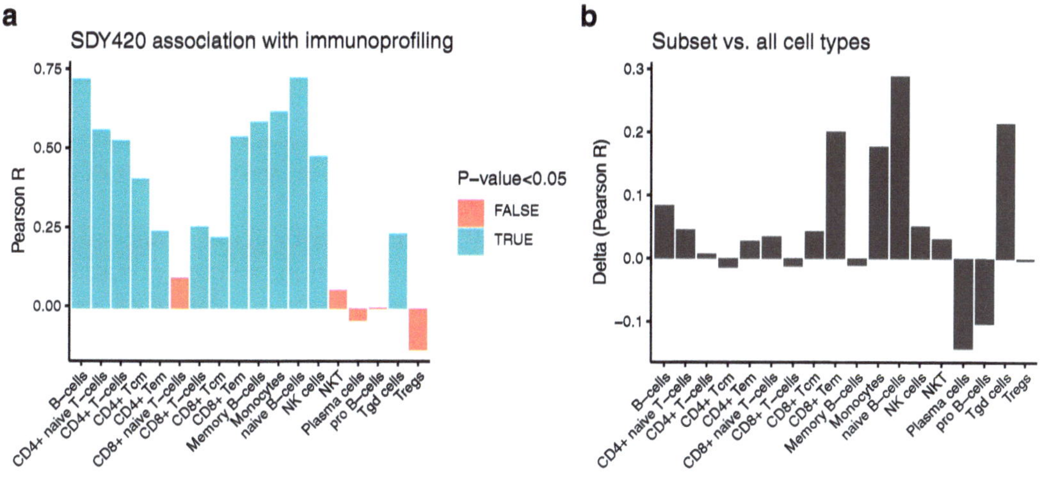

Fig. 2 Accuracy of xCell scores. (**a**) Pearson correlation coefficients of xCell scores and the CyTOF immunoprofiling counts for the 18 cell types, where both measurements are available in SDY420. Thirteen of the correlations have a *p*-value <0.05. (**b**) Differences in the correlation coefficients between using a subset of cell types in xCell and using all 64 cell types

```
scores.all = xCellAnalysis(sdy$expr, rnaseq=F)
res.all = corr.test(t(scores.all),t(fcs),adjust='none')
A = intersect(rownames(res$r),rownames(res.all$r))
qplot(x=A,y=diag(res$r[A,A])-diag(res.all$r[A,A]), geom='-
col',
    main= Subset vs. all cell types', xlab=",
    ylab='Delta (Pearson R)') + theme_classic() +
    theme(axis.text.x = element_text(angle = 45, hjust = 1))
```

This code generates a bar plot with the difference in correlation coefficients between the running xCell with a subset of cell types (as above) and running xCell with all cell types (Fig. 2b). We can see that for most cell types, there is no major difference between analysis with a subset of cell types or all cell types; however, major improvement is observed in monocytes (0.158) and naïve B cells (0.254).

3.6.4 Downstream Analysis with xCell Scores

The ImmPort study contains basic demographics for each of the subjects. We perform simple analyses to test for association between the xCell scores and these parameters. Age is a continuous variable; thus, we use Spearman correlation to analyze its association with xCell scores:

```
age = corr.test(t(scores), sdy$metadata$AGE_REPORTED,
      method='spearman', adjust = 'fdr')
qplot(x=rownames(age$r), y=age$r, fill=age$p<0.05,geom='col',
 main='Association with age', xlab=",ylab='Spearman R') +
      labs(fill = "p-value<0.05") + theme_classic() +
      theme(axis.text.x = element_text(angle = 45, hjust = 1))
```

As in the previous analysis, we can present a bar plot with the coefficients of the associations (Fig. 3a). The only significant correlation (FDR < 0.05) is with plasma cells (Fig. 3b):

```
qplot(x=scores['Plasma cells',], y=sdy$metadata$AGE_REPORTED,
    xlab='Plasma cells (xCell)', ylab='Age') + theme_classic()
```

We can also perform association with gender. Since the xCell scores do not follow a normal distribution, we use a nonparametric test (Wilcoxon rank-sum test). Here, we calculate *p*-values for all cell types association with gender, and then adjust the *p*-values with FDR correction:

```
p = p.adjust(apply(scores,1,FUN=function(x)
          wilcox.test(x~sdy$metadata$GENDER)$p.value),meth-
od='fdr')
```

One way to present this analysis is in the same manner as we presented above, using a bar plot, for example, presenting the $-\log10$ of the adjusted *p*-values (Fig. 3c):

```
qplot(x=names(p), y=-log10(p), fill=p, geom='col',
    main='Association with gender', xlab='',
    ylab='-log10(adjusted p-val') + theme_classic() +
    theme(axis.text.x = element_text(angle = 45, hjust = 1)) +
    theme(legend.position = "none")
```

The most significant association with gender is of regulatory T cells (Tregs) (adjusted *p*-value $= 0.0897$). Here, the best way to present such association is with box plots (Fig. 3d):

```
df = data.frame(Tregs = scores['Tregs',],Gender = sdy$metadata
$GENDER, Cells = colnames(scores))
qplot(data=df,y=Tregs,x=Gender,geom='boxplot', fill=Gender,
alpha=0.5, ylab='Tregs (xCell)', xlab='Gender') +
geom_point(position=position_jitterdodge()) +
  theme_classic() + theme(legend.position = "none")
```

Another way to present analysis between two groups is using a volcano plot, which presents the log fold-change vs. the $-\log10$ *p*-value. We can use the following code to create such a plot (Fig. 3e):

```
library(EnhancedVolcano)
df = data.frame(p = p, fc = log2(apply(scores,1,
      FUN=function(x) mean(x[sdy$metadata$GENDER=='Female'])/
      mean(x[sdy$metadata$GENDER=='Male']))))
EnhancedVolcano(df,lab = rownames(df),x='fc',y='p',
      pCutoff = 0.2,FCcutoff = 0.25, ylim=c(0,1.5),
      transcriptLabSize=4, transcriptPointSize=3,
      title = 'xCell scores: Female vs. males')
```

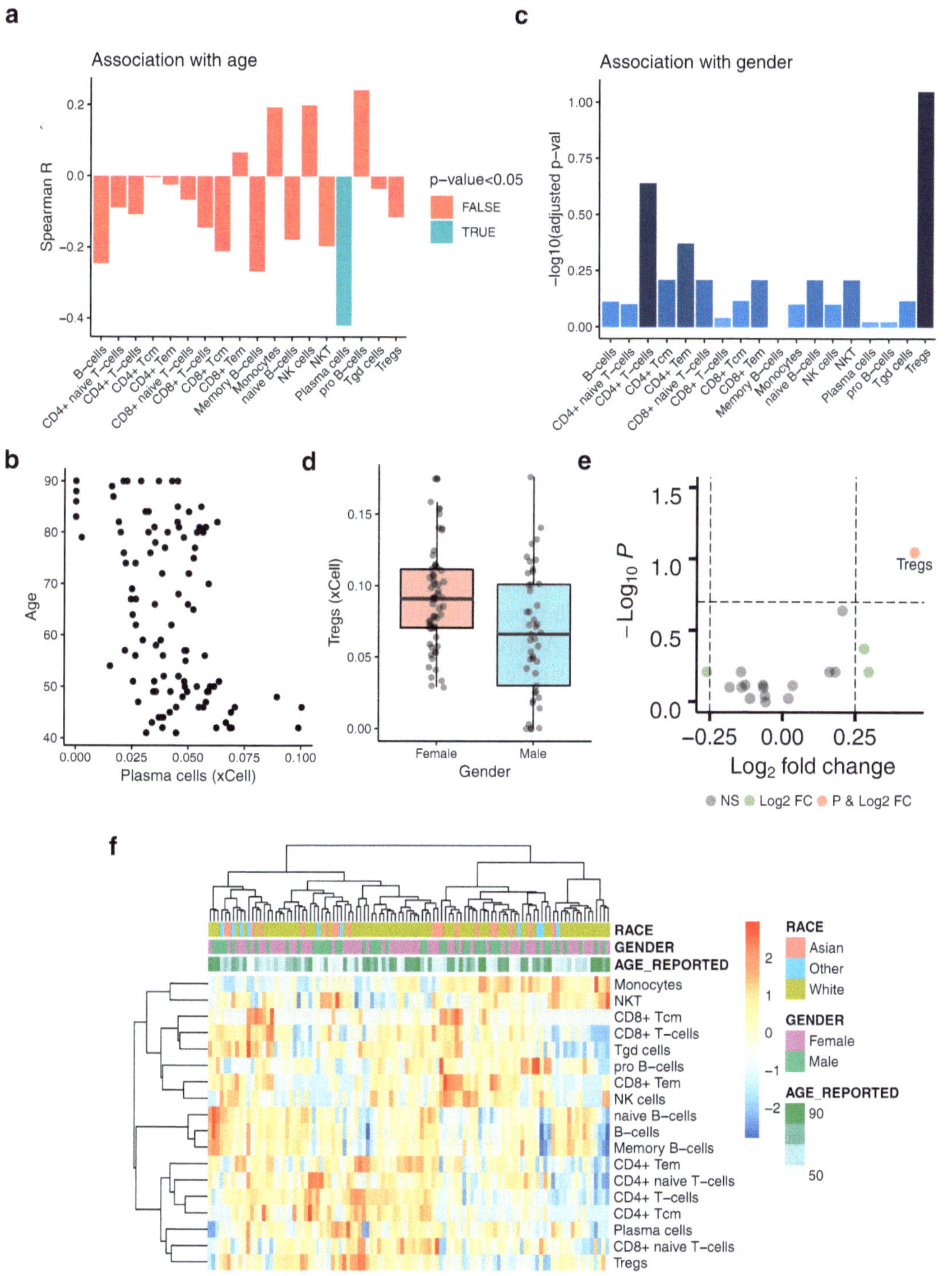

Fig. 3 Examples for downstream analyses with xCell scores. (**a**) Example for association with continuous variable. Spearman correlation coefficients of xCell scores with age in the 104 samples available in SDY420. Only plasma cells scores present a false discovery rate (FDR) < 0.05 (**b**) For continuous variables, it is best to present the scatter plot of particular associations. Here, we present a scatter plot of xCell scores for plasma

Of course, in this analysis, there are no strong associations, thus this plot is not useful.

Finally, we can present a summary of all the scores in a heatmap (Fig. 3f). Since some cells have low scores and some have high scores, we scale the rows to an average of zero and a standard deviation of 1.

```
library(pheatmap)
scaled.scores = t(scale(t(scores)))
scaled.scores[scaled.scores > 3] = 2
scaled.scores[scaled.scores < -3] = -2
pheatmap(scaled.scores,show_colnames = F,
      annotation_col=sdy$metadata[,
      c('AGE_REPORTED','GENDER','RACE'),drop=F],
      clustering_method = 'ward.D2')
```

It is important to note that this kind of visualization of the score may sometime be misleading as it can enhance small differences that are most probably noise. In addition, as we noted, it is inadvisable to use xCell for comparison across cell types (compare scores of cell types in a sample) and only compare a cell type across individuals. As a result, it is inadvisable to use xCell for determining cell type in single-cell RNA-seq data (*see* **Note 5**).

4 Notes

1. *Normalization*: xCell uses ssGSEA to calculate scores for gene signatures. This method is not affected by the actual expression values, just by the order of the genes according to the expression values. Thus, normalization or log transformation have no effect on the results. However, it is highly important to use as input for xCell expression values that are normalized to gene length, such as TPM or FPKM. It is highly important to note that xCell should not be sued with raw features, counts per

Fig. 3 (continued) cells (*x*-axis) and age (*y*-axis). (**c**) Example for association with binary variable. Wilcoxon rank-sum test *p*-values (in –log10 scale) of xCell scores and gender in the SDY420 cohort. Regulatory T cells (Tregs) are the most significant (FDR adjusted *p*-value = 0.0897). (**d**) For binary variables a box plot (or violin plot) is the best way to present particular associations. Here, we present box plots of the xCell scores for Tregs by gender. (**e**) Another recommended way to present associations with binary variables is a volcano plot (alternative to C). The *x*-axis shows the log fold-change of each cell type scores between the two groups, and the y-axis shows the –log10 adjusted *p*-values. (**f**) A heatmap, with hierarchical clusterings, is a recommended way to present unsupervised analyses. Since xCell scores should not be compared across columns, as the scores are not directly translated to percentages, we normalize the rows. However, it is important to note that this kind of visualization of the scores may be misleading, as it can enhance small differences that are most probably noise

million (CPM) or RSEM data, which are all expression values that are not normalized to gene length.

2. *Variability*: The 0 score in xCell is defined as the lowest score in the given dataset for the given cell type. Thus, to resemble the real percentages, the input data must be heterogenous. Moreover, if there is no sufficient variability in the cellular composition among the input samples, xCell will not be able to identify any signal. It is therefore highly recommended to use all data combined in one run. Failing to do so will again inevitably make the xCell results inaccurate.
3. *Calibration parameters*: The linear transformation in xCell uses calibration parameters to make the scores resemble percentages. There parameters may not be accurate for the specific experiment (*see* **Note 4**). If the analysis produces high scores for a cell type which are clearly false, it is possible to manually tweak the calibration parameters. The relevant parameters can be found in *xCell.data$spill$fv$V3*. Currently, the parameters are based on learning from the reference datasets that were used to train xCell; however, prior knowledge about the expected composition of the mixture can be helpful for adjusting these parameters.
4. *Enrichment scores*: xCell is based on gene signatures and thus produces enrichment scores, and not fractions as in methods that are based on regression. In the original publication, we discuss why we believe this is a better approach for this problem [6]. The implication of this distinction is that xCell scores should be used for comparisons across samples, not across cell types. In the transformation step, xCell does an attempt to make the scores resemble percentages; however, converting enrichment scores to percentages is a hard problem and is very platform and experiment specific.
5. *Cell of origin*: xCell is a method for detecting cell type's enrichments from mixed samples, not to detect the cell of origin. xCell is still able to recognize the cell of origin many times, but it was not trained for this problem, and there is no expectation that it will perform better than other methods for that problem. In accordance, it is strongly not advised to be performed on single-cell data. We refer the users to SingleR, another method we developed specifically for this purpose [10].

References

1. Aran D, Sirota M, Butte AJ (2015) Systematic pan-cancer analysis of tumour purity. Nat Commun 6:8971. https://doi.org/10.1038/ncomms9971
2. Galon J, Costes A, Sanchez-Cabo F et al (2006) Type, density, and location of immune cells within human colorectal tumors predict clinical outcome. Sci (New York, NY)

313:1960–1964. https://doi.org/10.1126/science.1129139

3. Carr EJ, Dooley J, Garcia-Perez JE et al (2016) The cellular composition of the human immune system is shaped by age and cohabitation. Nat Immunol 17:461–468. https://doi.org/10.1038/ni.3371
4. Binnewies M, Roberts EW, Kersten K et al (2018) Understanding the tumor immune microenvironment (TIME) for effective therapy. Nat Med 24:541–550. https://doi.org/10.1038/s41591-018-0014-x
5. Aran D, Butte AJ (2016) Digitally deconvolving the tumor microenvironment. Genome Biol 17:175. https://doi.org/10.1186/s13059-016-1036-7
6. Aran D, Hu Z, Butte AJ (2017) xCell: digitally portraying the tissue cellular heterogeneity landscape. Genome Biol 18:220. https://doi.org/10.1186/s13059-017-1349-1
7. Barbie DA, Tamayo P, Boehm JS et al (2009) Systematic RNA interference reveals that oncogenic KRAS-driven cancers require TBK1. Nature 462:108–112. https://doi.org/10.1038/nature08460
8. Bhattacharya S, Andorf S, Gomes L et al (2014) ImmPort: disseminating data to the public for the future of immunology. Immunol Res 58:234–239. https://doi.org/10.1007/s12026-014-8516-1
9. Whiting CC, Siebert J, Newman AM et al (2015) Large-scale and comprehensive immune profiling and functional analysis of normal human aging. PLoS One 10:e0133627. https://doi.org/10.1371/journal.pone.0133627
10. Aran D, Looney AP, Liu L et al (2019) Reference-based analysis of lung single-cell sequencing reveals a transitional profibrotic macrophage. Nat Immunol 20:163–172. https://doi.org/10.1038/s41590-018-0276-y

Chapter 20

Cap Analysis of Gene Expression (CAGE): A Quantitative and Genome-Wide Assay of Transcription Start Sites

Masaki Suimye Morioka, Hideya Kawaji, Hiromi Nishiyori-Sueki, Mitsuyoshi Murata, Miki Kojima-Ishiyama, Piero Carninci, and Masayoshi Itoh

Abstract

Cap analysis of gene expression (CAGE) is an approach to identify and monitor the activity (transcription initiation frequency) of transcription start sites (TSSs) at single base-pair resolution across the genome. It has been effectively used to identify active promoter and enhancer regions in cancer cells, with potential utility to identify key factors to immunotherapy. Here, we overview a series of CAGE protocols and describe detailed experimental steps of the latest protocol based on the Illumina sequencing platform; both experimental steps (*see* Subheadings 3.1–3.11) and computational processing steps (*see* Subheadings 3.12–3.20) are described.

Key words CAGE, TSS, Transcription start site, Transcription initiation, Promoter-level expression analysis, Enhancer, eRNA, Promoter activity

1 Introduction

1.1 Background

Quantification of gene expression originally required the researcher to conduct individual reactions per gene, such as a hybridization-based assay (northern blot analysis) or PCR-based method (quantitative real-time PCR). The former is parallelized in the large-scale methods of microarray analysis, fluorescent labeling, and designed probe hybridization [1]. These methods require predesigned probes or primers, which limits their scope to targets that are known genes or RNA species. In contrast, DNA sequencing technologies have been used to determine nucleotide sequences of unknown genes and can be used to count frequencies of RNA molecules in random samples to quantify the abundance of all RNA species: that is, those of both known and unknown primary structure.

Sebastian Boegel (ed.), *Bioinformatics for Cancer Immunotherapy: Methods and Protocols*, Methods in Molecular Biology, vol. 2120, https://doi.org/10.1007/978-1-0716-0327-7_20,

The earliest study of sequencing-based gene expression analyses relied on expressed sequence tags (ESTs), which are partial cDNAs generated from transcribed RNAs [2]. EST sequencing uncovered sets of RNA species transcribed from the genome and provides the level of their abundance. However, both the coverage of targeted genes and the precision of quantification are limited by throughput to determine nucleotide sequences. To increase the number of observations of RNA molecules in sequencing, serial analysis of gene expression (SAGE) [3], which is based on the production of a large number of short tags (typically 14 base pairs [bp]) generated by digestion of cDNAs with a type IIS restriction enzyme, was developed. In SAGE, the short cDNA tags are concatenated, ligated into a plasmid vector, cloned into E. coli cells, and then sequenced. SAGE analysis can quantify RNA abundance with larger coverage of genes than that achieved with EST sequencing because multiple short tags can be determined in one sequencing reaction. However, these short tags do not have enough information to recover complex RNA structures, such as alternative isoforms. Both EST sequencing and SAGE technologies primarily rely on capillary sequencers, where Sanger sequencing reactions automated in capillary electrophoresis are conducted in parallel for 96–384 samples. The sequencing capacity of capillary sequencers is substantially larger than that of manual gel-based electrophoresis; however, it is quite limited compared with the technologies described below.

The emergence of next-generation sequencing (NGS) instruments [4], which achieved remarkable progress in sequencing capacity by conducting parallel sequencing reactions on the flow cell surface, drastically extended the limit of sequencing-based transcriptome analysis. NGS instruments have enabled us to determine nucleotide sequences with similar length to those obtained in the EST-based studies described above, while allowing us to count substantially larger number of molecules than possible with SAGE. One of the widely used technologies based on NGS is RNA-seq [5–7], which determines exon structures and quantifies their abundance by sequencing random fragments of cDNAs. Exon–exon junctions can be profiled in cases where the random fragments span the boundaries, and the whole primary structure of RNAs (gene models) can be estimated computationally based on read counts across exon–exon junctions. RNA-seq has the potential to recover an entire transcriptome without prior knowledge of the genes expressed; however, it has two limitations. The first limitation is that the coverage of both 5′-end and 3′-end mapped reads yielded from RNA-seq procedures are not enough to precisely identify both mRNA ends [8, 9]. The second limitation is in the accuracy of the recovered gene models, which largely depends on the computational algorithms used; development of improved algorithms is still an active field of research [10]. Nevertheless, compared with microarrays, RNA-seq is a more accurate and powerful tool for transcriptome profiling [11]; the sequenced fragments can be aligned with mRNA sequences from RefSeq [12] or GENCODE [13].

1.2 Overview of Cap Analysis of Gene Expression

Cap analysis of gene expression (CAGE) [14] is an alternative approach to monitor the transcriptome; this approach has been employed in the Functional Annotation of the Mammalian Genome (FANTOM) project [15–22] as well as the Encyclopedia of DNA Elements (ENCODE) project [23]. It enables us to identify transcription start sites (TSSs) and to monitor their activities at single-nucleotide resolution, relying on two technologies: NGS and full-length cDNA synthesis [24], where the 5′-ends of capped RNA molecules (or the 3′-ends of cDNA molecules) are sequenced, aligned with the genomic DNA to identify TSSs, and counted to quantify their abundance. The resulting data enable us to identify core promoters (i.e., proximal regulatory regions located immediately upstream of TSSs) and to quantify their transcription initiation activities. Furthermore, the activity of distal regulatory regions, called enhancers, is also effectively assessed by CAGE profiles, by monitoring enhancer RNAs (eRNAs) divergently transcribed from enhancers [16]. While eRNA itself was described initially by monitoring enhancer regions with RNA-seq [25], CAGE-based 1-bp resolution profiles enable us to identify divergently transcribed regions, rather than just monitoring predefined regions. The capture of capped RNA molecules is specific enough that CAGE protocols do not require polyA selection or ribosomal RNA depletion which are usually required for mRNA enrichment step in RNA-seq preparation. Another major advantage of CAGE is that it makes one observation (5′-capped site) per RNA molecule regardless of its length. In contrast, read counts of RNA-seq are biased toward long RNAs, and so need to undergo length-based correction as RPKM (reads per kilobase per million). The unbiased nature of CAGE has been effectively used to uncover the transcriptome landscape, including lowly expressed RNAs and unstable RNAs [26].

1.3 History of CAGE Protocols

One of the most fundamental components of CAGE is a method, called cap trapper, to prepare full-length cDNAs [16]. The first step is biotinylation of oxidized diol in the ribose at both termini (cap and 3′-end) of RNA after reverse transcription (RT). For elimination of RNA/cDNA hybrid consisting with 3′-end incomplete cDNA (i.e., 5′-end of RNA is not complemented by cDNA), single-strand RNA is digested with RNase I. The resultant RNA/cDNA complex biotinylated in the cap structure is recovered by binding to streptavidin beads. The cDNA is then eluted by denaturation and processed depending on subsequent applications, such as sequencing platforms.

The first generation of CAGE was developed for capillary-based Sanger sequencing equipment, before the emergence of NGS [27]. In this system, 20–30 bp segments at the 5′-end of full-length cDNAs (CAGE tags) are cleaved from the cDNAs by digestion with a type IIS restriction enzyme and then concatenated to make a long template for sequencing. The templates are subsequently cloned into *E. coli* cells, amplified, and subjected to capillary-based Sanger sequencing; ~800-bp read lengths from 96 to 384 samples are sequenced in parallel. Because each template consists of multiple CAGE tags, it is processed computationally to reconstruct the original CAGE tags representing the 5′-ends of capped RNAs. This version of the CAGE protocol was effectively used in FANTOM3 to uncover the transcriptome landscape [8, 19].

A modified version of the first-generation CAGE protocol was used with the earliest commercially available NGS instrument, the 454 sequencer [20], which determines sequences up to 400 bases in length. By bypassing the labor-intensive step of template cloning into *E. coli*, this modified protocol was quite efficient, resulting in increased sequencing capacity (i.e., number of CAGE tags sequenced). This protocol was employed in FANTOM4, which addressed the gene regulatory network during a monocytic differentiation in addition to the transcriptome landscape [21, 28–30]. However, regardless of the use of capillary-based Sanger sequencing or 454 sequencing, read counts obtained from the first-generation CAGE protocol were largely reproducible but biased from theoretical random sampling, probably because the concatenation process consisted of a series of operations requiring multiple PCR amplification steps.

The second generation of CAGE, which does not require concatenation of CAGE tags, was developed for the Illumina Genome Analyzer [22], which produces 35–50 base short reads but larger amounts of read counts than the 454 genome sequencer. With the concatenation-free protocol and increased capacity of the Illumina Genome Analyzer, quantitation was improved and this protocol was employed in the ENCODE project [23] to identify more than 60,000 TSSs. However, several challenges remained in the amount of starting RNA required (microgram level) and in the quantification bias caused by generating 5′-end short tags with restriction enzymes and the need for a few PCR amplification steps.

To address the problem of the amount of starting RNA, an alternative protocol called nanoCAGE, which requires submicrogram amounts of total RNA, was developed for the Illumina sequencer [31]. This protocol is based on template switching, which relies on the terminal transferase activity of reverse transcriptase (RTase) [19, 32] to obtain 5′-end sequences of RNA molecules, and thus does not use cap trapping. The template switching chemistry is optimized to increase the specificity toward the

5′-terminal end of the RNA molecule with efficient addition of an artificial adapter at the 3′-end of the cDNA, with some preference to the cap structure [33, 34]. Since this alternative protocol does not require cleavage of the 5′-end from the cDNA as short tags, it enables us to identify the internal portion of the RNA and the priming site of RT by sequencing paired ends of cDNAs. The concept of sequencing the 5′-end and internal portion of an RNA molecule at the same time is termed CAGEscan and has been effectively used to identify longer 5′-ends of RNA than the other CAGE protocols relying on single-end sequencing [23].

The third generation of CAGE was developed to address the remaining challenge, the read count bias due to PCR amplification steps. The third-generation CAGE protocol bypasses PCR amplification by using a "third-generation" NGS instrument, a single-molecule sequencer called a HeliScope Genetic Analysis System, produced by Helicos Bioscience [34]. The unique feature of this sequencer is that it determines the nucleotide sequence of any template length in a single molecule, with the number of read counts comparable to that of the Illumina Genome Analyzer. Single-strand 5′-end complete cDNA, obtained by cap trapping, is sequenced after addition of poly-dA tail with a 3′-end block to prevent extension. It enables the protocol to bypass PCR amplification before loading material to the sequencer. Furthermore, the single-molecule sequencing reaction does not require PCR amplification, making the protocol completely PCR-free. Performance of the protocol is highly reproducible and quantitative [35, 36], and it has been employed in the FANTOM5 project, which aims to obtain a complete catalogue of TSSs in mammalian multiple cell types in their normal state. It has been applied to more than 3000 mammalian samples, and ~200,000 TSS regions and ~65,000 enhancers have been defined in these samples across the human genome [18, 37].

The third-generation CAGE protocol has now been applied to the Illumina sequencing platform because the HeliScope instrument and reagents have become unavailable on the market. By optimization of the cDNA recovery rate and careful control of template length, the protocol has been applied to the Illumina sequencing platform without either PCR amplification or digestion by restriction enzyme before loading sequencing materials on flow cells. Given that the sequencing platform requires clonal PCR for sequencing reactions in its flow cells, this application cannot be completely PCR-free. However, this protocol, which consists only of RT, cap trapping, and adapter ligations at both termini [38], is the one with the least bias due to PCR among the currently available protocols.

Hereafter, we describe the latest, the third-generation CAGE protocol for the Illumina sequencing platform, which is performed without PCR amplification in the library preparation (*see*

Subheadings 3.1–3.11). We also describe the computational processes required for interpretation (*see* Subheadings 3.12–3.19). Lastly, we provide an overview of the application of CAGE to investigations of biological systems, including gene regulation (*see* Subheadings 3.18–3.20).

2 Materials

2.1 Equipment

1. Agilent 2100 Bioanalyzer.
2. 1.5 mL SnapLock Microtube, Non-sterile, Max Clear, Maximum Recovery.
3. 16-well micro PCR plate, clear.
4. 0.2 mL 8-strip PCR Dome Tube.
5. 0.2 mL Thin Wall Clear PCR Strip Tubes and Clear Strip Caps.
6. Micropipettes.
7. Multichannel pipetters, 8-channel.
8. Low-binding filter tip.
9. Dynal magnetic bar (Invitrogen).
10. Dynal magnetic stand (Invitrogen).
11. Thermal cycler, 96-well format.
12. 96-well plate and tube centrifuge.
13. Centrifugal concentrator.
14. ABI PRISM 7900HT Sequence detection system.
15. ABI PRISM 384-well plate.
16. ABI PRISM Optical Adhesive Covers.

2.2 Reagents

1. RNase Zap (Ambion).
2. Quant-iT Oligreen ssDNA Reagent and Kit (Invitrogen).
3. 10 mM dNTP mix.
4. SuperScript III RNase H^- Reverse Transcriptase (Invitrogen).
5. Sodium periodate ACS Reagent Grade.
6. Biotin (Long Arm) Hydrazide.
7. Dynabeads M-270 Streptavidin (Invitrogen).
8. LiCl binding buffer (7 M LiCl, 20 mM Tris–HCl (pH 7.5), 0.1% Tween 20, 2 mM EDTA (pH 8.0)).
9. TE wash buffer (10 mM Tris–HCl (pH 7.5), 0.1% Tween 20, 1 mM EDTA(pH 8.0)).
10. Release buffer (1× RNase ONE buffer, 0.01% Tween 20).
11. RNase ONE nuclease.

12. Ethanol, ≥99.5%.
13. 2 M NaOH.
14. LCL-LB1 buffer (200 mM Tris–HCl (pH 7.0), 1 M NaCl).
15. CAGE library preparation kit (Dnaform) [39].

3 Methods

3.1 Outline of the Experimental Process

Production of a CAGE library follows the following steps: reverse transcription (RT), oxidation, biotinylation, RNase I digestion, cap trapping, releasing cDNA, and sequencing library preparation. In principle, the diol of ribose sugars at both ends of RNA is oxidized with sodium periodate, generating dialdehyde by cleavage of the sugar ring. The dialdehyde reacts with biotin hydrazide generating a covalent bond between ribose and biotin. The biotinylated riboses are at both termini. The single-strand RNA portion is digested with RNase I. This process makes it possible to select fully reverse-transcribed cDNA (which reaches the 5′-end of the RNA) from among incomplete cDNA or uncapped RNA. In the following sections, the key points of each step are described. The detailed practical protocol is described in Chapter 7 of Methods in Mol. Biol. vol. 1164 [38].

3.2 RNA Quality and Quantity

For CAGE library preparation, 5 μg of total RNA is required for the standard protocol. The RNA should be prepared by using a commercially available kit such as RNeasy or miRNeasy mini or micro kits, or equivalent. The RNA quality is important and should satisfy the criteria of OD260/230 ratio of ≥1.8, OD260/280 ratio of ≥1.8, and RNA Integrity Number (RIN) value of ≥7 as determined by BioAnalyzer RNA nano chip analysis.

3.3 RT

1. Mix 5 μg of total RNA and 2.5 nmol of RT primer in 10 μL water: RNA/primer mix (*see* **Note 1**).
2. Incubate at 65 °C for 5 min.
3. Chill on ice.
4. Mix the following reagents: Enzyme mix.

Regents	Volume	Final concentration
First-strand buffer, 5×	4 μL	1×
0.1 M DTT	1 μL	5 mM
dNTPs, 10 mM each	1 μL	500 μM each
SuperScriptIII RT (200 U/μL)	2 μL	400 U
Water	2 μL	
Total volume	10 μL	

5. Add 10 μL of the enzyme mix solution to RNA/primer mix solution.
6. Incubate at 25 °C for 30 s, and then at 50 °C for 30 min.
7. Chill on ice.
8. Purify the RNA/cDNA complexes by using 1.8× volume of Purification beads 1 (or Agencourt RNAClean XP kit) and elute with 40 μL of water.

In RT reaction, important factors affecting the length of the resultant cDNA are reaction temperature and primer concentration (*see* **Note 2**).

3.4 Diol Oxidation with Sodium Periodate

1. Mix the following reagents:

Regents	Volume
RNA/cDNA from RT	40 μL
1 M sodium acetate (pH 4.5)	2 μL
250 mM sodium periodate	2 μL

2. Chill on ice in the dark for 5 min.
3. Add 16 μL of 1 M Tris–HCl (pH 8.5) to adjust pH.
4. Purify the resultant oxidized RNA/cDNA complexes by using 1.8× volume of Purification beads 1 (or Agencourt RNAClean XP kit) and elute with 40 μL of water.

In this chemical reaction, prepared solution should be dispensed in aliquots to avoid repeat freeze-and-thaw cycles, and then stored in a −80 °C freezer (*see* **Note 3**).

3.5 Biotinylation

1. Mix the following reagents:

Regents	Volume
RNA/cDNA after oxidation	40 μL
1 M sodium acetate (pH 6.0)	4 μL
100 mM biotin hydrazide	4 μL

2. Incubate at 40 °C for 30 min.
3. Purify the resultant biotinylated nucleic acids by using 1.8× volume of Purification beads 1 (or Agencourt RNAClean XP kit) and elute with 40 μL of water.

Biotin hydrazide should be dissolved at 100 mM in DMSO, dispensed in aliquots to avoid repeat freeze-and-thaw cycles, and then stored in a −80 °C freezer (*see* **Note 4**).

3.6 RNase ONE Digestion (See Note 5)

1. Mix the following reagents:

Regents	Volume
Biotinylated RNA/cDNA	40 μL
RNase ONE buffer, 10×	4.5 μL
RNase ONE (10 U/μL)	0.5 μL

2. 37 °C for 30 min.
3. Purify the resultant RNase One digested sample by using 1.8× volume of Purification beads 1 (or Agencourt RNAClean XP kit) and elute with 40 μL of water.

3.7 Cap Trapping

1. Dispense 30 μL × sample number of M-270 streptavidin beads slurry in a new tube.
2. Place on the magnetic stand and remove supernatant.
3. Wash the beads with 30 μL × sample number of LiCl binding buffer.
4. Place on the magnetic stand and remove supernatant.
5. Repeat wash three times.
6. Resuspend 95 μL × sample number of LiCl binding buffer.
7. Mix RNase I-treated RNA/cDNA complexes with streptavidin beads:

Regents	Volume
RNase ONE-treated RNA/cDNA	40 μL
M270 streptavidin beads slurry	95 μL

8. Incubate at 37 °C for 15 min with interval pipetting for 7 min.
9. Place in a magnetic stand for 2 min and remove the supernatant.
10. Wash the beads with 150 μL of TE wash buffer for three times (removing the supernatant each time).
11. Suspend beads in 35 μL of release buffer.
12. Incubate at 95 °C for 5 min.
13. Chill on ice for 2 min or more.
14. Place in a magnetic stand for 2 min and recover the supernatant into a new tube.
15. Resuspend beads in 30 μL of release buffer.
16. Place in a magnetic stand for 2 min and recover the supernatant into the recovery tube.

17. Mix released cDNA with RNase H and RNase I:

Regents	Volume
Released cDNA	65 μL
RNase H (2 U/μL)	3 μL
RNase I (10 U/μL)	2 μL

18. Incubate at 37 °C for 30 min.
19. Purify the released cDNA by using 1.8× volume of Purification beads 2 (or Agencourt AMPure XP kit) and elute with 48 μL of water.

The interaction between biotin and streptavidin is dependent on the collision rates among the molecules (*see* **Note 6**). Therefore, the beads should be resuspended at 7 min intervals for 15 min at 37 °C. After the supernatant is completely removed from the beads, and the beads have undergone a series of washes, the beads are suspended in release buffer, heated at 95 °C for 5 min, and then chilled on ice to denature the RNA/cDNA double-stranded molecules. The released single-stranded cDNA molecules are recovered in the supernatant. The eluent contains not only cDNA but also pieces of RNAs, and so needs to be digested with RNase H and RNase ONE.

3.8 Quality Check of cDNA

1. Remove 8 μL of eluted cDNA for quality check (4 μL for quantification of cDNA; 4 μL for qPCR.
2. Dry the released cDNA by using a SpeedVac at 37 °C.
3. Dissolve the cDNA in 4 μL water.

The recovered "cap-trapped" cDNA is the single-stranded DNA that is used to prepare the sequencing library. Before the library construction, the cDNA yield and quality need to be determined. First, the cDNA yield is measured by using fluorescence dye for single-stranded nucleic acids, such as that provided in the Oli-Green fluorescence measurement kit. The standard curve range should be between 10 pg/μL and 1 ng/μL. The expected yield of cDNA is 5–20 ng. Second, the contents of the cDNA can be measured by qRT-PCR (*see* **Note 7**).

3.9 Adapter Ligation

Single-Strand Linker Ligation (SSLL) performs first with 5′-linker and then independently with 3′-linker (*see* **Note 8**).

1. Preheat 5 μL of 2.5 μM 5′-linker at 55 °C for 5 min, and then chill on ice for 2 min.
2. Preheat released cDNA at 95 °C for 5 min, and then chill on ice for 2 min.

3. Mix 4 μL of linker and 4 μL of released cDNA.
4. Add 16 μL of DNA Ligation Kit (Mighty Mix).
5. Incubate at 16 °C for 16 h.
6. Add 46 μL of Nuclease-free water.
7. Purify the ligated cDNA by using 1.8× volume of Purification beads 2 (or Agencourt AMPure XP kit) and elute with 40 μL of water.
8. Heat at 95 °C for 5 min and chill on ice immediately.
9. Purify the ligated cDNA by using 1.2× volume of Purification beads 2 (or Agencourt AMPure XP kit).
10. Dry the ligated cDNA in a SpeedVac at 37 °C.
11. Dissolve the ligated cDNA in 4 μL of water.
12. Preheat 5 μL of 2.5 μM 3′-linker at 65 °C for 5 min, and then chill on ice for 2 min.
13. Preheat released cDNA at 95 °C for 5 min, and then chill on ice for 2 min.
14. Mix 4 μL of linker and 4 μL of released cDNA.
15. Add 16 μL of DNA Ligation Kit (Mighty Mix).
16. Incubate at 16 °C for 16 hrs.
17. Add 46 μL of Nuclease-free water.
18. Purify the ligated cDNA by using 1.8× volume of Purification beads 2 (or Agencourt AMPure XP kit) and elute with 40 μL of water.
19. Heat at 95 °C for 5 min and chill on ice immediately.
20. Purify the ligated cDNA by using 1.2× volume of Purification beads 2 (or Agencourt AMPure XP kit) and elute with 40 μL of water.

Ligation of adapters to 5′- and 3′-ends should be done independently to avoid generating artifacts by ligation of adapters together. After every ligation reaction, the nucleic acids should be purified by Purification beads 2 (or Agencourt AMPure XP kit) to remove excess free adapters (*see* **Note 9**).

3.10 Quality Check of Sequencing Library

Unlike other NGS sequencing libraries, it is difficult to measure the molar concentration of CAGE sequencing libraries by calculation from molecular weight and the recovery mass. This is because CAGE preparation cannot employ size control. The qPCR measurement kit such as KAPA Library Quantification kits is available for this purpose.

3.11 Sequencing on Illumina HiSeq2500 by One-Shot Loading

To load the cDNA libraries onto the cBot, we recommend the use of a special loading method ("one-shot loading") described here rather than the standard procedure provided by Illumina because the library yield is too low when following the manufacturer's instructions:

1. Dilute 320 amol of cDNA library in 19 μL of water.
2. Add 1 μL of 2 M sodium hydroxide.
3. Incubate at room temperature for 5 min.
4. Add 60 μL of LCL-LB1 buffer.
5. Load 75 μL of sample onto the cluster generation system (cBot).

3.12 Outline of Computational Processing

The CAGE protocol produces a large amount of short reads. Interpretation of the data as TSS activities or promoter-level expression requires a series of computational processes: preprocessing to filter out contaminated signals, alignment with the reference genome sequence, counting the 5′-ends of the alignments as frequency of transcription initiation at 1-bp resolution (called CTSS, CAGE tag start sites), and aggregation of the neighboring signals as promoter-level expression (Fig. 1). A series of CAGE data sets have been published and maintained through large consortia, in particular, FANTOM and ENCODE. Comparison of newly obtained data with these reference data sets will help to check data quality and facilitate the interpretation of data in biological systems.

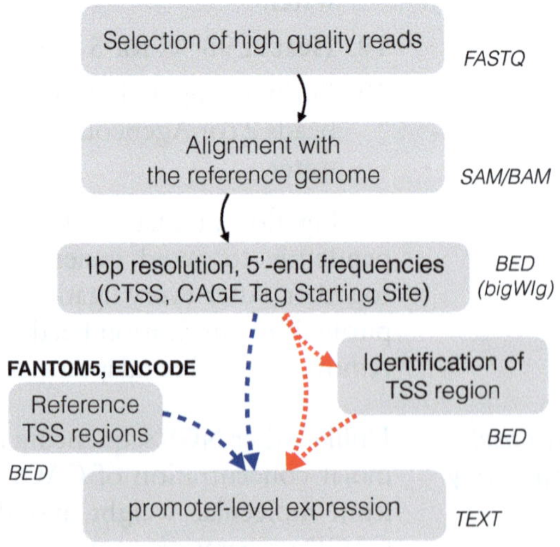

Fig. 1 Overview of the computational steps. Gray boxes indicate individual computational steps; italic font indicates the data format used to store the resulting data. Blue and red arrows indicate two alternative approaches to monitor promoter-level expression: assaying of a predefined reference set of transcription start site (TSS) regions (blue) and assaying of TSS regions identified based on the newly obtained data (red)

The most straightforward way to conduct the series of data processing steps described below would be to work with a command line interface; however, this would not be suitable for wet-lab researchers. Therefore, we routinely use a workflow system called MOIRAI [40], where whole processing steps are graphically represented, while keeping all parameters and intermediate files as traceable.

3.13 Raw Read Preprocessing

Raw data produced by the Illumina sequencer (in vendor-specific Binary base call (BCL) format) is converted to Sanger FASTQ format by the vendor's software, bcl2fastq. The reads, 5′ barcode (linker) and 3′ equivocal (1 bp) sequences are trimmed by fastx_trimmer, a program in the FASTX-toolkit [39]. When multiple samples are mixed into one library after indexing with sample-specific linkers with unique barcode sequences (*see* Subheading 3.9), one FASTQ-formatted file corresponding to one lane is split into multiple files according to the unique barcode sequences. The result is that each data file represents one sample.

All reads containing ambiguous bases (N) are discarded because sequencing reactions producing such reads are likely unreliable. Reads with low complexity or known contaminants such as primer dimers are discarded by using TagDust [41], and reads matching the rDNA repeat unit (U13369.1), which is not assembled in the reference genome, are discarded by using the rRNAdust program if there are no more than two mismatches, insertions, or deletions [42].

3.14 Alignment of CAGE Reads

The remaining reads are aligned with the reference genome sequences in a process called genome mapping to identify the genomic location from which the sequenced transcripts are derived. For short reads (such as <50 bases), Burrows-Wheeler Aligner (BWA, [44]) can be used effectively. For longer reads (such as 250 bases), the use of spliced read aligners (such as TopHat [44] and STAR [45]) has to be considered because these reads may span across exon-exon junctions. Note that BWA is not suitable for HeliScope-derived reads, which range in length from 20 to 50 bp and have a higher error rate, including insertions and deletions, than Illumina-derived reads; for HeliScope-derived reads, DELVE [46] can be used effectively. Typically, alignments are stored in SAM (Sequence Alignment/Map) format or its binary version, BAM.

An example of command lines to align CAGE reads to the reference genome and to store the result in BAM format:

```
% bwa index REFERENCE.fa # only once
% bwa aln REFERENCE.fa CAGE_READS.fa > aln_sa.sai
% bwa samse REFERENCE.fa aln_sa.sai CAGE_READS.fa > CAGE_-
MAPPED.sam
```

```
% samtools view -Shb CAGE_MAPPED.sam > CAGE_MAPPED.bam
% samtools sort CAGE_MAPPED.bam CAGE_MAPPED.sorted
```

Following genome mapping, alignments are assessed for their suitability for inclusion in downstream analysis according to two properties: goodness of the alignment itself (i.e., alignment score and sequence identity) and mapping quality (i.e., the probability that a read is misplaced). When using BWA, alignments with mapping quality (Phred scale) of 20 or more are selected, which means that the accuracy of the mapping is more than 99% (probability of misplacement is less than 1%) (*see* **Note 10**). Selection of suitable alignments can be achieved by using SAMtools [47], a set of utilities to process SAM or BAM formatted files, with appropriate options such as "–q 20" (*see an example of command lines in* Subheading 3.17).

3.15 Quantification of Transcription Initiation Activities

The selected alignments are counted by their 5′-ends (CTSSs) to give the observed number of transcription initiation events. The result, TSS activities at single-nucleotide resolution, can be stored in a data file in Browser Extensible Data (BED) format, which is the most basic data structure of CAGE profiles. To generate the CTSS profiles in BED format, selected alignments from the SAM or BAM file are converted into BED and counted by the genomeCoverageBed program in BEDtools, a toolset for processing BED-formatted data files [48]. With appropriate genomeCoverageBed options, that is, "-5" for counting 5′-end positions, "-strand" for strand-specific processing, and "-bg" for printing the result in bedGraph format, CTSS profiles are generated.

3.16 Visualization and Quality Assessment

The resulting data, CTSS profiles, can be visualized via genome browsers. The bedGraph format generated above can be directly subjected to widely used genome browsers, such as IGV (Integrative Genomics Viewer) [49] and the UCSC (University of California, Santa Cruz) Genome Browser [50]. Note that bigWig is an efficient format to query and retrieve data points across the genome [51]; bigWig can also effectively handle a large number of CAGE profiles. To convert bedGraph to bigWig, the bedGraphToBigWig program in Kent source utilities can be used [52], (*see* **Note 11**). Alternatively, ZENBU [53] can be used to visualize the data in SAM, BAM, or BED.

Here is an example of command lines to count 5′-ends of individual alignments and generate a CTSS file:

```
#Counting the number of 5′-ends of CAGE read alignments as
bedGraph
% samtools view -F 4 -u -q 20 CAGE_MAPPED.sorted.bam \
  | genomeCoverageBed -g ChromInfo.txt -ibam /dev/stdin -5 -bg
-strand + \
```

```
  | sort -k1,1 -k2,2n > CAGE.ctss.fwd.bedGraph
% samtools view -F 4 -u -q 20 CAGE_MAPPED.sorted.bam \
  | genomeCoverageBed -g ChromInfo.txt -ibam /dev/stdin -5 -bg
-strand - \
  | sort -k1,1 -k2,2n > CAGE.ctss.rev.bedGraph
# Convert the bedGraph files into a BED file
% cat CAGE.ctss.fwd.bedGraph \
  | awk 'BEGIN{OFS="\t"}{print $1,$2,$3,".",$4,"+"}' > CAGE.
ctss.bed
% cat CAGE.ctss.rev.bedGraph \
  | awk 'BEGIN{OFS="\t"}{print $1,$2,$3,".",$4,"-"}' \
  >> CAGE.ctss.bed
% sort -k1,1 -k2,2n CAGE.ctss.bed > CAGE.ctss.sort.bed

# Convert the bedGraph files into bigWig files
% bedGraphToBigWig CAGE.ctss.fwd.bedGraph ChromInfo.txt CAGE.
ctss.fwd.bw
% bedGraphToBigWig CAGE,ctss.rev.bedGraph ChromInfo.txt CAGE.
ctss.rev.bw
```

The most basic way to assess data quality is to examine whether the signals (or read counts) are accumulated in TSSs of genes expressed in the target cells since CAGE protocols are developed to quantify TSS activities. Data visualization of CAGE reads on the reference genome is an effective way to manually inspect data from this perspective. Alternatively, the data can be inspected quantitatively by assessing the "promoter rate", defined as the number of CAGE reads starting from promoter regions within mapped reads on the genome as a proportion of total CAGE reads (*see* **Note 12**). CAGE peaks produced in the FANTOM5 project [16] or proximal regions (e.g., +/−500 bp) relative to the 5′-ends of gene models, such as those in RefSeq [12] and GENCODE [13] gene sets, can be used as reference promoter sets.

3.17 Reference Sets of TSS Regions in Human and Mouse

A promoter is a region of genomic DNA that initiates transcription, which means that adjacent TSSs share almost the same promoter sequence. Since CAGE protocols essentially capture transcription initiation events triggered by promoter activation, identification and monitoring of a group of such adjacent TSSs, called a TSS cluster or TSS region, is used to monitor promoter activities. We would like to note that the set of TSS regions defined in FANTOM5 can be used as a comprehensive reference set for expression analysis of newly obtained CTSS data (Fig. 1, blue line). The reference regions are defined based on a wide range of samples, and most CAGE signals fall within these regions, unless the profiled cells or tissues are extremely different from the FANTOM5 samples. While this approach is limited to studies on human and mouse, it reduces the substantial effort that would be required to identify and annotate TSS regions from scratch, and it facilitates data interpretation because the expression patterns have been profiled across a series of samples.

```
# Counting the number of CAGE reads
% sum_fwd=$(bigWigToBedGraph CAGE.ctss.fwd.bw /dev/stdout \
  | awk '{ sum = sum + $4 * ($3 - $2) }END{print sum}' )
% sum_rev=$(bigWigToBedGraph CAGE.ctss.rev.bw /dev/stdout \
  | awk '{ sum = sum + $4 * ($3 - $2) }END{print sum}' )
% sum=$(echo | awk -v fw=$sum_fwd -v rev=$sum_rev '{sum=fw +
rev; print sum}')
# Making a header line
% printf "01STAT:MAPPED\t$sum" > CAGE.count.txt

# Counting the CAGE tags on forward strand
% zcat hg19.cage_peak_phase1and2combined_coord.bed.gz \
  | awk '{if($6 == "+"){print}}' \
  | bigWigAverageOverBed CAGE_MAPPED.ctss.fwd.bw /dev/stdin /
dev/stdout \
  | cut -f 1,4 >> CAGE.count.txt

# Counting the CAGE tags on reverse strand
% zcat hg19.cage_peak_phase1and2combined_coord.bed.gz \
  | awk '{if($6 == "-"){print}}'\
  | bigWigAverageOverBed CAGE.ctss.rev.bw /dev/stdin /dev/stdout \
  | cut -f 1,4 >> CAGE.count.txt
```

3.18 Approaches to Identifying TSS Regions

In the analysis of the first-generation CAGE data, a simple approach to define TSS regions is employed: neighboring CTSSs within 20 bp of each other are aggregated by a single-linkage approach into a unit called a "tag cluster," which can be used effectively to delineate promoter shape as well as serving a unit to define promoters [31]. In parallel, a method called Paraclu [54] has been developed for multiscale clustering, in which several small clusters may be included in a large cluster. With the remarkable increase in sequencing capacity gained by employment of NGS in the second- and third-generation CAGE protocols, baseline signal around primary TSSs and exon signals become evident, while some of the exon signals are explained by recapping of fragmented RNA [55]. Increase of baseline signals connects neighboring tag clusters when using the single-linkage approach [56]. To select peaks against the baseline signals, a method called RECLU [57] selects regions that are reproducibly identified by Paraclu based on assessment of IDR ("irreproducible discovery rate") across replicates. CAGEr [58], a Bioconductor R package, consists of several programs to process CAGE data, such as quality filtering, trimming of CAGE-specific G nucleotide additions at the 5′-end of the read (*see*

Note 13), normalization of CAGE tag counts based on power-law distribution, and detection of TSS shift (in comparisons of different tissues or conditions). Additionally, CAGEr implements three clustering methods: disticlu, which is based on the distance between neighboring TSSs, a "custom" method that is based on a user-defined window size, and Paraclu. The FANTOM5 project, which aimed to produce a comprehensive catalogue of promoter-level expression in various cell types, had an additional challenge—to identify TSS clusters across very heterogenous cellular states in which several of neighboring TSSs are under different regulations. A method called "decomposition peak identification" [16] was developed to cope with such heterogenous TSS profiles with the use of independent component analysis [59]; application of this method resulted in the selection of nearly 200,000 CAGE peaks in human and 240,000 peaks in mouse that represent robustly expressed RNAs [27].

3.19 Promoter-Level Expression Analysis

Promoter-level expression profiles are obtained by counting CAGE tags starting from each of the TSS regions defined by FANTOM5 or newly obtained data. The resultant read counts are then subjected to a digital gene expression analysis method, such as edgeR [60] or DEseq2 [61], which normalizes the expression intensities and performs differential analysis. Notably, quantification of CAGE does not require consideration of exon length since only one 5′-end is determined per RNA molecule; in contrast, read counts in RNA-seq require scaling according to exon length, with the total number of mapped reads being expressed as RPKM or FPKM (reads or fragments per kilobase of transcript per million mapped reads) [37]. Thus, promoter-level expression measured by CAGE is primarily assessed based on counts per million (CPM), a simple scaling to adjust library sizes. Additional normalization scheme, such as relative log expression [60], can be effectively used.

```
#Normalization steps in R environment
#Start with R
library(edgeR)

# Creating an object of class "DGEList" from table of CTSS
counts.
d = DGEList(counts=data[-1,], group=factor(c(1,1,2,2)), lib.
size=data[1,])

# Calculation of normalization factors and estimating disper-
sions for statistical tests (e.g., Relative Log Expression)
dgeL = calcNormFactors(d, method="RLE")
dgeL = estimateCommonDisp(dgeL)
dgeL = estimateTagwiseDisp(dgeL)
```

```
# Statistical test (e.g., exactTest function)
dep = exactTest(dgeL,pair=c("2","1"))
# statistical test result of all promoters
exp = as.data.frame(topTags(dep, n = nrow(data[-1,])))
```

3.20 Applications of CAGE to Study Cancers and Cancer Immunotherapy

Monitoring transcription initiation activities has a wide range of applications, and here we briefly overview a limited number of examples. Transcription of retrotransposons can be quantified by careful interpretation of CAGE reads, and their unique pattern of expression across tissues [21], induced pluripotent stem cells, embryonic stem cells [62], and liver cancer [63] has been uncovered. Application of CAGE to time series samples during cellular differentiation generates information on the dynamic behavior of promoter activities, and its combination with genome scanning of DNA motifs recognized by transcription factors can be used to estimate the level of transcriptional control by regulatory motifs, termed "motif activities" [30]. Combination with cellular perturbation, for example, siRNA-based knock down, can be effectively used to predict DNA motifs [64]. Surprisingly, distal regulatory elements, enhancers, can be identified by recognizing their unique pattern of CAGE profiles, divergent transcription [65]. By counting CAGE reads obtained from the enhancer region, we can monitor the abundance of eRNAs, which likely corresponds to enhancer activity. A series of time course profiles indicate that change in the activity of enhancers happens much earlier than that of promoters [17].

CAGE was further used to study cancers, including identification of diagnostic markers [66–68]. Novel biomarkers discriminating squamous cell carcinoma (SCC) from adenocarcinoma (AD) in lung cancers were identified through application of CAGE to primary lung cancer tissues in nearly 100 cases [68]. CAGE profiles obtained from primary lesions of endometrial cancers, not lymph node, uncovered molecular markers indicating presence or absence of lymph node metastases [67]. In-depth sequencing of CAGE from five melanoma cell lines that are directly derived from patient biopsies revealed melanoma specific promoter of Prominin-1 (PROM1), a cell surface marker used to isolate cancer stem cells [68]. Alternative promoters in cancers were also studied by using exon arrays [69] and RNA-seq [70], and the impact of TSS heterogeneities on translation initiation event was further revealed by reporter assays [71].

Treatment of DNA methyltransferase and histone deacetylase inhibitors are used as epigenomic therapies for the treatment of several hematopoietic malignancies [72–76]. The genome-wide TSS profiles of CAGE revealed that these epigenomic agents induce cryptic transcription of thousands of treatment-induced nonannotates TSS (TINATs) which frequently spliced into protein-coding exons and encoded truncated or 5′-chimeric open reading frames in cancer cells [73]. Notably, some of the chimeric peptides translated from TINAT fusion transcripts present on MHC class I molecules in cancer cells, suggesting that cryptic transcription of TINAT fusion are a potentially immunogenic. CAGE can be effectively used to identify such immunogenic factors crucial to cancer immunotherapy, as well as exploration of cancer biomarkers and transcriptional dysregulation at promoter level.

4 Notes

1. RT primer (cartridge purified): 5′-Phos-TCTNNNNNN-3′.
2. RT of total RNA should be performed with random primers to recover cDNAs from every RNA molecule including non-polyA+ RNA. Important factors affecting the length of the resultant cDNA are reaction temperature and primer concentration. High reaction temperature is required for an enzyme to overcome the strong secondary structure to reach the 5′-end of the RNA. Recent commercially available RTases can be used at relatively high temperatures, such as 50 °C or more. Illumina NGS employs cluster generation for the amplification of sequencing templates. The template length must be controlled for the bridge PCR on the flow cell surface. However, CAGE cDNA is different from other NGS libraries in which the length of template was controlled; it is difficult to control its length only by RT reaction. Therefore, a high concentration of primer enabled to shorten the cDNA length by increasing the collision rate of primer against RNA molecules.
3. In this protocol, RNA/cDNA complexes are oxidized with sodium periodate. This chemical reaction is rapid. Sodium periodate solution is stable, so the solution should be prepared in a large volume under RNase-free conditions. After the reaction, the nucleic acids must be purified by using a Purification beads 1 (or Agencourt RNAClean XP kit) because the oxidized complexes still contain RNA.
4. Biotin hydrazide has a long alkyl arm between the biotin and hydrazide moieties to reduce steric hindrance. The biotinylation reaction is not rapid. The optimized reaction condition is 23 °C for 2 h; shorter times will decrease the yield of trapped RNA/cDNA.
5. The biotinylated RNA/cDNA complexes are digested with RNase I to remove the biotinylated ribose at the both 5′- and 3′-end of incompleted RNAs. RNase I digests the single-strand RNA portion, but not the RNA/cDNA double-strand portion of the complex. Activity of RNase I from different manufacturers varies, and weak RNase I reactions should be avoided

because they cause high ribosomal content in the CAGE library. We consider that RNase ONE ribonuclease (Promega) is one of the best enzymes for this reaction.

6. Cap trapping is used to select biotinylated capped RNA/cDNA complexes among other RNA/cDNA molecules. The biotin moiety is trapped by streptavidin attached on the solid surface of the bead. Streptavidin magnetic beads are convenient for this step because they are easy to wash by dispersal in buffers and can be simply harvested by magnet. In the early development version, we employed multiporous glass beads to ensure high capacity. However, the excess streptavidin capacity caused a high ribosomal content in the library, and the porous structure caused nonspecific adsorption of nucleic acids. Thereafter, we changed to using magnetic beads that have streptavidin immobilized only on the surface, M270 streptavidin beads from Invitrogen. Streptavidin beads from other manufacturers might be available for this step, but it is important to only use bead materials that show low nonspecific adsorption of nucleic acids.
7. Routinely, we use primer sets from near the 5′-end of the transcribed RNA molecules of the 18S ribosomal RNA and beta actin genes of human and mouse. Higher cDNA recovery sometimes correlates with the higher ribosomal content. Therefore, if the cDNA yield is higher than expected, qRT-PCR for ribosomal RNA should be conducted to check the ribosomal content prior to sequencing.

   Primer set for qPCR:

   5′ Human ACTB (Fwd: GGCATGGGTCAGAAGGATT, Rev.: AGGTGTGGTGCCAGATTTTC), 5′ Human 18S rRNA (Fwd: CTGGTTGATCCTGCCAGTAG, Rev.: TCTAGAGTC ACCAAAGCCGC).

   5′ Mouse ACTB (Fwd: TATCGCTGCGCTGGTCGTCG, Rev.: TAGGGCGGCCCACGATGGAG), 5′ Mouse 18S rRNA (Fwd: GCCATGCATGTCTAAGTACGCACG, Rev.: TCAGC GCCCGTCGGCATGTA)
8. 5′linker & 3′linker oligos (HPLC purified)

   5′linker_N6up: 5′-AATGATACGGCGACCACCGAGA TCTACACTCTTTCCCTACACGACGCTCTTCCGATCTN NNNNN-Phos-3′.

   5′linker_GN5up: 5′-AATGATACGGCGACCACCGAGA TCTACACTCTTTCCCTACACGACGCTCTTCCGATCTG NNNNN-Phos-3′.

   5′linker_down: 5′-Phos- AGATCGGAAGAGCGTCGTG TAGGGAAAGAGTGTAGATCTCGGTGGTCGCCGTAT-CATT-Phos-3′.

3′linker_up: 5′- NNNAGATCGGAAGAGCACACGTCTGAACTCCAGTCACXXXXXXATCTCGTATGCCGTCTTCTGCTTG-Phos-3′.

3′linker_down: 5′- CAAGCAGAAGACGGCATACGAGATxxxxxxGTGACTGGAGTTCAGACGTGTGCTCTTCCGA-3′.

XXXXXX: is the index sequence of Illumina sequence (xxxxxx is a compliment sequence of XXXXXX).

The "up" and "down" oligo are mixed (final 10 μM, 0.1 M NaCl). They are annealed by following condition: 95 °C, 5 min—gradient −0.1 °C/s to 83 °C—83 °C, 5 min—gradient −0.1 °C/s to 71 °C—71 °C, 5 min—gradient −0.1 °C/s to 59 °C—59 °C, 5 min—gradient −0.1 °C/s to 47 °C—47 °C, 5 min—gradient −0.1 °C/s to 35 °C—35 °C, 5 min—gradient −0.1 °C/s to 23 °C—23 °C, 5 min—gradient −0.1 °C/s to 11 °C—11 °C, forever.

In case of 5′linker, mix 5′linker GN5 and N6 at 4:1 ratio. 5′linker and 3′linker should be diluted to 2.5 μM with 0.1 M NaCl.

9. This ligation step is the most critical step through entire CAGE library preparation process because the SSLL efficiency affects library recovery. While the target molecule is single stranded rather than double stranded, conventional A/T tailed adapters cannot be used. In SSLL, the adapter is specifically ligated at the 5′- or 3′-end of the cDNA, by using a 6-base 5′- or 3-′-random overhang, respectively. We employ DNA Ligation Kit (Mighty Mix) to maximize the ligation yield at every step. The manufacturer recommends reaction conditions of 30 min at 16 °C or 5 min at 25 °C, but the reaction condition in our protocol is 16 h at 16 °C. By using SSLL under the latter condition, it is possible to achieve a ~80% ligation yield. Another important issue for adapter ligation efficiency is the terminal transferase activity of RTase. Most commercially available RTases (from moloney murine leukemia virus or avian myeloblastosis virus) have terminal transferase activity: in a template-independent way, they add one or more bases at the 3′-end of cDNA (i.e., when the enzyme reaches the 5′-end of template). The base(s) added are almost always C or CC, but in rare cases, A, G, T, CA, CT, or others are added. This means the majority of cap-trapped cDNA carries a C at the 3′-end. This would affect adapter ligation efficiency if the first overhang base were random because three-quarters of adapter molecules would not be able to anneal with C-tailed cDNA. Therefore, we mixed GNNNNN and NNNNNN overhang adapters at a ratio of 4:1 to maximize the adapter ligation recovery. In the adapter ligation step, it is possible to install the sequencing index for each sample to enable samples to be combined in a

sequencing lane. Various indexing methods are available, but Illumina indexing is most widely used.

10. When using DELVE, a percent identity threshold of 85% or more is combined with the above mapping quality threshold because the mapping quality threshold does not eliminate low-identity alignments. For both TopHat and STAR, criteria for suitable alignments have to be examined carefully since mapping quality indicates only the number of multiple hits across the search space.
11. bedGraph and bigWig formats do not store strand information, so two data files (one for each individual strand, forward [Watson], or reverse [Crick]) are required to represent one profile (e.g., suffixes of bigWig will be ∗.ctss.fw.bw and ∗.ctss.rev.bw, respectively).
12. When the promoter rate is over 70%, the data quality is considered to be reasonably good.
13. Majority of CAGE reads have an additional nucleotide G to the start position of RNA, owing to the terminal transferase activity of reverse transcriptase, without RNA template. It is not possible to discriminate whether G at the first base of CAGE read represents genuine 5′-end of RNA or nontemplated nucleotide when the upstream neighbor of the TSS on the genome is G nucleotide. CAGEr has a method to estimate frequency of genuine 5′-end of RNA with a probabilistic model, while it has to be noted that the alignment algorithm may influence the correction process. Notably, CAGE reads produced by HeliScope sequencer do not require such correction, as the instrument is designed to sequence only from the second position of the template DNA, rather than the first position.

References

1. Schena M, Shalon D, Davis RW et al (1995) Quantitative monitoring of gene expression patterns with a complementary DNA microarray. Science 270:467–470
2. Adams MD, Kelley JM, Gocayne JD et al (1991) Complementary DNA sequencing: expressed sequence tags and human genome project. Science 252:1651–1656
3. Velculescu VE, Zhang L, Vogelstein B et al (1995) Serial analysis of gene expression. Science 270:484–487
4. Margulies M, Egholm M, Altman WE et al (2005) Genome sequencing in microfabricated high-density picolitre reactors. Nature 437:376–380. https://doi.org/10.1038/nature03959
5. Morin R, Bainbridge M, Fejes A et al (2008) Profiling the HeLa S3 transcriptome using randomly primed cDNA and massively parallel short-read sequencing. BioTechniques 45:81–94. https://doi.org/10.2144/000112900
6. Mortazavi A, Williams BA, McCue K et al (2008) Mapping and quantifying mammalian transcriptomes by RNA-Seq. Nat Methods 5:621–628. https://doi.org/10.1038/nmeth.1226
7. Cloonan N, Forrest ARR, Kolle G et al (2008) Stem cell transcriptome profiling via massive-scale mRNA sequencing. Nat Methods 5:613–619. https://doi.org/10.1038/nmeth.1223

8. Carninci P, Kasukawa T, Katayama S et al (2005) The transcriptional landscape of the mammalian genome. Science 309:1559–1563. https://doi.org/10.1126/science.1112014
9. Kawamoto S, Yoshii J, Mizuno K et al (2000) BodyMap: a collection of 3′ ESTs for analysis of human gene expression information. Genome Res 10:1817–1827
10. Martin JA, Wang Z (2011) Next-generation transcriptome assembly. Nat Rev Genet 12:671–682. https://doi.org/10.1038/nrg3068
11. Marioni JC, Mason CE, Mane SM et al (2008) RNA-seq: an assessment of technical reproducibility and comparison with gene expression arrays. Genome Res 18:1509–1517. https://doi.org/10.1101/gr.079558.108
12. Pruitt KD, Tatusova T, Maglott DR (2005) NCBI reference sequence (RefSeq): a curated non-redundant sequence database of genomes, transcripts and proteins. Nucleic Acids Res 33: D501–D504. https://doi.org/10.1093/nar/gki025
13. Harrow J, Frankish A, Gonzalez JM et al (2012) GENCODE: the reference human genome annotation for the ENCODE project. Genome Res 22:1760–1774. https://doi.org/10.1101/gr.135350.111
14. Shiraki T, Kondo S, Katayama S et al (2003) Cap analysis gene expression for high-throughput analysis of transcriptional starting point and identification of promoter usage. Proc Natl Acad Sci U S A 100:15776–15781. https://doi.org/10.1073/pnas.2136655100
15. FANTOM Consortium and the RIKEN PMI and CLST (DGT), Forrest ARR, Kawaji H et al (2014) A promoter-level mammalian expression atlas. Nature 507:462–470. https://doi.org/10.1038/nature13182
16. Andersson R, Gebhard C, Miguel-Escalada I et al (2014) An atlas of active enhancers across human cell types and tissues. Nature 507:455–461. https://doi.org/10.1038/nature12787
17. Arner E, Daub CO, Vitting-Seerup K et al (2015) Transcribed enhancers lead waves of coordinated transcription in transitioning mammalian cells. Science 347:1010–1014. https://doi.org/10.1126/science.1259418
18. Dunham I, Kundaje A, Aldred SF et al (2012) An integrated encyclopedia of DNA elements in the human genome. Nature 489:57–74. https://doi.org/10.1038/nature11247
19. Carninci P, Sandelin A, Lenhard B et al (2006) Genome-wide analysis of mammalian promoter architecture and evolution. Nat Genet 38:626–635. https://doi.org/10.1038/ng1789
20. Valen E, Pascarella G, Chalk A et al (2009) Genome-wide detection and analysis of hippocampus core promoters using DeepCAGE. Genome Res 19:255–265. https://doi.org/10.1101/gr.084541.108
21. Faulkner GJ, Kimura Y, Daub CO et al (2009) The regulated retrotransposon transcriptome of mammalian cells. Nat Genet 41:563–571. https://doi.org/10.1038/ng.368
22. Takahashi H, Lassmann T, Murata M et al (2012) 5′ end-centered expression profiling using cap-analysis gene expression and next-generation sequencing. Nat Protoc 7:542–561. https://doi.org/10.1038/nprot.2012.005
23. Djebali S, Davis CA, Merkel A et al (2012) Landscape of transcription in human cells. Nature 489:101–108. https://doi.org/10.1038/nature11233
24. Carninci P, Kvam C, Kitamura A et al (1996) High-efficiency full-length cDNA cloning by biotinylated CAP trapper. Genomics 37:327–336. https://doi.org/10.1006/geno.1996.0567
25. Kim T-K, Hemberg M, Gray JM et al (2010) Widespread transcription at neuronal activity-regulated enhancers. Nature 465:182–187. https://doi.org/10.1038/nature09033
26. de Hoon M, Shin JW, Carninci P (2015) Paradigm shifts in genomics through the FANTOM projects. Mamm Genome 26:391–402. https://doi.org/10.1007/s00335-015-9593-8
27. Kodzius R, Kojima M, Nishiyori H et al (2006) CAGE: cap analysis of gene expression. Nat Methods 3:211–222. https://doi.org/10.1038/nmeth0306-211
28. Ravasi T, Suzuki H, Cannistraci CV et al (2010) An atlas of combinatorial transcriptional regulation in mouse and man. Cell 140:744–752. https://doi.org/10.1016/j.cell.2010.01.044
29. FANTOM Consortium, Suzuki H, Forrest ARR, van Nimwegen E et al (2009) The transcriptional network that controls growth arrest and differentiation in a human myeloid leukemia cell line. Nat Genet 41:553–562. https://doi.org/10.1038/ng.375
30. Taft RJ, Glazov EA, Cloonan N et al (2009) Tiny RNAs associated with transcription start sites in animals. Nat Genet 41:572–578. https://doi.org/10.1038/ng.312
31. Plessy C, Bertin N, Takahashi H et al (2010) Linking promoters to functional transcripts in small samples with nanoCAGE and CAGEscan.

Nat Methods 7:528–534. https://doi.org/10.1038/nmeth.1470

32. Zhu YY, Machleder EM, Chenchik A et al (2001) Reverse transcriptase template switching: a SMART approach for full-length cDNA library construction. BioTechniques 30:892–897
33. Ohtake H, Ohtoko K, Ishimaru Y et al (2004) Determination of the capped site sequence of mRNA based on the detection of cap-dependent nucleotide addition using an anchor ligation method. DNA Res 11:305–309
34. Harris TD, Buzby PR, Babcock H et al (2008) Single-molecule DNA sequencing of a viral genome. Science 320:106–109. https://doi.org/10.1126/science.1150427
35. Kawaji H, Lizio M, Itoh M et al (2014) Comparison of CAGE and RNA-seq transcriptome profiling using clonally amplified and single-molecule next-generation sequencing. Genome Res 24:708–717. https://doi.org/10.1101/gr.156232.113
36. Itoh M, Kojima M, Nagao-Sato S et al (2012) Automated workflow for preparation of cDNA for cap analysis of gene expression on a single molecule sequencer. PLoS One 7:e30809. https://doi.org/10.1371/journal.pone.0030809
37. Kanamori-Katayama M, Itoh M, Kawaji H et al (2011) Unamplified cap analysis of gene expression on a single-molecule sequencer. Genome Res 21:1150–1159. https://doi.org/10.1101/gr.115469.110
38. Murata M, Nishiyori-Sueki H, Kojima-Ishiyama M et al (2014) Detecting expressed genes using CAGE. Methods Mol Biol 1164:67–85. https://doi.org/10.1007/978-1-4939-0805-9_7
39. Hasegawa A, Daub C, Carninci P et al (2014) MOIRAI: a compact workflow system for CAGE analysis. BMC Bioinformatics 15:144. https://doi.org/10.1186/1471-2105-15-144
40. Gordon A, Hannon GJ (2010) Fastx-toolkit. FASTQ/A short-reads pre-processing tools. http://hannonlab.cshl.edu/fastx_toolkit. Accessed 19 Jul 2019
41. Lassmann T (2015) TagDust2: a generic method to extract reads from sequencing data. BMC Bioinformatics 16:24. https://doi.org/10.1186/s12859-015-0454-y
42. FANTOM Consortium (2014) rRNAdust program. http://fantom.gsc.riken.jp/5/suppl/rRNAdust/. Accessed 19 Jul 2019
43. Li H, Durbin R (2009) Fast and accurate short read alignment with Burrows-Wheeler transform. Bioinformatics 25:1754–1760. https://doi.org/10.1093/bioinformatics/btp324
44. Trapnell C, Pachter L, Salzberg SL (2009) TopHat: discovering splice junctions with RNA-Seq. Bioinformatics 25:1105–1111. https://doi.org/10.1093/bioinformatics/btp120
45. Dobin A, Gingeras TR (2015) Mapping RNA-seq Reads with STAR. Curr Protoc Bioinformatics 51:11.14.1–11.14.19. https://doi.org/10.1002/0471250953.bi1114s51
46. Lassmann T (2011) DELVE: a probabilistic short read aligner used in FANTOM5 and ENCODE. http://fantom.gsc.riken.jp/5/suppl/delve/delve.tgz. Accessed 19 Jul 2019
47. Li H, Handsaker B, Wysoker A et al (2009) The sequence alignment/map format and SAMtools. Bioinformatics 25:2078–2079. https://doi.org/10.1093/bioinformatics/btp352
48. Quinlan AR (2014) BEDTools: The Swiss-army tool for genome feature analysis. Curr Protoc Bioinformatics 47:11.12.1–11.12.34. https://doi.org/10.1002/0471250953.bi1112s47
49. Robinson JT, Thorvaldsdóttir H, Winckler W et al (2011) Integrative genomics viewer. Nat Biotechnol 29:24–26. https://doi.org/10.1038/nbt.1754
50. Rosenbloom KR, Armstrong J, Barber GP et al (2015) The UCSC genome browser database: 2015 update. Nucleic Acids Res 43:D670–D681. https://doi.org/10.1093/nar/gku1177
51. Kent WJ, Zweig AS, Barber G et al (2010) BigWig and BigBed: enabling browsing of large distributed datasets. Bioinformatics 26:2204–2207. https://doi.org/10.1093/bioinformatics/btq351
52. UCSC Kent source utilities. http://hgdownload.soe.ucsc.edu/admin/exe/. Accessed 19 Jul 2019
53. Severin J, Lizio M, Harshbarger J et al (2014) Interactive visualization and analysis of large-scale sequencing datasets using ZENBU. Nat Biotechnol 32:217–219. https://doi.org/10.1038/nbt.2840
54. Frith MC, Valen E, Krogh A et al (2008) A code for transcription initiation in mammalian genomes. Genome Res 18:1–12. https://doi.org/10.1101/gr.6831208
55. Fejes-Toth K, Sotirova V, Sachidanandam R et al (2009) Post-transcriptional processing generates a diversity of 5′-modified long and short RNAs. Nature 457:1028–1032. https://doi.org/10.1038/nature07759

56. Hirzmann J, Luo D, Hahnen J et al (1993) Determination of messenger RNA 5′-ends by reverse transcription of the cap structure. Nucleic Acids Res 21:3597–3598
57. Ohmiya H, Vitezic M, Frith MC et al (2014) RECLU: a pipeline to discover reproducible transcriptional start sites and their alternative regulation using capped analysis of gene expression (CAGE). BMC Genomics 15:269. https://doi.org/10.1186/1471-2164-15-269
58. Haberle V, Forrest ARR, Hayashizaki Y et al (2015) CAGEr: precise TSS data retrieval and high-resolution promoterome mining for integrative analyses. Nucleic Acids Res 43:e51. https://doi.org/10.1093/nar/gkv054
59. Hyvärinen A, Oja E (1997) A fast fixed-point algorithm for independent component analysis. Neural Comput 9(7):1483–1492
60. Robinson MD, McCarthy DJ, Smyth GK (2010) edgeR: a bioconductor package for differential expression analysis of digital gene expression data. Bioinformatics 26:139–140. https://doi.org/10.1093/bioinformatics/btp616
61. Love MI, Huber W, Anders S (2014) Moderated estimation of fold change and dispersion for RNA-seq data with DESeq2. Genome Biol 15:550. https://doi.org/10.1186/s13059-014-0550-8
62. Fort A, Hashimoto K, Yamada D et al (2014) Deep transcriptome profiling of mammalian stem cells supports a regulatory role for retrotransposons in pluripotency maintenance. Nat Genet 46:558–566. https://doi.org/10.1038/ng.2965
63. Hashimoto K, Suzuki AM, Dos Santos A et al (2015) CAGE profiling of ncRNAs in hepatocellular carcinoma reveals widespread activation of retroviral LTR promoters in virus-induced tumors. Genome Res 25:1812–1824. https://doi.org/10.1101/gr.191031.115
64. Vitezic M, Lassmann T, Forrest ARR et al (2010) Building promoter aware transcriptional regulatory networks using siRNA perturbation and deepCAGE. Nucleic Acids Res 38:8141–8148. https://doi.org/10.1093/nar/gkq729
65. Lizio M, Harshbarger J, Shimoji H et al (2015) Gateways to the FANTOM5 promoter level mammalian expression atlas. Genome Biol 16:22. https://doi.org/10.1186/s13059-014-0560-6
66. Takamochi K, Ohmiya H, Itoh M et al (2016) Novel biomarkers that assist in accurate discrimination of squamous cell carcinoma from adenocarcinoma of the lung. BMC Cancer 16 (1):760
67. Yoshida E, Terao Y, Hayashi N et al (2017) Promoter-level transcriptome in primary lesions of endometrial cancer identified biomarkers associated with lymph node metastasis. Sci Rep 7(1):14160. https://doi.org/10.1038/s41598-017-14418-5
68. Sompallae R, Hofmann O, Maher CA et al (2013) A comprehensive promoter landscape identifies a novel promoter for CD133 in restricted tissues, cancers, and stem cells. Front Genet 4:209. https://doi.org/10.3389/fgene.2013.00209
69. Thorsen K, Schepeler T, Øster B et al (2011) Tumor-specific usage of alternative transcription start sites in colorectal cancer identified by genome-wide exon array analysis. BMC Genomics 12:505. https://doi.org/10.1186/1471-2164-12-505
70. Demircioğlu D, Kindermans M, Nandi T et al (2017) A pan cancer analysis of promoter activity highlights the regulatory role of alternative transcription start sites and their association with noncoding mutations. bioRxiv. https://doi.org/10.1101/176487
71. Dieudonné FX, O'Connor PB, Gubler-Jaquier P et al (2015) The effect of heterogeneous Transcription Start Sites (TSS) on the translatome: implications for the mammalian cellular phenotype. BMC Genomics 16:986. https://doi.org/10.1186/s12864-015-2179-8
72. Conte M, De Palma R, Altucci L (2018) HDAC inhibitors as epigenetic regulators for cancer immunotherapy. Int J Biochem Cell Biol 98:65–74. https://doi.org/10.1016/j.biocel.2018.03.004
73. Brocks D, Schmidt CR, Daskalakis M et al (2017) DNMT and HDAC inhibitors induce cryptic transcription start sites encoded in long terminal repeats. Nat Genet 49(7):1052–1060. https://doi.org/10.1038/ng.3889
74. Navada SC, Steinmann J, Lübbert M et al (2014) J Clin Invest 124(1):40–46. https://doi.org/10.1172/JCI69739
75. Pan T, Qi J, You T et al (2018) Addition of histone deacetylase inhibitors does not improve prognosis in patients with myelodysplastic syndrome and acute myeloid leukemia compared with hypomethylating agents alone: a systematic review and meta-analysis of seven prospective cohort studies. Leuk Res 71:13–24. https://doi.org/10.1016/j.leukres
76. Pleyer L, Greil R (2015) Digging deep into "dirty" drugs—modulation of the methylation machinery. Drug Metab Rev 47(2):252–279. https://doi.org/10.3109/03602532.2014.995379

Index

A

Allele-specific prediction 113, 118, 119, 122
arcasHLA 74–79, 81–84, 87, 89–91
Artificial neural networks (ANNs) 113, 114, 169
Athlon 94–98

B

Benchmark.. 38, 45, 73, 75, 78, 91, 117, 198, 203, 211, 218, 219, 230, 232, 235, 236
Bioinformatics .... v, 1–7, 13, 17, 94, 102, 113, 120, 147, 165
Bulk RNA-Seq R package.................................. 223–246

C

Cancer genomics 12, 234
Cancer immunotherapy .... v, 1–7, 93, 121, 129, 197, 213–220, 294
Cancers .... v, vi, 1–7, 11–13, 37–45, 47, 69, 71–91, 93, 102, 110, 129, 130, 132, 138, 139, 141, 143, 147–159, 164, 165, 167, 183, 184, 186, 187, 197–199, 201, 203, 213–220, 230, 231, 233, 234, 236–238, 242, 245, 246, 249–252, 254–260, 294, 295
Cap analysis of gene expression (CAGE) 277–298
CD8+ T cells 162, 168, 173, 213, 242, 245, 246, 253, 254, 257, 271, 273
Cell fraction predictions 219, 235
Cell type deconvolution..............................213–220, 230
Comparative analyses 224, 226–230
Computational biology...................................... 234–244
Computational proteomics...185

D

Database 4, 20, 25, 28, 73, 75–79, 82, 94, 104, 106, 107, 110, 116, 117, 129, 132, 133, 135–137, 142–144, 162–169, 174, 177–179, 184, 187, 198–201, 203, 204, 208, 211, 255, 259
Data repository...173–180, 234
Deconvolution 5, 6, 213–220, 224–226, 228, 230, 232, 234, 235, 237, 245, 246, 249–261, 264
Docker 48, 49, 61–66, 130, 132, 141, 142, 148, 150, 151, 154, 158, 159

E

Enhancer RNAs (eRNA) 276, 294
Enhancers, v 279, 281, 294
Ensemble method 37–45, 59–69, 133
Epitope predictions 4, 110, 187
Epitope-specific T cells ....v, 183–195
Estimating the Proportions of Immune and Cancer cells (EPIC)...214, 216–220, 224–228, 231, 233–246
Estimations 18, 103, 109, 110, 131, 144, 148, 162, 169, 178, 193, 214, 218, 226, 242, 252–255, 258
Expression v4, 6, 74, 81, 89, 91, 101–111, 129–131, 136, 137, 140, 141, 143, 144, 148, 155, 157, 164, 206, 214, 215, 217, 218, 226, 227, 231–246, 249–261, 263, 264, 266, 267, 269, 270, 274, 275, 277–298

G

Gene expression analysis 233, 240, 259, 293
Gene signatures 216, 218, 265, 275
Genotyping..................................... 4, 56, 76, 78, 81, 83, 84, 87, 89, 91, 104–107, 110

H

HLA expression........................ v, 81, 105, 107, 109, 110
HLA ligandome 163, 164, 167
HLApers 101–111
HLA type ....v, 4, 131, 135, 138, 139
Human leukocyte antigens (HLA) .... v, 4, 6, 71–91, 93–98, 101–111, 115, 116, 120–122, 124, 129–132, 135–140, 152, 161, 163–166, 168, 169, 173, 174, 177

I

Immune cell quantification..5, 6
Immune checkpoint blockade 2, 129
Immunedeconv 223–232
Immunoinformatics 5
Immunopeptidome 173, 174, 178
Immunopeptidomic analyses 173–180

Sebastian Boegel (ed.), *Bioinformatics for Cancer Immunotherapy: Methods and Protocols*, Methods in Molecular Biology, vol. 2120, https://doi.org/10.1007/978-1-0716-0327-7, © Springer Science+Business Media, LLC, part of Springer Nature 2020

Immunotherapy .... v, 1–7, 93, 110, 213–220, 294
Infiltrating Immune Cells 249, 255
Interactive Website 187

K

Konda, P. 161–169

L

Ligands .... v, 113–127, 157, 161–169, 197

M

Machine learning 38, 47–69, 184, 264
Major histocompatibility complex (MHC) 2, 4, 6, 7, 71–74, 101, 102, 105, 106, 110, 113–127, 145, 147, 148, 153, 156, 159, 161–169, 173, 174, 190, 197, 198, 200, 201, 203, 295
Mass spectrometry 4, 129, 161–169
MHCflurry 113–127
Modeling 6, 197–212
Mutations .... v, 1–4, 6, 11–13, 18, 20, 23, 28, 37–45, 47–69, 75, 89, 102, 110, 130–132, 134, 135, 137–139, 147, 149, 157, 162, 164, 167, 255, 258, 259

N

Nanopore sequencing 93–98
Neoantigens .... v, 3, 6, 75, 87, 93, 110, 129–145, 147–160, 204
Next generation sequencing (NGS) 2–5, 18, 61, 73, 78, 93, 102, 110, 131, 167, 278–281, 287, 292, 295

O

Open-source 147–160, 272
OpenVax 147–160
Oxford nanopore 94

P

Personalized cancer vaccines 129
Personalized index 101–111
Personalized mutanome vaccines 2
Personalized reference 12, 24, 26, 28–31, 104
Personalized Reference Editor for SomaticMutation (PRESM) 12–14, 17, 19, 23, 24, 26–34
Promoter activity 291, 294
Promoter-level expression analysis 293

R

Random forest 38, 185
RNA-seq deconvolution 5, 217, 218, 220, 235, 245, 246
RNA sequencing (RNA-seq) 5, 12, 71–91, 102, 104–106, 109, 110, 129–132, 134, 135, 139, 143, 148, 158, 165, 167, 168, 214, 217–220, 223–232, 234–238, 241, 242, 244–246, 250, 254, 264, 274, 278, 279, 293, 294
Rosetta 198–200, 206, 209–211

S

Single nucleotide variations (SNVs) 3, 6, 38–41, 43, 44, 51, 52, 54, 55, 59, 60, 62, 66, 258
Small insertions/deletions (indels) 14
Somatic Mutation calling method using a Random Forest (SMuRF) 38–45
Somatic mutations .... v, 2, 11–13, 18, 20, 37–45, 47–69, 75, 102, 130, 132, 164
SomaticSeq 47–69
Somatic variant discovery 11–34
Somatic variants 11–34, 47, 148, 149, 153–156
Spectra 162, 167, 169, 178
Spill-over 267, 269, 270
SysteMHC 117, 173–180

T

Targeted sequencing 62, 163–168, 179, 278
T-cell-based immunotherapies 129
T cell receptors (TCRs) .... v, 2, 4–7, 162, 183–189, 191–194, 197–212
T cells .... v, 1, 2, 4–6, 71, 72, 93, 102, 113, 124, 129, 140, 147, 161–163, 168, 173, 183–195, 197–213, 218, 227–229, 235, 242, 244–246, 249, 253, 254, 257, 258, 260, 271, 273, 274
TCR3d 197–212
TCRex 184–194
TCRmodel 197–212
TCR repertoire analysis 4, 5, 184–187, 189, 195
TCR repertoire sequencing 4, 5, 184, 185
3D structures 197–211
TIminer 129–145
Transcription start sites (TSS) .... v, 277–298
Tumor IMmune Estimation Resource (TIMER) 6, 214, 216, 217, 224, 226, 228, 230, 235, 251, 253–261
Tumor-immune interactions 249, 254
Tumor immune microenvironment 5, 213, 223, 233, 249, 250

X

XCell 6, 215–217, 219, 220, 224–226, 234, 263–275

MIX
Papier aus verantwortungsvollen Quellen
Paper from responsible sources
FSC® C105338

If you have any concerns about our products,
you can contact us on
ProductSafety@springernature.com

In case Publisher is established outside the EU,
the EU authorized representative is:
Springer Nature Customer Service Center GmbH
Europaplatz 3, 69115 Heidelberg, Germany

Printed by Libri Plureos GmbH
in Hamburg, Germany